Towards The Death Of Humanity

DEHUMANIZATION: THE AFFLICTION DESTROYING MANKIND AND MODERN SOCIETY

Immunologist and Emeritus Professor

By

Dr. Gilles Lamoureux M.D., Ph.D.

authorHOUSE™

1663 LIBERTY DRIVE, SUITE 200
BLOOMINGTON, INDIANA 47403
(800) 839-8640
WWW.AUTHORHOUSE.COM

First published by AuthorHouse 09/10/04

ISBN: 1-4184-8027-4 (sc)

Printed in the United States of America
Bloomington, Indiana

This book is printed on acid-free paper.

TO ALL MY CHILDREN
AND GRANDCHILDREN

With special thanks to my daughters Marie-Andrée and Marie-France.

I also wish to thank Mrs Jacqueline Gauthier and Mrs Monique Cadieu for their valuable proofreading work.

My thanks also go to Mr Michel Campeau and Mrs Nicole Henri for their precious help in correcting this text, and to Mrs June Hacala for the translation of the original French document.

Men accept with great difficulty
The things they do not wish to change
From the bottom of their hearts
For it would oblige them to question absolutely everything

Alexis Carrel

TABLE OF CONTENTS

CHAPTER 1
The benefits of science and technology 1

Introduction 1

The technological revolution and its benefits 2

The acquired knowledge 2

The acceleration of technological applications 3

Changes in the range of diseases 4

Life expectancy 4

Disease prevention 5

Life balance 5

Food 6

Regression of poverty 6

Communications 6

An already present future 7

In physics 8

In biology 8

Genetic cartography 10

In biogenetics 11

Artificial limbs 11

In robotics 12

In nanotechnology 12

In technological applications 13

The car industry 14

In ballistics 14

In data processing 14

Knowledge 14

Proof of the evolution since a century 15

Warning 16

Thoughts on the implementation of all this science 18

CHAPTER 2
The endless spiral of human aspirations 23

Introduction 23

The situation today 24

The migrating of populations 25

The credit 26

Chaos is here and problems accumulate 27

The exhaustion 28

The abuse 28

The debts 29

Globalization 30

The program 31

CHAPTER 3
The negative effects of technological applications on the environment 35

Introduction 35

A. The elements of the life 36

Life 38

Water, the problem of the twenty-first century 39

Drinking water 41

On the wasting of water 41

The pollutants mainly of water 42

Industrial pollutants and others 43

Acid rains 43

Herbicides 43

Medication 44

Genetically modified organisms 46

Sewage water 46

The pollutants found in human tissue 47

The effects of certain pollutants on life 48

Ground water 52

B.   Deterioration of the air and of the earth.....................................53

The sources of atmospheric pollution..56

The tendency of atmospheric pollutants 56

What are the potential effects on health?....................................57

What is Health Canada doing?...57

What *you* can do 58

Health Canada 59

The Sword of Damocles 59

Global warming 59

Warming of the air and climatic changes....................................60

We take very lightly the commitments against pollution...........62

The extent of the current destruction processes........................64

Rating the ecosystem pollution..65

The decline of the ecosystems 65

C.   Expansion and thoughts on the processes of environmental
destruction...66

Considerations on the physical effects of technological
applications 69

D.   Diseases caused by the effects of technological applications on
life...70

Diseases directly linked to environmental pollution.................73

Diseases associated to air pollution 77

Impacts of certain technological applications on the future of
botanical and animal species...81

Medicine and GMOs...82

The GMO is not the first intervention of man on nature 89

Could scientists have unknowingly messed around with the
genes?..89

Indirect proof that infections and epidemics are natural
selectors 90

Let us compare the data with those acquired after the
interventions of man on natural selection 92

The phenomenon of mass immunity...95

Individuals responding and non-responding to infection 95

Proof by the common sense 96

Who are these less resistant individuals? 96

Internationalization of danger 99

Regarding the GMOs 99

Risky medical technologies 101

From the ethical point of view 101

E. Discussion and reflections 104

CHAPTER 4
The dehumanising effects of technological applications on man and society

The dehumanising effects of technological applications on man and society **107**

Introduction 107

Plan 108

A. Technologies dictate their own laws and materialize 109

B. With the help of technology, the authorities rob the citizens from their individual rights and liberties 110

Proofs of violation of our rights and liberties 113

Governments supercomputers 113

A "unique government file" on each citizen 114

More demonstrations of power abuse 117

Encoding 118

A law to locate individuals 120

The White House and the secret of identity 121

Social repercussions of these amoral and inhuman measures . 122

A major problem created by this culture 122

Other technologies, other violations of our personal life 123

Private information and insurance companies 124

After September 11, 2001 124

Thoughts and comments about people disgusted by the technologies which allow people in power to abuse them 126

C. Examples of administrative technologies propelling
    materialism.. 128

    Examples of materialization of administrative decisions 131

    The materialistic culture and the irresponsibility of the
    authorities... 132

    Can we return to the past?.. 133

D. Technological applications mind-destroying to man 134

    Same phenomenon for administrative techniques applied in
    private institutions... 138

    Implementing computer techniques in administration............ 139

    The weight of administrative techniques in the exercise of a
    profession.. 140

    Other characteristics of administrative technologies.............. 143

E. When techniques create disease....................................... 143

    The "stress diseases"... 145

    Stress and human behaviour 146

    Behavioural diseases.. 148

    Stress and human errors 149

    Linking diseases... 150

F. When the materialistic methodology of the search for truth takes
    charge of education... 151

    The first changes provoked by these education technologies .. 153

    The people in power at the Ministry of Education warp
    education 156

    The atrocities of the powers.. 157

    Let us think about the dehumanizing depth of technological
    applications 159

G. Summary.. 159

CHAPTER 5
Repercussions and impacts of materialism on the physical world,
on man and on society... 163

Introduction 163

A. The benefits and setbacks of scientific education 164

Materialism carried through education becomes the motor..... 171

Examples of failures in education.. 174

The powers in the universities 175

The violation of human rights across the planet...................... 177

The decline of human values 177

Materialism modifies the expression of morality 178

Sharing the wealth – let's talk about it 179

The influence of materialism on communication between
people.. 179

B. The materialism of technological applications deeply modifies
human behaviour.. 182

The materialism which transforms student mentality, even in
universities 184

The role of materialism in the induction of crime 186

C. Materialism and deterioration of the exercise of power 189

The technologies that make mistakes acceptable in powers and
stimulate violence 190

Materialism and arrogance of the powers 192

Materialism and families 195

The materialism of Daniel Kemp's Teflon child...................... 197

Materialism and religions 198

Materialism and international powers 200

Materialism and the confidence of citizens............................. 203

Governments: omnipresence and incompetence....................... 205

D. The influence of materialism on the degrading of man's
work 208

The desire of freedom in employees and employers 208

Stupidity.. 210

E. Materialism is invading the professions 211

Example of the materialization in the medical profession....... 214

An unhealthy medical conservatism 216

The gratuitous assertions of medicine 216

The example of Jenner's vaccination 218

Consequence: medicine misses the boat of scientific
revolution 219

The absence of medical researchers... 219

Lines of conduct.. 220

Experimentation victimized.. 220

Patients are quick to understand 222

The socialist speeches of the powers infiltrate medicine......... 222

F. The influence of materialism on other professions and
    institutions 226

    Justice... 226

    Correction institutions.. 232

    The Police system 232

    When the materialistic culture influences sports and State
    actions 234

    Influence on people in power.. 234

G. The influence of scientific and social materialism on poverty,
    criminality, violence.. 239

    Materialism and poverty 239

    Materialism and suicide 241

    Abuses leading to violence 242

    Defence mechanisms of individuals assaulted by their
    leaders 243

    Materialism and violence in parents 246

    Materialism and street gangs 248

    Materialism and violence... 249

    Materialism and criminology... 250

    Materialism, violence and overpopulation................................ 252

    Materiality and other activators of violence 252

    The materialism of violence in children and teenagers 253

    The explanation of the psychologist 256

H. Attempts at restraining violence 256

Material techniques of evasion 256

The "comfort" techniques 257

The "restaurant" techniques.. 257

The "shopping" techniques 257

Techniques "to change the vocabulary".................................. 258

Are there human solutions? 259

I.   Conclusion 259

CHAPTER 6

**How do materialized people manage to live in their soulless
world?**.. **263**

Introduction 263

A. Let us examine how the people and the powers of our societies
    live this materialism and this lack of humanity in their everyday
    life. 264

B. Materialism has destroyed the human being........................... 267

C. Something, somewhere, has been broken 271

D. Materialism disillusions man 273

E. Conclusion 276

CHAPTER 7

**The history, the interest and the ideological impact of scientific
and social materialism**.. **279**

Introduction 279

Plan 279

A. New ways of thinking that give rise to scientific and social
    materialism 280

The desire to monopolize the power of the Church................. 280

One day Hegel came 282

Again the new philosophy 286

The choice of a methodology to search for the truth 287

Societies are quickly becoming saturated with this scientific
materialism... 289

Craze without limits for science 289

B. Some of the great materialistic ideological tendencies 291

Science becomes the fashion and the screen through which
everything must pass 294

Creation of education fields on a scientific basis 295

The explosion of materialistic and socialistic ideologies 296

Major currents of political thoughts that continue to divide
societies today 297

Countries governed by liberal or neo-liberal ideologies 298

Countries governed by Marxist or socialistic ideologies 299

Vices of the scientific and social materialism 302

Adapting our ways of thinking to the new ideologies 304

C. The role played by the teaching of sciences in the expansion of
materialism 305

Where are we at with this education of scientific materialism at
the beginning of this twenty-first century? 307

The consequences of a scientific education without
humanism 309

D. Matter is not eternal and the incomplete scientific methodology
must be rethought to take into account new and non tangible
facts 313

Matter is not eternal 314

The limits of the method 317

What has become of man and society? 318

Will the twenty-first century be religious or not? 321

E. Conclusion 322

CHAPTER 8
Materialism and the downfall into despair 325

Introduction 325

A century of "reason" or of nonsense...? 326

How will tomorrow be if nothing changes? 328

Conclusions 331

CHAPTER 9

Are solutions beginnings to emerge? 333

Introduction 333

    Possible solutions.. 335

    Will we succeed in extricating ourselves from all of this? 342

    The solution of the conscience.. 343

SUMMARY

"Towards the Death of Humanity" is the endless demonstration of the **disastrous side effects** left on our **environment**, on **life** on this planet, on **health and most of all on human dehumanization** by a century of tremendous scientific and technological realizations and their material values.

It illustrates how these **unhealthy side effects** are highly linked to the hasty and thoughtless decisions of scientists, intellectuals and governments to replace the **humanities and the traditional methods of teaching with their own methods of science, technology and their essentially material values**. Unfortunately totally deprived of the smallest bit of humanity, this method of education has **killed the rudiments of humanity acquired during the course of history, has dehumanized the people, their institutions and their societies.**

These side effects originate in the age-old desire of intellectuals and scientists **to take hold of the power of religion over man and to satisfy their morbid need to build a new world,** re-designed following Hegel' belief that the matter is eternal and that God is eternal. Although these **ideologies have lost all their legitimacy today** because science has recently demonstrated that the matter is not eternal, scientists and intellectuals keep on teaching the falseness of their **materialistic social system as if the logic of their reasoning were the only truth.**

This work promotes certain guidelines that individuals must follow to make the lucid and effective compromises needed to try and solve these destructive problems affecting man and society before the end of History.

OBJECTIVES

The main objectives of this work are to draw the attention of all the people of this planet to the beneficial as well as the harmful side effects that science and technological applications have spread everywhere along their path.

"**Towards the Death of Humanity**" claims that these paradoxical side effects stem from a science that gives a lot and at the same time destroys a lot from the environment and life on earth, up to the **deprivation of men of their humanities and of their human values**. This lack of humanity is linked to the system of education with science and technologies which greatly help to materialize and dehumanize individuals and their societies, while enabling scientists and intellectuals to control the people, their thoughts and their actions, as religions formerly did.

Even though the comments may sometimes seem alarmist, the intention of "**Towards the Death of Humanity**" is to send out a **cry of alarm** for each individual to become aware of the dangers that threaten the environment and the alarming materializing effects of science and technology on the disappearance of men' humanity. A cry of distress in the name of all the men and all the women of this planet who suffer the materialization of their world and the sick deceits of their "reasonable" scientific governments that take pleasure in undermining what little humanity is left in man.

It intends to protest against a socialist point of view presently discussed in current pseudo scientific movements, university faculties, as well as by scientists and intellectuals that it belong to the society and government to control men' thoughts, rights and liberties, including their diseases, their education and all aspects of people' lives by using their power and the technology.

It claims high and loud **that every individual on the planet must rapidly become aware of this deep evil of materialism which suffocates him and eats him away and** that this planet is in danger, whatever his race, religion, country, culture, the ideologies he shares or the excessive use of the soulless technologies.

Let us hope that the actual degree of destruction of our material and dehumanized world and the gigantic mess in which our civilization is now locked, will rapidly **touch the conscious mind** of most individuals of this planet as well as that of the leaders of our governments.

It is not intended to offer all the possible solutions that will have to be taken one day only to begin the necessary corrections urgently needed and to simply control the scientific ideologies that have driven men in this abyss, but to suggest some lines of thought.

THE AUTHOR

Doctor Gilles Lamoureux is a medical doctor, a researcher and a laboratory man specialized in clinical immunology. Doctor Lamoureux is a retired professor from the *Université de Montréal* and the *Université du Québec* in Montreal. He guided dozens of students in their extracurricular activities during their MSc, their PhD and their post-doctoral research work on multiple sclerosis, cancer immunology, allergy, clinical immunology and environmental immunology. He published over a hundred international scientific papers, presented over a hundred scientific conferences on his research work, wrote a book on BCG in cancer immunotherapy and produced a film entitled "A Question of Immunity" with colleges, etc. He just finishes writing his last book "**Towards the death of Humanity**".

Doctor Lamoureux considers himself as an ordinary citizen and gives himself the right to wonder as he watches his fellowmen **and observes the regrettable dehumanization phenomenon taking place all around him.** He is very much aware of the colossal movement that is hindering the people, the modern world, society, environment and the very existence of life on earth, and has always denied having the scientific authority to write this book or to blame anyone.

His knowledge comes from his education, his readings and his life spent between suffering individuals and a laboratory, where he tries to find solutions to relieve men of their diseases. At the school of science, he learned to watch life evolve around him and has always conserved the faculty to be amazed by the unlimited forms of life found on our planet. However far he looks, he always sees life and is constantly impressed by the biodiversity of forms of life in our world: human beings, animals, insects, trees and plants, earth itself with its seas, its rivers and its brooks which are all filled with all this life. **He is completely conscious that life is the outcome of our universe** even though scientists have not yet been able to prove this with scientific evidence. As a professor, he considers it his task to become conscious of all this and to make others aware of it.

CHAPTER 1

The benefits of science and technology

Introduction

For over a century, our world has been engulfed in a whirlwind of scientific and technological revolutions impossible to avoid. The knowledge acquired on our physical and material universe has provided endless possibilities in the development of numerous technological applications. During the last century, these applications have also released waves of deep changes which relentlessly surge and eventually flood man and society.

Multiple technological changes have and will continue to influence the way people live. Modern technologies have enabled people to lead easier and more enjoyable lives. Men and societies have been undergoing these transformations for decades now. With every five year cycle, for example, communications, transportation, cities, household commodities, trips, trades, education, entertainment, life styles, styles in clothing, ways of thinking, etc., change so quickly that we are left with the impression that each cycle is so different from the previous one that it seems to have become a new generation.

The explosion of data accumulated on the subject enables us to foresee that this stream of material changes will continue to surge over our world at the same staggering rhythm as the arrival of the newly acquired knowledge. In fact, never before in the history of humanity has a century been as productive in accumulating scientific knowledge and applying new technologies as the twentieth century. It is also the first time that people will have gained so much from such a great quantity of technological applications. It is without a doubt to science and technology that we owe this incredible revolution of knowledge in our material world.

The era of Knowledge and Science now walks hand in hand with the new Industrial era and we now witness revolutions in computer science, in robotics, in space science and so much more... The enthusiasm shown by everyone for these novelties, revolutions and technologies is not likely to fade away.

Each day, new technologies, biotechnologies and/or genetic manipulations, progresses in telecommunications, robotics, physics, genetics and so many others, continue to emerge and to noticeably transform the material life of man and society. Scientists have been able to

1

clone at will, be it animals or humans, in fact any live entity. Over 4 000 scientific journals of all sorts regularly publish either weekly or monthly results of their most recent findings.

A simple overview of the scientific and technological revolutions pouring their overflow over the world and mankind makes us realize the enormous quantity of material benefits that do not cease to amaze us.

The technological revolution and its benefits

Not even for a moment could we do without all the material possessions that have become part of our existence. Could we, for example, do without the electronic devices that make our every day modern life easier? Would we accept going back to the way people lived in the early nineteenth century, in homes without running water, electricity, outhouses without baths or showers, and without the numerous domestic appliances that have become part of our daily lives? Can we even begin to imagine life without a car, with no modern transportation, no telephone, no computer, no Internet, to name but a few? All the driving forces of our consumer societies, our industries, our production companies providing these goods and possessions, are intimately bound to this scientific knowledge and its technological applications.

Much like children in awe, such amazing and remarkable progress has made our need for new technologies insatiable. The modern man, much like his counterpart at the beginning of the technological revolution, is forever grateful for this new knowledge because it makes his every day life so much easier by satisfying his material needs. We have quickly taken for granted that all the benefits gained through science and new technologies are essential to our very existence.

The acquired knowledge

Never before has scientific knowledge progressed as quickly as it does today. The knowledge acquired since the beginning of conscious humanity until the dawn of the twentieth century has doubled in the first fifty years of the twentieth century. It doubled once more between the fifties and the mid-seventies, which at the time seemed quite fantastic. Since the eighties, it has been doubling almost every year.

Thousands of researchers and research labs are working to discover the secrets of the matter, using computers 50,000 times more powerful than the first ones back in the sixties. Internet and numerous technological tools from all branches of science promote the acquisition of new technologies so much that, should a researcher now prove them to be doubling every week if not every day, we would not be in the least bit surprised.

With the speed of communications, radio, television, telephone, Internet and its incredible contents of brain-storming ideas of all kinds, the new data on research is immediately available in all the homes everywhere on earth. New knowledge will continue to flow into the future in a way yet impossible to conceive. With each passing day, each passing hour if not more quickly, humanity's knowledge will continue to double.

The acceleration of technological applications

It all begins with Denis Papin's discovery of the steam engine, at the end of the seventeenth century. The first real technological discoveries begin in the nineteenth century and the twentieth century literally explodes in all directions and in all areas of human material activities. Gas engines will come soon after the steam engines, which will then be followed by electric motors. Production of goods increases at the speed of lightning to meet the constant and growing demands of the consumer society. The tremendous energy required to meet such a demand in production leads to the development of new sources of energy. Means of travelling have evolved greatly in the past century. Space carriers use new kinds of fuels. The most recent engines are of an exceptionally scientific refinement. At the beginning of the twenty-first century, scientists already used ionic fuel for space travel. Fuels of a completely different nature will soon enable us to travel in space, using energy that even the most brilliant scientists could not imagine just a few decades ago.

From one discovery to the next, knowledge is accumulated at an accelerated rhythm for our greatest well-being. Today's houses are more comfortable than yesterday's castles, and their functional structures cannot begin to compare with those of just a century ago. New factories continually modernize to produce more goods and make our every day life easier. New professions and trades contribute to the acquisition of new knowledge. With such unprecedented human activity, knowledge progresses at an incredible speed, a speed unthinkable to man only half a century ago.

Men from the post-industrial era need not spend their days searching for food or walking to the well for water, as their ancestors did. Powerful engines used in modern cities activate aqueducts and distribute drinking water to each home. Thanks to the income from their work, men can easily purchase the essential goods using only a fraction of their wages, and thus enjoy more free time and vacations. They can choose to work a certain number of years and then enjoy an early and fruitful retirement, in comparison to their parents and grandparents. By choosing to be well prepared by an education available to all, they can get financially rewarding jobs and be protected by syndicates. All this magic is part of the era of the new knowledge.

Changes in the range of diseases

Epidemics of smallpox, diphtheria, whooping-cough, poliomyelitis, scarlet fever, plague, which in the past decimated entire populations, have almost completely disappeared from the surface of the earth. In every country, education with science and technology is more and more available. Schools and universities are also very good tools of education. Although occidental countries have been the first to profit from therapies in disease prevention, other countries are quickly catching up.

Infant mortality rate (before the age of five), which just fifty years ago decimated over 40% to 50% of the African populations as well as many other countries, has been greatly reduced in almost all countries of the world. Smallpox has been officially eradicated from the surface of the earth since 1977. Several diseases which were at one time deadly are disappearing and no longer a threat. Plague, scurvy, poliomyelitis, scarlet fever and other infectious diseases are now extremely well controlled almost all over the world. Vaccine discovery, good hygiene and the use of antibiotics have greatly contributed to a greater life expectancy in all humans on the planet in the twentieth century.

Life expectancy

Although the increase in life expectancy may be lesser in certain countries, it is nonetheless noticeable. Neonatal and infant mortality have decreased very significantly all over the world. Life expectancy at the end of the twentieth century in America exceeds 78 years in men and 82 years in women, in comparison to 50 and 55 years at the beginning of the twentieth century.

More adequate control of cardio-vascular diseases and cancers will soon increase life expectancy by an additional five years. Scientists foresee that babies born at the beginning of the twenty first century will live for an average of 100 to 110 years and possibly more, due to the accumulated knowledge in the medical field.

Thanks to the accumulated data on the prevention of certain diseases, scientists can already evaluate the number of years of life we can expect to gain. Changes in eating habits, life hygiene, the efficiency and specificity of new medication to control protein in urine, diabetes, heart diseases, hypertension and even certain cancers, should soon contribute to raise life expectancy and allow us to enjoy those extra years in good health. These diseases are obviously more evident today because man lives more than 50 or 55 years, as was the case in the beginning of the twentieth century. Heart diseases, strokes and cancers are the main causes of death. Scientists foresee that today's diseases will soon be a thing of the past, considering our new means of prevention and treatment.

Heart disease and cancer are much better controlled today than they were just ten years ago. Knowledge acquired in that specific field has increased tenfold. People accept to let go of their bad habits and opt for a more balanced lifestyle. Research done in numerous sectors related to medicine has accomplished wonders in this area. When we say "to live a more balanced life", we think exercise, adding certain trace-elements, vitamins, food containing more fibre, reducing animal fat and better controlling the necessary intake of elements essential to the cellular metabolism, such as fruits and vegetables.

Disease prevention

A healthy life style coupled with regular exercise is already a great start. Reducing alcohol intake, tobacco, more or less potent drugs, a well-balanced diet, new forms of therapies based on the prevention of diseases already in our genes instead of their treatment, the choice of foods and nutriments that we know are essential to our good health and to the prevention of disease, will soon make all these preventions well worth our while.

The use of natural antibiotics – by this I mean substances which possess antibiotic potentialities acquired during millions of years on earth – must be taken into consideration. Let us take, for example, the lacto-ferritine and the lacto-peroxidase in milk, the antibodies of the egg yolk, several other natural antimicrobial substances present in certain foods or the micronutrients capable of stimulating the immune system. The better informed and more advised persons already use these ingredients.

Life balance

We are beginning to understand a forgotten ancient rule of life which commands the respect of a certain balance that men of more ancient societies mastered more naturally than we do. The respect of that balance is a necessary condition to our survival. We must be careful not to break this fragile balance and transgress the rules of nature. Man will live longer and healthier only if he learns to respect nature and its balance.

We can observe that it is often worthwhile to modify our habits if we want to, and in so doing, return to the core of a more natural way of living in greater harmony with nature. "Transgression of nature's rules leads us inevitably, more often than not, to chaos", wrote Alexis Carrel in "*L'homme, cet Inconnu*".

The considerable increase in life expectancy over this past century is attributed to the acquisition of knowledge, the decrease in the spreading of infectious diseases and the prevention of diseases such as scurvy and goitre, by the simple use of vitamin C or in the iodine found in food. Prevention of tobacco's disastrous effects on lung cancer could serve as an example for learning how to live a more balanced life.

Man must now learn to use this newly acquired knowledge to better preserve his environment and his life.

Food

We no longer need to search for food for hours on end every day, as did men in prehistoric times, nor do we need to produce it ourselves as did our ancestors. Today, thanks to new means of transportation, all the food and micro nutriments from all over the world are available, at the corner store, at least in industrialized countries.

Regression of poverty

In spite of regional disparities where poverty might seem predominant, and although this is by no means an equal panacea in all countries of the world, one can think if not assert that, in spite of several incontrollable factors (ideological quarrels, religion, the disastrous domineering instincts of certain twentieth century men…), the material life conditions of most people on earth have greatly improved over the past fifty years. We can foresee an even greater improvement in the coming century.

Communications

Communication techniques have made possible extraordinary exchanges between the people. We are all familiar with the potential of radio, telephone, telecopy, cinema, video, television and especially Internet in this field. New technologies have made it possible to store, on data tape or information chips, an infinite quantity of information on the newly acquired knowledge in all possible areas of human activity and science.

This information is available to the whole world by the phenomenal technology of the Internet, and to each individual from the comfort of his own home. We no longer need to go to public or regional libraries to have access to this information: the information is here, it is available, in many homes on this planet. The availability and the broadcasting of this information to the whole world are ferments which increase, as never before, the need to acquire even more knowledge. Examples of this are already numerous. Let us remember that is has been less than ten years since the Internet has become involved in this revolution of knowledge.

One can imagine that these endless exchanges between individuals of all ages, all races, all cultures, everywhere, at the same time, all over the planet, could become a tremendous source of knowledge for mankind. This boundless culture is bubbling and ready to explode. Tomorrow's world will never be the same as yesterday's because of the forever easy access we now have to humanity's knowledge and heritage.

An already present future

New technologies are developed every day. Specialties, unknown just a few years ago, are already taking shape all over because of the new developments available to mankind. They are available to all human beings on this planet. Employment in telecommunications, in aeronautics, in pharmacology, in biology, in the media, in photonic optic, in robotics to name but a few, is here to stay and succeeds the industrial revolution that still marks the twentieth and twenty first century.

Teaching institutions still functioning as in the old days will soon be forced to adapt to the new methods or they will die. Too bad for the governments that delight in pedantic expressions and that are unable to make and take good decisions. Very soon, the free man will finally take charge of his own training and it will be a plus for mankind. Even the illiterates, who for some reason did not have access to such training, will benefit from it. All the knowledge in the world is now available to all individuals on earth without exception. Can we name just one area of human activity that cannot be found on the Internet?

Let us hope that certain narrow-minded and retarded leaders will quickly understand that their country is in urgent need of solid and well-formed minds for teaching. For education to become a cultural value, and for a country to become a centre of education able to overcome the enormity of the scientific innovations overcoming today's world, we need more qualified professors than incompetent civil servants, sticking their dead wood in the wheels of progress.

He who thinks that new knowledge and technological applications have come to a standstill is very wrong. The future is at our doorstep. Open your eyes wide and look…

In physics

A few decades ago, Soljenitsyne, the Russian physicist, found a way to store atoms in a magnetic field. Today, Californian scientists have discovered ways to carry protons across their laboratories. Others use magnetic fields to develop spaceships, thus accelerating the speed in space by an unconceivable multiple.

Scientists have already succeeded in carrying molecules. They are now working on the tele-transportation of human beings while others work on the antigravity belt.

Hydrogen batteries are recharged by solar energy. This research is being made at the Fraunhofer Institute. It is a technology that could considerably prolong battery life. The regular recharge of batteries in cellular telephones and portable computers will soon become obsolete. Casio and Siemens, notably, are working of the first prototypes of these chargers.

New fuels are also being studied, certain of which will enable us to use photonic energy to travel in space by using the sun as a source of photons.

In pure physics, ultra-fast lasers, made of ultra-short X-Ray pulsations, enabled a research team directed by Markus Drersscher, of Bielefed University (Germany), and Ferenc Krausz, of the Vienna University of Technology (Austria), to take photos of electrons of krypton atoms circulating around the nucleus and travelling at the speed of light.

Astro-physicians discover, identify, measure and survey the depths of the black holes within the galaxies. They have identified them as immense star gobblers and gas charged with incredible energy. When telescopes look far into the past, they can find the oldest star in the universe and can go even further back in time. Researchers in this field even believe that there may be several universes such as ours.

In biology

Since 1990, a public consortium on gene composition in humans, financed by eighteen countries, has the mission to establish, free of charge, the genetic mapping of the human genome. All the experiences acquired during the evolution of life on earth, be it pre-human or actual, are coded in humans in the genes of all body cells. Gathered in the form of three billion pairs of nucleotidic bases, this information is composed of deoxyribonucleic acid: the human and animal DNA. **This work presents the history of the great book of genetics on the secrets of life in both man and animal, and even in the history of life on earth**. It is the most imposing scientific piece of work ever to be done. The genetic mapping of the human genome is the first stage of this colossal project. It ended at the beginning of 2001.

In the following decades and centuries, researchers will identify some 150,000 groups of genes, which represent approximately 20% of the DNA found in humans.

During the coming decades and centuries, researchers will try to identify the functions of these 150,000 groups of genes, which represent approximately 20 % of the DNA ribbon in humans. These genes code day after day the millions of products and enzymes necessary to assure the multiple cellular functions in the bodies of living creatures. Scientists are already working at identifying the genes which code the production of the billions of enzymes necessary to ensure the good "behavior" of those organisms and of the protein synthesis, called the genomic of proteins.

The human genome and that of several other animal species are almost completely known today. One will have to wait to find out if the new knowledge will enable, in the near future, the cure of certain human genetic diseases. In spite of the hypotheses which will certainly not be proven before several decades, the "blind" economic forces of the stock markets world wide try to convince investors that these realizations will happen tomorrow. This knowledge, associated to the technologies already on the way, but will only open the way to new avenues in a quite distant future.

The magazine *Science* of January 25th 2002 showed that scientists from the National Centre for Scientific Research (NCSR) (Centre National pour la Recherche Scientifique) of France and from Harvard University in Maryland, U.S.A., are building a laboratory which is stocked on a computer chip. They have managed to set up an effective method for mixing substances and reagents at microscopic scales. This laboratory chip consists of micro canals in which are found the products intended to react. These techniques will soon allow the analysis of tiny quantities of products in any given sample. The abolition of turbulence, in such a small scale, is a problem which seems to have been solved by Armand Adjari, of the NCSR team of the Laboratory of Theoretic Physico-Chemistry, and Abraham Stroock, of Harvard University. These researchers have managed to engrave small grooves on one face of the canals, which induces a helical drainage enabling the mixture. This microscopic laboratory can easily be made in series (so it seems) and would save, according to the researchers, a surface equal to one or two rooms of a laboratory.

Genetic cartography

Sheep, cows, monkeys, mice and rats have already been cloned. It is naïve to think that human beings have not yet been cloned, even though governments forbid it and that fundamental ethical problems remain to be resolved. Several countries, such as Great Britain to name but one, have already accepted that their researchers work on cloning human cells at least for medical purposes. A Montreal researcher has even identified such stem cells in the skin of individuals, while another found similar stem cells in the dentine of the teeth, thus creating a fountain of youth, without having to use ovules and embryos!

The cartography of the genetic mapping and its identification to certain diseases offer little chance of a cure for human genetic diseases in the near future. However, it can already be established that human genes are 98% similar to those of apes and 70% to those of mice. For those who believe in creationism (the theory that God created man as he is today), this is an insult. But the evolution of species is now here, proven a million times over. About 98 % of the three billion human genes are exactly the same as the billions of genes found in apes. Whether we like it or not, this is indisputable proof that man is a descendent of the ape. Creationists must face the fact that their thinking must be corrected in the face of these new realities. Evidence now proves universally in thousands of different ways the falseness of this type of education. One must find the maturity to realize and admit it. It is also necessary for us to admit that, somewhere in the history of our development, the human branch has evolved alongside the branch of the other monkeys and even of that of the mouse and of the rat…

The identification of the millions of genetic codes, exactly similar, which links millions of pairs of nucleotide acids, in the same order, in different animal species, provide an incredibly strong argument in favor of the theory of evolution of species of Lamarque and Darwin. Paleontologic discoveries have placed at some 75 million years back in time, the branch which divides and separates rodent mammals from the main trunk which will become man, and the rodents possess a 70 % genetic resemblance to man…For the benefit of creationism, it is not compulsory to reject the fact that God has created life with all its possibilities some three or four billion years ago, and that certain forms of this life have developed into a human being who will one day become conscious of the good and of the evil, and we will call him "man".

In biogenetics

The sequencing of three billion chemical nucleotides forming the codes of the human genome represents a scientific breakthrough as important as the theory of the relativity, the domestication of the atom and the landing of man on the moon in 1969. Even if since his quest for fire man has not yet been destroyed by his creatures, numerous are those who think that deciphering human patrimony as well as human cloning could bring forth biblical images where God's wrath would descend upon us like the ten plagues of Egypt.

Man has not forgotten the terrible technological disasters caused by the bursting of the Banquio and Shimantan's dams in China in 1975, or the three thousand deaths caused by the toxic gas leaks in Bhopal, India, in 1984. The nuclear accident in Chernobyl, Russia in 1986 resulted in thousands of deaths. The nuclear power station was afterwards definitely closed in December of 2000 after (or so we were told) 300,000 persons had died… Man is quickly loosing faith in the sole rational intelligence. The possibilities of discriminating people because of their genetic profile or by the simple marketing process of genetic therapy which could follow the deciphering of certain sequences of the DNA of human genetic coding are frightening to many individuals.

Manipulating cells, chromosomes and particles of matter is no longer an obstacle. Scientists now use high technology microscopes connected to computerized systems to guide them, in order to manipulate the particles of matter. At this point, tiny particles of matter are moved by using special pipettes.

Artificial limbs

Today, marvels of medical engineering enable human beings to move with artificial limbs so perfect that even the "Six Million Dollar Man" would envy them. The only body parts that cannot be replaced as of yet are the brain and the spine. We are told that Christopher Reeves who played *Superman* on the screen, as well as many researchers, are working on the subject …

Technologies applied in medicine, in diagnosing diseases and in surgery have allowed the development of superbly sophisticated instruments combining the knowledge of several scientific fields.

In robotics

In 1999, the Japanese firm *Sony* has marketed 5,000 robotized metallic "AIBO" dogs. These robots are equipped with sophisticated computers able to recognize all obstacles in the home and in the environment, and also differentiate and remember voices. They also have the capacity to learn their owner's human feelings, to react like their owners and finally to store this information on a computer chip for further use. The knowledge of human reactions to such and such an event can be analyzed, synthesized and transferred to extremely more intelligent second generation robots. By integrating such quantities of human feelings, these robots could one day behave in a more human way than many humans…In May of 2002, *Sony* announced on the Internet that AIBO software would be available to those interested in adding their own programming to the dog's. In June 1999, when *Sony* launched its first generation of AIBOs in Tokyo, 3,000 of these "electronic animals" were sold in 20 minutes. Since then, *Sony* has sold over 100,000 AIBOs in Japan and abroad, according to Mr. Obana. In March of 2003, the company announced on Internet that the dogs are now equipped with cameras and can take photos or films and deliver them instantly through the Internet.

Bonzi Body, a small intelligent monkey on Internet, can talk, communicate with you and learn from you. For quite awhile now, this robot has been able to guide you through the Net, download your programs, execute them, check novelties and do your research … Incredible!

In 2003, portable videophones appeared on the market. Equipped with video cameras, they can <u>indiscreetly violate</u>, without their consent, the intimacy of the people wherever they are and immediately show these pictures on the Internet. Millions of copies were sold in the days following their appearance on the market. These portable photo-telephones, equipped with a camera and a video, can immediately send through the Internet the pictures taken by amateurs.

By learning human reactions and feelings and in a given situation, these robots will soon be able to react as we do in that same situation. "They" never forget and, even more, the broad range of reactions integrated tomorrow will represent a synthesis taken from thousands of human beings... Very soon, robots will be more sophisticated and closer to humans than many individuals today!

In nanotechnology

Nanotechnology already develops at a molecular scale, hard-wearing paints, extremely reaction resistant fibers, implants which restore hearing and sight, molecular computers, made-to-measure medications, nerves that self-repair, robots that constantly patrol the human body in search of diseased cells …

In technological applications

Robotized road sweepers and lawn mowers are already being used on our lawns and sold in stores.

The Japanese people are now sporting Dick Tracy watches. With these watches, the fictional characters of the fifties can talk and see each other from a distance. With these technologies, we can now immediately speak to and/or see each other, at any given time, both on earth and in space.

Paper-thin televisions that can be rolled under the arm have already come out of laboratories and are being sold in stores, for the moderate sum of 15 to 18 000$ (Can.)!

English and Israeli technicians have perfected a miniature video-camera one centimeter long equipped with a luminous source. When swallowed, as a tablet of aspirin, this miniaturized camera can film all the areas of the body where it travels. This information is then transmitted to a monitor where it can be read. "The Fantastic Journey", a film made in the sixties in which miniaturized doctors traveled in the blood flow to repair defective organs, has today become reality! Surgeries at a distance, using robots, are now done between countries.

For example, supports of one thousandth of a billionth of a meter (a nanometer) are presently built with chemical molecules on which it is possible to "attach" medications that can be injected into the body, penetrate a predetermined cell and deposit, with the utmost precision, a medication on specific cellular receivers. Experiments on these new particle carriers are patented and going at a good pace in research labs.

An IBM research team managed to make nano-tubes with atoms of carbon, offering a conductivity equivalent to that of a silicone transistor. This biomolecular transistor is much smaller than a traditional transistor and also less energy-consuming. This same company announced in October of 2002, that it had developed a computer "chip" on a carbon molecule 260,000 times smaller and equally successful as the silicon support.

Lasers and sonars, hauled by boats, can measure the degree of reflection of sound waves on the ocean bed. Coupled with seismic detection apparatus, these sonars are capable of mapping the ocean beds without using cameras. Seismic remote detection are already localizing with precision deeper and deeper oil spills in the North Sea and in the Gulf of Mexico using these sonars.

Sonars could soon be replaced by GPS transmitters connected to special software. Captains will view, on 3D screens, their ship and the state of the sea beds in which they navigate. The System of Location by Satellite, known as the GPS, will determine the exact latitude and longitude of a ship.

Soon, these technologies will enable us to discover the age and the species of trees that grow in our forests.

The car industry

Today's economical cars will average 60 kilometers per liter (200 miles per imperial gallon) of gas. Cars using hydrogen molecules are now appearing in the big cities. They can be seen in dealer shop windows and some of them are even used for public transportation in certain big cities of the world …

Certain busses and cars are already equipped with cells using hydrogen as fuel. These engines do not produce polluting substances: almost pure water comes out of the exhaust system.

In ballistics

An ultra light high technology material known as "Kevlar" can stop bullets.

The precision of rockets and spaceships no longer cause problems.

In data processing

In 1978, *Intel* launched the processor 8086 containing 29,000 transistors and able to execute instructions at the incredible speed of 5 megahertz (MHZ). In 2003, the Pentium 4 contains 55 million transistors and works at frequencies reaching 3000 MHZ.

Knowledge

Knowledge is an individual possession that can travel with us the world over. Scientists are overwhelmed with job applications. There are no limits in the field of science. The people in power are still unable to realize the full potential of this precious possession. And so, thousands of scientists from all fields leave each year the countries that formed and educated them at the cost of several thousands of dollars, to find work in other countries where private companies much better equipped with sophisticated labs will pay them what they are worth.

Germany did not want to be left behind while new technologies developed at the speed of light, so Chancellor Gerhard Schroeder announced a few years ago that his country now emits five-year work permits, known as "red-green" cards, which is the color of the social democratic environmentalist coalition. With this, they plan to attract at least 20,000 technicians of all sorts.

A growing number of sophisticated technological applications are unfortunately used to produce war weapons by countries whose scientists have the appropriate knowledge. History shows that a number of old technological applications, from the Greek fire to cannon powder and gun powder and then to the atomic bomb, were first used by the leaders of these countries as weapons to destroy human beings and conquer other countries.

Proof of the evolution since a century

It would be interesting to know if all that knowledge has really brought a "measurable" evolution to our world. Scientific works have enabled us to establish that in certain areas at least, there was in fact evolution for the best.

Calculations on the gross domestic product (GDP) made in constant dollars, show that between 1890 and 2000, the American GDP *per capita*, has effectively increased from 13 700$ (US current dollars) to 65 540$. These data indicate that the country is five times wealthier than a century ago (from the French daily newspaper *La Presse* of March 25th, 2000). Bradford DeLong, a Berkeley University economist, demonstrates with figures to prove it, that never before had humanity known such a period of growth as in the twentieth century. DeLong's data indicate that the United States is five times wealthier than a century ago.

This phenomenon is not unique to America. Inhabitants from Thailand or Tunisia also enjoy, according to DeLong, a productive potential three times as large as the average American did in 1900. The same conclusion applies to most developing countries. Although in the poorer countries these statistics don't make much sense because of social disparities, there is indisputable proof that, at least on the material plan, there has been evolution and progress in productivity during the last century.

Everything is relative, of course... In the case of a rich manufacturer of the nineteenth century, as J. P. Morgan, for example, DeLong indicates that, at that time, he could not go to the cinema, watch a football match on television, enjoy a movie on video, and that he had to travel for a week to go from New York to Italy, rather than one night.

In 1895, a one-speed bicycle required 260 hours of work against 7,2 hours today; a desk chair 24 hours against only 2 hours today; a tableware set, 44 hours against 3,6 hours; a hairbrush 16 hours against 2 hours; Encyclopedia Britannica 140 hours against 33,8 hours and a Steinway piano 2,400 hours against 1 107,6 hours today. All proportions kept, these calculations seem to demonstrate that there has been progress in certain areas, even though the poorest citizens of the rich countries are, in more ways than one, quite wealthier than the middle-class individuals of the nineteenth century.

Today, many middle-class individuals are forced to live in filthy and cold apartments. They live in worse conditions than farmers did a century ago. On the other hand, the poor now have access to much better health care. There are still regions, even in the United States, where infant mortality rate (an indication used to measure the quality of life), is still as high as 50 out of 1 000 births. From the 100 deaths for 1 000 births at the end of the nineteenth century, it has gone down to 2 to 3 per 1 000 in the majority of North America and in industrialized countries today.

"Everything is relative" concluded the article by exaggerating, "because Occidentals complain on a full stomach". Thomas Jefferson, as did Emperor Tiberius, found it normal to eat raspberries only in the summertime. Today, we complain when raspberries imported from Mexico in the winter are not sweet enough!

Warning

A growing number of technologies, already in place or about to be, will be used by individuals as a means of dominating others. Radioactive isotopes could contaminate, for hundreds of years, the ground of one country or the oil wells of another. Microwave rifles can already destroy any computer in an instant. Certain micro and nanotechnologies, much smaller than a speck of dust, can be used as a microscopic tool to destroy specific groups of individuals, oil fields, entire populations and even contaminate them with viruses, toxic particles or other specific molecules.

Bill Joy, one of Silicon Valley's renowned scientists and Director of research at Sun Microsystems, believes that the twenty-first century technologies, such as genetics, nanotechnology and robotics are so powerful that they can very well generate abuse and accidents *(Wired, American magazine, March 13th, 2000)*.

Scientists forecast such possibilities. Today's technologies are indeed more and more within everyone's reach, including terrorists and scientists without a conscience. Even without sufficient means or complex raw material, it is now possible, with only minimal knowledge on how to operate technologies, for an individual or a dishonest government to appease his needs for personal vengeance.

"By 2030", says the scientist, "it will be possible to build computerized machines by the lot and make them a million times more powerful than today's personal computes". And to think these are already 50,000 times more powerful than the first large computer systems of the sixties!

We are on the same wavelength as Joy when he writes that the enormous computer power, added to the progress in other areas of science such as physics, for example, will soon enable us to redesign the world for the best or for the worst...

Will September 11th 2001 be the beginning of this new era? **When we think of the sick madness with which scientists already use their knowledge to make viruses able to destroy computers, block Internet access to powerful navigation engines, or make sophisticated weapons of destruction, we can fear the worst**. By presuming the that past is guarantor of the future and that the human beings will be tomorrow what they are today, we can certainly assume that he who has such a mind will continue to use his evil spirit to destroy his fellow man.

And so we must, as of this day, worry that knowledge, added to more and more intelligent robots capable of learning and of storing man's knowledge and using it, will soon give these machines inconceivable powers. Man is already surpassed by his own intelligent machines. It would not be impossible to build, using these tools, for instance, genetically modified machines that could accidentally or deliberately manufacture epidemic weapons or specifically destroy certain human beings.

People being what they are since the beginning of times, it is certain that at this very moment, in some great big sophisticated military or private subsidized laboratory on the planet, dishonest scientists are already at work. These heads filled with knowledge are already planning microscopic war machines to destroy particular human beings or to eliminate predetermined geographic zones. How do you think the knowledge to guide destructive missiles, such as the "Cruise" missile, or any of the numerous data engineered with artificial intelligence was acquired? Important research army grants in computer engineering and in artificial intelligence were given to the researchers of our magnificent modern universities.

When I read these lines to a precious friend, she told me that if I could not publish these lines by tomorrow, they will already be outdated, because of the speed of evolution in all technologies. And man is just beginning to know how to control some rudiments of his material universe! No one will stop the development of knowledge. It will be necessary to either learn to use it or to die with it or because of it.

Thoughts on the implementation of all this science

Especially during the twentieth century, humanity witnessed much tangible progress for better or for worse. Accustomed for thousands of years to a peaceful way of life, man was suddenly projected into a dizzying physical modernization. He benefited from it even before learning how to use all the new technological tools. Numerous technological applications were developed because of this scientific knowledge, and the numerous changes which quickly took place in the world have transported man, as a leaf on a tree is transported by a spring brook... "A pupil in elementary school today knows certain truths about the material world for which Archimedes would have given his life", said Ernest Renant some decades ago in the foreword of " *Souvenirs d'Enfance et de Jeunesse" (Recollections of Childhood and of Youth)*.

With all this knowledge, children evolve more quickly than we did. And we evolved more quickly than our parents and grandparents. Why then do our children have more difficulty growing vegetables, recognizing a fruit tree as our grandparents did almost automatically? Why do they not know where eggs and ham come from? Why do the youngest in our modern societies think that tap water always poured from the faucets in their house? It is because history is not always taught according to the book, if taught at all! Learning and work have considerably been modified during this evolution. The knowledge that our children learn to master is different from what we learned at their age. Today, the young learn to master the sophisticated techniques they will need in their future profession. And although they are skillful at practicing these new techniques, our children have drifted away from nature which had always served man and that he knew so well.

Yesterday, we mastered the art of living with nature and little scientific knowledge. **Yesterday, schools taught the art of thinking and techniques that are no longer used today**. These "experiments" were probably necessary for the development of the world and the evolution of the people. Today, the young do not learn to think freely; they mostly learn the special techniques of the evolution of science and how to live in their

18

world of technologies which is in perpetual movement. Their thoughts amount to the scientific way of "reasoning". This is their specialty. They learn to master the large number of techniques needed for them to earn their living. The authorities made them understand that they had no time to waste on a general formation or to study man' less tangible thoughts and human values.

From this new knowledge emerged thousands of new professions and new openings in the work force. Children don't necessarily follow their fathers' footsteps, as they almost automatically did in the past, at least where work is concerned. They are more attracted today by the new technical and scientific professions. The young have learned to use computers in a much more efficient way than their parents ever did. They know that they will have to use this tool in their work, as their fathers used a hammer and a set square to build a house. The professions people practice today are less and less traditional. Their knowledge is not less practical: it is different and more intimately linked to the newly acquired knowledge.

Young people must study much longer that their parents to assimilate only a small part of the current knowledge. They will have absorbed more new knowledge than their ancestors or their parents, but these represent only a small percentage of the knowledge acquired to this day. The young are less fortunate than their parents in this respect, because **it is more difficult for them to have a general view of their universe**. Many teachers have little knowledge of the complexity of the world around them, so they can hardy communicate an accurate overall picture to their pupils. The world is becoming more and more complex. Some can feel it but cannot explain why. There are no landmarks because everything moves too fast. Nowadays, no one can know everything about his universe. We evolve in a universe of techniques very so different one from the other. Although we are compartmentalized and specialized, in order to understand, it is essential that we have a general macro vision of the world we live in, even though it is not perfect.

Although school is compulsory until the age of 15 or 16, it only teaches the basic rudiments of reading, writing, mathematics and some techniques necessary in our modern society. Trades acquired in secondary school enable young people earn their lives quite well, but very few possess a general view on life.

The first university cycle, or bachelor's degree, is hardly sufficient to allow an individual to acquire enough scientific and technical baggage to make a living, except in the professions related to specialized technical groups. In certain traditional and more and more restricted fields, such as general medicine, law and engineering or in more complex special techniques such as administration, data processing or psychology, students acquire enough knowledge to practice their profession. The art of practicing medicine or law where doctors and lawyers were favorably distinguished is today replaced by the study of technical applications in more restricted fields. Practitioners must more and more conform to the rules dictated by their peers and that are in no way based on science.

Because of the fast evolution of knowledge and consequently of new professions appearing everywhere, students know that they will need to work in two or three different professions during their lifetime. **If professions change as fast as knowledge is acquired, students should have access to a more general training before learning about sciences and all the techniques related to these given professions**. But current school programming does not allow students to acquire a general formation which would open doors to the future and give them greater autonomy in making decisions about their lives and their professions. **It seems to me that only a more general formation could allow individuals to make an easier transition in their professions, should this become necessary**. Unfortunately, we are so far away from this concept with the education by sciences today, because often the student chooses himself the techniques of his training. There is no more general training in schools. Young people arrive on the work force, heads filled with the sophisticated techniques required to practice their first profession. But they are insufficiently formed to face the problem of a change in professions, because they do not possess this general formation needed to quickly get back on their feet or to embrace another profession.

The people in power at the Departments of Education of Occidental governments have not yet really understood all the implications related to knowledge. Due to the speed of the changes and their slowness to understand and to integrate the new knowledge, the programs they put forward in education are already out-of-date when time comes to teach them, plus the fact that they are inadequately adapted to the changing needs of consumer societies.

It is rather easy to understand that under these conditions the education system is late in teaching new knowledge. Curriculum, planned by the people in power, is so far behind that it will always have a hard time keeping up the pace. The administrators are worlds away from education, and it is they who decide! Professors and **truly competent career advisors** should be in the forefront of the new knowledge in order to advise students more adequately. They should decide of the programs they want to teach. Professors waste precious years trying to convince incompetent authorities to adapt programs and implement more adequate strategies of orientation and formation for the students.

Learning how to think is not taught anymore. Only sciences and technology according to the scientific methods of thinking used by scientists are now valid, because the young people are formed to learn through science. Some 180 days of annual schooling may not be sufficient to learn modern knowledge and there is no time left for adequate personal forming. This lack of time for teaching is spreading everywhere.

There are numerous and measurable consequences of this in our western societies. The eldest as well as the youngest individuals have no time to adapt to this accelerated rhythm of evolution in acquiring the knowledge. And without sufficient training, it is even more difficult and eventually we find them dropping out of school…

The young are unable to follow the rhythm of technology, which seems to cause numerous psychological disorders, side effects and affect the people and the populations subjected to this accelerated modernism: dropouts, suicides, behavioral problems can most certainly be counted among these side-effects...

We shall see in the following lines, **how the physical as well as the spiritual internal universe of man has been deeply disrupted by being in contact with all this science and technology without a soul, and, paradoxically, by all the changes caused by knowledge**. These changes present themselves as plausible causes of the new and sometimes distressing human problems.

First of all, if all these technological novelties have greatly facilitated the life of man and society, they have also modified the delicate balance of man in his environment and have left tangible impacts on man and society. Secondly, in grips with the difficulties of adapting to all these rapid changes, man has not even had the time to really think about the repercussions of those technologies on his environment or on himself, and even less time to reflect on everything that destroys his environment and dehumanizes him.

And so, **when one analyzes the side effects of this prodigious material scientific evolution on the environment and the individuals, he is frightened to notice that so much material progress has not developed the human side of man himself**. Knowledge and technological applications have even destroyed what was deep within him, his humanity. It is indeed sad to notice that amidst all of these changes, individuals did not have the time needed to evolve of a single iota on the human plan. Consequently, the scientific and technical evolution regrettably became established, causing wounds in the balance between life, environment, man, his freedom and his society.

Let us see together how all this knowledge and technology have badly affected and continue to destroy our environment, life on earth and the deep humanity in man.

CHAPTER 2

The endless spiral of human aspirations

Introduction

The profusion of scientific knowledge has lead man to believe that science and technology would constantly improve his life. But it was not meant to be, for most things have a bad side as well as a good side. Scientific promises have only been partially kept, and those that seem successful often bring frustration and false hope to man and society.

The generations following World War II were the first to benefit from science, knowledge and technological applications. The people more aware of this conflict were too busy surviving to really appreciate and participate in this progress. The "baby boomer" generation, which followed the war, was one of great political and economical freedom. Nothing got in the way of those wanting to work at "fixing" the world destroyed by this conflict. **Nevertheless, the tremendous post-war expansion has simultaneously been beneficial and detrimental to the economy, to man and to society**. The not so attentive observer was simultaneously witnessing the progressive deterioration of the environment, the pollution of drinking water, the quality of the air, the fast degradation of human values and of humanity in man and society.

As a colleague states it in *The Medical Post (Canada) of June 11th, 2002,* "Science", writes Doctor Bill Lancashire, "really contributed in making life easier and more comfortable. More accessible means of transportation made it much easier for us to travel. And so, I left my family and emigrated from England to work in Australia. Communication with my family was broken. Even though my childhood is filled with beautiful memories, I lost contact with my family which I have not seen in twenty years! I tried to compensate this loss with new friendships, but they will never compare with or replace what I shared with my family. Scientific progress and smaller families also caused much distress in my community. New technologies forced the closure of small factories in my childhood village. To survive, people had to find work elsewhere, and this caused the destruction of my small community. It was a disaster. But at the same time, the pressures of modern technology (computers, cellular telephones, etc.) ironically made life easier".

About the effects of technology on medicine, Doctor Lancashire mentions that if scientific evidence is essential to good medical practice, it does not apply to alternative medicines which accept facts on a rational basis. The English scientists have convinced, for example, the National Health Service of Great Britain to authorize the use of alternative medicines which do not require the scientific evidence of conventional medicine. "Protest is good", he writes, "but it should not be made in a passive way to the detriment of science by rejecting progress and knowledge. Science should be active and implicit in working against injustice, poverty, violence and child abuse".

We have used Doctor Lancashire's comments regarding certain side effects of science on his life to introduce this following chapter, **for he describes well how knowledge and technological applications stemming from science have also shattered the environment, man and his way of life and man and his society**.

New knowledge is in logarithmic expansion since the mid nineteenth century. The new technological applications have no limits and have created incredible expectations. Factories for products of all kinds are using this scientific knowledge and technology. Business is globalizing and telecommunications are now computerized. Carried by the pressures of all these novelties, men and societies are changing before our very eyes. This rapid evolution of science and technology has deeply transformed life habits in man and society. Let us take a look at how this evolution in science and knowledge has affected the environment and left its deep mark on the life of man and society.

The situation today

At the beginning of the twentieth century, the washboard is replaced by the electric washer. Houses now have electricity. As World War II begins, rural Quebec in Canada completes its electrical installations and simultaneously, electricity becomes available all over the world. Electrification brings tap water into our homes, driven by pumps and electric engines. Electric refrigerators replace iceboxes. Electricity brings light and replaces candles, while several electrical devices appear in the homes... Bathrooms with toilets are built indoors. Hot water tanks are warmed by electricity, gas or oil. It is no longer necessary to heat our bath water over the wood stove. Toilet water can be flushed...no more running to the outhouse!

Just imagine, for the first time since the beginning of the world, water pours from faucets in all the homes in cities, countries and even in the stables in America and in Europe just a few years later. For the first time since the beginning of the world, no more nauseous odors from the outhouses dug in back yards; they are now filled with earth and have become a distant memory... Chamber pots are a thing of the past and to some, collection pieces... Today's houses and even attics are more modern than the castles in which kings lived just a few decades ago. The motors of electric pumps have forever modified the habits of families and society.

On the farm, tractors replace horses. Cars and steam locomotives now travel countrywide. Planes fly over our heads. Engines have improved, are more powerful and transportation is possible over very long distances.

Using new sources of energy, boats and ships are bigger and are more powerful. Trains and locomotives have increased in number since the mid-nineteenth century and are now found in all countries. Cars, trains, roads and highways are more and more sophisticated and are seen all over the Occidental world as everywhere else. At the end of the third decade of the twentieth century, planes can fly across seas and continents.

Even with all the improvements brought on by this fantastic physical evolution, people are not satisfied. Like children, the more we have, the more we want. More factories are built to meet these demands. Cities grow and people leave the country to work closer to these factories. Several new products can easily be found on the shelves of our shops. But, and there is a but, at the same time, **more and more toxic substances are used or ejected by these factories and added to the herbicides, pesticides and chemical fertilizers used in large quantities on private farms and by important producers,** all these substances further destroy the environment and intoxicate the people...

The migrating of populations

In the early twentieth century entire populations left the country where they had lived for centuries, and began migrating towards industrialized cities in order to find work in factories where all the goods people needed were being produced. This migration from the country to the city went on for decades, emptying the countryside of its populations, destroying several families and many small villages. Large cities became crowded with people looking for work. Both the environment and the former ways of life were destroyed, increasing the demands in sanitary accommodations and water supplies, as well as extra food for such large populations. Science and technology are greatly responsible for all these changes...

New professions and trades appear in the workplace. Even education methods are changing to better serve these new consumer societies, but this was too often done, as we will explain later on, to the detriment of the humanity of man. Farmers, short of hand power, modify their food producing techniques and factories modernize their production techniques to keep up with the demand. Except for the Great Depression, which remains a period of adaptation to this great production period and until the eighties, the world experiences little unemployment. But, with the improvement of technology and the machinery that eventually becomes robotized, unemployment begins to show its head at the horizon, **while water, air and earth begin to show the first signs of pollution; by-products of consumer societies change, and polluting products of all sorts invade the air, the water and the earth. And simultaneously to the destruction of the environment we witness the loss of humanity in individuals**.

Influenced by the new technologies which influence methods of marketing and advertising and which are repeated and repeated *ad nauseam* on television, in newspapers, on billboards, everywhere including universities walls, people keep buying new products, even those they never thought they would ever want or need. To satisfy never ending desires for these new products, industries must increase their needs in energy and its toxic by-products. Everything is set up to exploit human weakness, to create debts and keep the economy running, and in so doing, further destroys the environment with new toxic products.

The credit

People's enthusiasm and their desire to acquire new products are limitless. To enable them to buy what they cannot afford and still keep companies in business, financial institutions have created new forms of credit. Credit becomes more and more available to everyone at the end of the sixties. Credit keeps consumer societies going. Without credit, society would run at a very slow pace.

Credit cards then make their appearance, replacing cash. Banks, gas stations, chain stores, all go out of their way to make it easy for anyone to obtain these cards. Interest rates on credit cards vary between 18 and 30 % a year, sometimes more, and people easily forget that one day soon they will need to repay this money. Credit is easily available to everyone. We use it to buy car(s), television set(s), audio-video system(s), household appliances, furniture, swimming pools, etc… People travel on credit. They eat on credit. They get all the products they want on credit… In 2003, it was reported that 28% of American children are not aware that a credit card is some kind of loan which will one day soon have to be repaid.

To give their citizens the services they often do not need but keep on demanding, governments also depend on credit and must borrow billions of dollars. Without this credit which suffocates people and will soon choke the governments, consumer societies would grow much more slowly. Individuals, cities, institutions and governments, all are in debt. In no time, individuals and governments are saturated with huge debts. For years, nobody thought they would have to repay these debts! Certain not so conscious scientists in economy even told governments and people not to worry, that debts need not be reimbursed right away, because inflation and devaluation of money would eventually erase the debts... But the debts continued to increase and many were never able to free themselves without declaring bankruptcy. The European countries have established that the debts could not exceed 1, 3% of the gross domestic product (GPD) without facing bankruptcy. In the United States, the excess of the GPD is 3, 5%, which means that this country is spending more than three times its production income each year. They are well on their way to bankruptcy...

Governments use the money from our highly taxed services to repay their debts. In some countries, governments spend between 30 and 40% of the income tax money to pay back their debts. Governments only have to increase our taxes to survive, whereas individuals only have their salary to lighten their credit burden...

Chaos is here and problems accumulate

People wear themselves out at work to buy new products and to pay all their debts. Some work at two jobs to make ends meet. Many kill themselves at work and become overly stressed. Values and rules of conduct tend to disappear. The social wheel turns in circles and amplifies the chaos in individuals as well as in families and society. In spite of this, Union employees require less working hours, more leisure activities and always more money. Both the workers' and the governments' problems accumulate. Created decades ago to bring workers together and help them negotiate better working conditions, Unions finally went too far and kept asking full fringe benefits and larger and larger salaries. **Often incapable to increase salaries and social benefits, factories closed their doors and moved to other countries looking for cheaper ways to produce their goods.** This is what is happening in USA in 2003. And jobs were lost for good while companies moved to less demanding countries...

During the second half of the past century, for instance, governments in occidental countries sink more and more into debt. **The social measures put forward to satisfy the demand for services become overpriced.** Income taxes were on the rise to pay for the implementation of expensive social programs. In many occidental capitalist or socialist countries, over

70 % of the money earned by the employees is deducted in various forms of taxes. In certain socialist countries such as Sweden, which socialist occidental governments imitated during years, the governmental powers monopolize up to 90 % of the workers' salaries to offer various social programs. Even though these governments take care of many aspects of the people's lives (health, schooling, transportation, etc.), these people have lost their freedom and their right to oppose their governments' extravagances.

The last two decades of the twentieth century are characterized by several personal, private or public bankruptcies collapsing under their debts. Several countries even went bankrupt. Countries supporting these enormous debts become the new owners of these countries which lost the power over their own countries. Many African countries and some South American countries are in such a situation today.

The twentieth century ends in challenging trade-union fights, in associations of syndicates with governments, and in attempts to make compromises to prevent such disastrous situations. At the end of the last century, the industrial revolution, the knowledge revolution and the science revolution added to the numerous changes in society created social conflicts between employees, families, municipalities, regions, central governments as well as between other poorer countries.

The exhaustion

All these changes tire out the citizens as well as the governmental powers. To get away from the people's demands, governments start building intermediate structures between them and the people. These intermediate structures are not empowered to make decisions. They are still there after 20 years, with twice as many civil servants as before. Incapable of keeping up with the new technology, these useless intermediate structures block everything. Their jobs are protected by powerful conventions and so they remain in place. Governments implement everywhere linear and logical economic plans, apply administrative techniques without human sense, and turn into chaos all the problems they attempt to solve.

The abuse

Power abuse is found everywhere in governments. Costly inappropriate structures exist in all areas of state administrations, where dozens of employees continue to receive their full salary without having to work, because their employers are subcontracting to independent companies. All these structures increase the debt of governmental administration.

And many countries, states, cities and institutions continue for years to get into debt and to borrow to repay the day to day administration which we commonly call "the grocery bill".

In overly indebted countries, money looses its value. Economists are working at creating a more reliable accounting system to calculate the worth of the monetary unit. Private and governmental debts reach proportions well over the paying capacity of the citizens, and in certain countries, beyond the production capacity of their GDP (gross domestic product). To survive, many companies falsify their accounting procedures such as at "Enron / Anderson", Sprint PCS, JDS Unisphase, Adelphia Communication and dozens of other large companies in the Occidental world, undermining the investors' confidence and thus contributing at further destroying the capitalist as well as the socialist systems.

The debts

During the last two decades of the twentieth century, many countries went into monstrous debts exceeding more than two to three time their GDP, indicating that their debts exceeded their production capacity by 2 to 3 times. These debts burden the future generations and discourage the younger generation who will have to pay them back. What is most discouraging to the young people is that their governments continue to spend without paying off these monstrous debts, this being added to the cost of holding these debts. At the beginning of 2003, the debt of the American government reached trillions of dollars or thousand of billions (1,800,000,000,000.00$) of dollars totaling tens of thousands of dollars per individual, including the children and the elderly. These debts exclude those of individuals, cities, states... and represent the largest debt ever incurred by a country in the history of humanity.

At the end of the century, more than 95 % of the people found themselves in unthinkable personal debt, some owing more than their annual salary to banks or credit companies. Combined with municipal, state and national government debts, the all-around American debt amounted to quite a bit of money: a trillion dollars altogether, which is roughly 100 000, 00$ per individual, in that country alone!

In several other countries, the debts also exceed the production capacity of the country as well as the capacity of its citizens to repay. Many countries and companies were or are on the verge of bankruptcy and many hundreds of such companies are in that condition in United States and abroad...

Banks lend more money than what they have, also risking bankruptcy (12 times more in the case of the J.P.Morgan American bank). This is several times the bankruptcy limit for this American bank. In Argentina and in other countries, governments must use their citizens' savings to finance their own administration. They oblige banks, by law, to refuse to citizens the access to their own savings account for a while. Not being able to use the money from their own bank account has ruined many individuals, reducing them to begging or to simply leaving their country for more favorable skies.

Government loans are often owned by groupings of pension plans, banks and foreigners. And so, financially, these countries eventually belong to the foreigners who own more than half of their debts, and those creditors use their new power to control these deeply indebted countries.

This stratagem has been used for decades in certain less fortunate countries. As I write these lines, Japan, an extremely prosperous country less than two decades ago, has been in permanent recession for the last 13 years. The economy is literally collapsing under the enormous debts of banks and companies. The country itself could even go bankrupt if it became clear that it could not repay its debts. Many economists took a long time to realize that the monetary system of the world will have to be rethought, because the leaders can print money as they want.

Globalization

During the last decade of the twentieth century, economists imagined stratagems to pay off these enormous deficits by diluting the debts. They imagined the world as "a global village" without borders, in which products and savings could be exchanged. And here we are, at the beginning of the twenty-first century, on the edge of ruin, in the global village. And we looked at this "drunken ship" coming at us without even reacting!

After the incredible effervescence in the last decade of the twentieth century, stock markets of industrialized countries are on the verge of bankruptcy. Nevertheless, they carry on their crazy momentum into the new century, drawing other countries into this whirlwind. Stock-exchange securities reach 500 – 1,000 times, and sometimes even 1,500 times their company profit. Investors live in this bubble threatening to burst at any time. Unfortunately, few persons realize that they are living the lies of companies and of their government... Large companies have tilted big stock-exchange indications, confusing economists, bookkeepers and young crooks who know nothing of history. Several countries find themselves on the verge of an economic crisis, and the world economy is tottering under the burden of the debts of governments, states, cities and people throughout the occidental world. There is no sane driver to control this mad train of debt. During this period of stock-exchange madness, people keep on gambling their savings, as their grandfathers did before 1929 (the Great Depression), and lose their lifetime earnings in no time. A recent example of this occurred in 2002 when 3, 8 trillion dollars was suddenly removed from the NASDAC New York Stock Exchange within a few weeks, leaving thousand of individuals, institutions and private companies losing billions of dollars in pension funds.

The administrative, economic and sociological sciences which are not, strictly speaking, sciences, but rather a cluster of techniques, try to use these dubious scientific techniques which are in fact only based on human behavior, to steer economy. These techniques propose weird solutions to very real administrative problems. Even though history shows that there is nothing less scientific than the human behavior in a state of panic, governments cling to these desperate solutions and try to save themselves from disastrous economical situations. Some believe this, others do not.

The program

And so the twentieth century ends in an apotheosis of unprecedented materialism. Individuals, societies and industries collapse under the goods produced for the whole world. The "baby boomer" generation sees "BIG". **Always in exponential growth, factories continue to automate and robotize. Pollution increases all over the world**. Norms, laws and governmental efforts to control and limit thousands of polluting agents poisoning the planet could not manage to control the mess, but who cares? Because of this limitless production of goods, pollution increases all over the world and this, in spite of all the political speeches. **We prefer closing our eyes in order not to see what does not affect us personally at this very moment.**

The leaders of democratic countries have found the solution: produce for the entire planet and develop the international exchange beyond what is humanly possible. And so Nortel, a gigantic technological Canadian company which almost went bankrupt at the beginning of the new millennium, as hundreds of others in the United States, built in China, in just a few weeks, a gigantic factory of more than one kilometer long. Often, these factories continue to pollute the planet, but not in their own country, so no one cares. The real reason is that the expansion of worldwide production brings a lot of money to the indebted countries that have their goods produced in poorer countries where salaries are less than nothing.

And the occidental governments go on spending as if they were still rich. And so life goes on… Societies are in full decline and on the edge of a worldwide recession, but drowned in their ocean of material assets, neither the people in power nor the people can see it.

All over the world, the air, the water, the earth and life on earth continue to deteriorate. Shops are loaded with goods. **New diseases, caused by the toxic pollutants released by these factories, start to appear**. Medical needs and medication are on the increase, men are poorer and, as the wheel continues to turn, the closer we are to the breaking point. Stress and depression have become routine, nervous breakdowns occur all the time. Medication against stress has become more and more necessary and represents over 30% of all prescriptions drugs in a year in America, and so we find extremely rich pharmaceutical companies and shareholders throughout the world.

The materialization of society reached summits at the beginning of the millennium. In occidental countries, the God that was ever present in schools, in classes, in municipal courts, in cities councils, in assemblies, in institutions and in government buildings, can no longer be seen. The secularization of schools, of governments and of institutions was one of the most outstanding phenomena in the materialization of individuals and in the dehumanization of man and his societies. Atheism flourishes every where. Associations were even formed to de-baptize those who were baptized at birth!

Flooded with science, knowledge and scientific applications, intellectuals and scientists keep on denying the human identity of man and to materialize him. Education is becoming so "scientized" and "technologized" that scientists, educated without the humanities, do not even know that there exists other ways of thinking and of teaching than their own, (*Can Science Save its Soul, by Mary Midgley, New Scientist, August 1ˢᵗ, 1992)*.

In the following pages of this work, <u>**we will brink proofs that science and technology does not only destroy the environment, but also the humanity in man and in his world**</u>. First, man lost his soul and now he is losing his mind. "Nevertheless", says Fukuyama in his book *The End of History and the Last Man,* "liberalism has clearly established its superiority by defining itself as the ultimate great find of humanity".

Let us see now how the disastrous side-effects of science and technological applications have succeeded in destroying our physical environment (Chapter 3), and have largely contributed to materializing and dehumanizing individuals and societies and their deep human spirituality (Chapters 4 and 5).

CHAPTER 3

The negative effects of technological applications on the environment

Introduction

Ever since man began to evolve from the animal state to the human being as we know him today, he is forever getting engulfed by the knowledge offered to him by science on a silver platter. He enjoyed distinguishing himself from animals because of his capacity to abstract and generalize things. He chose to subject all his actions to the uncompromising logic of his intelligence, losing the human side of his being that was filled with feelings and love. Unfortunately, when he **deliberately chose to accept as the only truth the inflexible logic of his intelligence, he rejected at the same time the truth about everything he could not feel with his senses** and so he lost the humanity he had acquired all along his history.

He put all of his faith in his god: science, a science promising to prevent and to cure all his diseases and to make him immortal. Science, indeed, had the mission to discover all the secrets of the matter to assure man of an easier life. Science would bring him happiness, time, comfort, reduce his workload, take care of his physical needs and give him his delay bread... **These promises were only partially kept, but at what cost to the environment, to life on earth and to humanity!**

Man made a god of science and technological applications because of the numerous benefits that improved his physical and material world. He does not dare blame this same science for the disastrous side effects to his environment and to his humanity. As geographer Paul Adam said one day, each technological progress, be it the telephone, the train, the plane, television, the Internet or the thousands of other scientific applications, **all have produced their share of pollution, modified man's relation to the environment, to other people and to society.**

One can hardly imagine to what degree technology has been both beneficial and harmful to life, after being repeatedly applied with no human conscience. Before elaborating on the damages caused to nature by these technological applications, let us briefly recall the elements necessary to life.

A. The elements of the life

We often forget that, in order to survive, men, other animal species and plants absolutely need the following essential elements: water, air and earth.

The wisdom of primitive men was expressed in this very simple way, and to this day it remains true: **if man contaminates water, he will die, poisoned by the polluted water. If he contaminates the air he breathes, he will die suffocated by breathing the air he himself polluted. If he contaminates the earth, he will also die, poisoned by the food produced in that earth.**

No one had ever even dared think that if man simultaneously poisoned the water, the air and the earth, he and other forms of life in his universe would be doomed to a certain death. It seemed so unthinkable just a few decades ago that no one even considered it. But today, with over 6 billion persons on this planet and the large production of goods and food needed for all these people, everything has changed, everything is off balance. **Man has infringed the limits of decency. Unthinkable errors leading to the destruction of the earth, the polluting of the air and of the drinking water, happen every day under our very eyes, as if it were normal**... Man has polluted his world so much during the industrial evolution that any form of life on earth is doomed to disappear, sooner or later, if nothing is done immediately to stop this polluting of all the elements necessary to life. Thousands of the more fragile plants and animal species have already disappeared from the surface of the earth and thousands of humans become sick and even die every year because of pollution. This chapter will be an eloquent testimony of the abuses of man on his environment.

We are just becoming aware of this phenomenon of the destruction of our physical environment as being connected to life itself. This link is more and more unstable and weak. And this fragile balance is intimately linked to man's interventions and to his inconsiderate application of technology. Some technological applications put at great risk the delicate balance between life and the environment. This is all taking place in front of our very eyes everyday and is in full expansion.

The governments' attempts as well as those of several groups trying since several decades to right these mistakes seem little more that drops of water in an ocean. In spite of the very few advantages of technology, the water we drink, the earth producing our food and the air we breathe continue to deteriorate day after day. The growing load of toxic and polluting products from factories producing goods, from agriculture producing food and by several unconscious interventions of man on the environment is forever increasing.

Over 100,000 products are used every day by factories to produce the variety of goods people need and require for their daily living. **Many toxic and non biodegradable substances that are needed to produce these goods are discharged in the environment.** New ones appear each year to swell the ranks. Thousand of these substances have already caused irreparable damage to the life of certain animals and plants of our planet. No governmental agency is able to regulate such large quantities of substance. Each one of us should be aware of this and act accordingly. Rivers, lakes, brooks and seas are filled today with loads of these polluting products and are dangerously increasing. The fauna of the rivers and lakes is considerably modified by many of these products. Different forms of life in these habitats are being transformed under these toxic polluting products and specific modifications are seen in their cells. The genes of these cells can be modified by all this pollution, which in turn change the body cells and cause the death of humans and animals in a more or less short period of time.

In spite of all the interventions and the insufficient and binding laws imposed on the industry, thousands of tons of toxic products are being poured in the environment every year, and each day the water, the air and the earth of the planet are more and more filled with these toxic matters. By contaminating the elements necessary to life they contribute at weakening a little more every day the delicate balance on our planet. **The air, the water and the earth contaminate today at an alarming rate**. This pollution creates conditions favorable to the development of new diseases in people, animals and plants.

The scientist is mostly to blame for all this pollution, because he sets up technological applications in the most inconsiderate way. The scientist is a special human being, believing only in tangible and palpable things. What he cannot see, touch or hear does not exist for him. The scientist works only with matter, which is his daily bread and that he worships as a god. He is the sorcerer's apprentice who gives us the material knowledge necessary to the proliferation of the technological applications that improve our lives. The following pages will illustrate how these high priests of "god the matter" are responsible for most of the observable destruction processes threatening life on our planet. They are also the ones who acquire the knowledge on our material environment and the ones who destroy it... Scientists act as the avenging arm of "god the matter" in the same way the great mullahs killed people without discrimination in the name of their god!

Let us first recall the usefulness of the elements to life, so that we can better understand the importance of their delicate physical balance in the survival of life on earth.

Life

Seen from space, the earth appears as a magnificent sphere speckled with blues and whites floating in the infinite sky. Forever in perfect order, it turns on its axis over every 24 hour period and goes around the sun in exactly 365 days and a quarter. The terrestrial biosphere travels in a very precise way in our galaxy which contains billions of stars floating amidst thousands of other galaxies and unknown worlds.

Life has been developing on earth for billions of years. "Everywhere I glance", wrote a wise man, "I see life". Life is everywhere, powerful and overwhelming: it is on the mountains, in forests, on plains, in the deepest ravines and in the strangest places... Life is there, in the layer of translucent gas surrounding earth a few kilometers above our heads, as well as in the deepest places on earth, in the craters of volcanoes and in the depths of the seas... Millions of forms of life are found everywhere. Every single square centimeter of the earth's surface is covered with multiple forms of life of great biodiversity; even in extreme conditions such as in active volcanoes live micro-organisms can be found. Life can survive at over 200° Celsius, as well as in the cold boreal seas, in the ice of the North or the South Pole and in the large and small equatorial or Scandinavian forests. All forms of life can be found in an incalculable number of animal species who manage to reproduce in such extreme conditions. Life is also present in the phytoplancton and the tiny zooplancton of the ocean, as well as in multiple living bodies who adapted to living in extreme conditions (the extremophiles), since over 4, 2 billion years. Kilometers below the surface of the sea, in the crevices of mountains, scientists find micro-organisms very much alive! Over 1, 5 million different species of life have so far been identified by biologists, and this would represent about only 10 % of the existing species, or so they think. This is the biodiversity of genes!

Paleontologists admit that during its evolution, the earth underwent six great periods of massive destruction, all caused by cataclysms, the last one in the Yucatan Peninsula, after the fall of a meteorite 65 million years ago. Biologists believe that this accident is probably responsible for the disappearance of dinosaurs. According to certain scientists, today's world is most probably in the midst of the seventh great period of destruction of life. But, unlike the previous periods, this one is not caused by a cataclysm or a volcanic eruption, **but by the numerous inconsiderate interventions of the people on their environment.**

The water, the air, the earth and the mineral salts they contain are essential elements to the continuance of all these of life. Contamination of these essential elements with toxic waste products discarded in the environment by man himself has already destroyed thousands of species of life and endangered thousands of others, such as our own. **Biologists estimate that human intervention on the environment is causing the destruction of species at a rhythm 100 to 1,000 times greater than before the beginning of humanity.**

With these species gone, billions of genes have disappeared, much of which we imagined could have been used to fight diseases such as cancer, to control pathogenic micro-organisms or to insure the biodiversity of various animal or plant species. The world thus became impoverished at the same rate as the destructions of these various forms of life. Man took hold of the earth by using only his intelligence, as our governments seized their citizens by using arguments based only on the intelligence of the brain without using the intelligence of the heart.

Water, the problem of the twenty-first century

Water is indispensable and necessary to life. Seventy percent (70 %) of the surface of the earth is made of water. This water is salted and only about 1% of it is fit to drink. Over 70 % of our body weight is water. Life began in water. This is why water is essential and intimately linked to life. It is also why cities all around the world are built near water sources to which people can have access and which is an essential source of life.

When we first buy a lot or start building a city, the first thing to do is to find the water supply. It is then compulsory to determine if this water is drinkable, if the source can reasonably meet the requirements of the population and prevent the possibility of contamination. In today's cities the situation is different. No one thinks about the water he uses every day. People take for granted that water will naturally be there and that they will always be able to use it...

Formerly, roman aqueducts brought water into the homes, but this was not the case everywhere. Today electric pumps push the water into the faucets. People do not have to pump water with their arms anymore when they need it. When wells dried out, they often had to go several kilometers from their homes to get their daily water supply. Certainly these people were more conscious of the value of water in their life than today's people who just have only to turn on the faucet to fill their glass with water.

Scientists have established that drinking water will be one of the major problems of the twenty- first century. No one thought a few decades ago that the world would be short of drinking water. Now, we are quickly running out of reserves which decrease more quickly than populations grow. From 1950 until 1999, the water supply for a family of three passed from 17,000m3 to 7,300m3 a year, and could become as low as 4,800m3 per person in 25 years.

The World Council on Water, made of international experts, reported in November 2001, (*cf. Le Journal de Montréal, (Daily Newspaper), November 12th, 2001*) that over than half of the five million lakes and reservoirs of the planet face important ecological threats. Besides pollution, lakes undergo excessive diversions of their waters to irrigate the ground. These combined effects present a major ecological threat and, because of the increasing population which is estimated to reach 8.5 billion individuals by the year 2025, this human pressure will be even more threatening. "The senseless use of natural resources increases at such a rhythm that it will kill the environment which supplies our resources and enables us to live" said William Cosgrove, Vice-President of a committee on the World Council on Water held in Montreal in 2001.

The final report of this congress described exactly which zones are most affected by a lack of water. The quality of the water of Victoria Lake, Africa's most important stretch of water, is degrading so rapidly that several species of fish have already completely disappeared. Lake Chad in Africa has reported serious subsiding of the water, while the water in the Amazon River is heavily taxed by agricultural activity. Between 1850 and 1980, over 534 lakes have disappeared in China only, because of the diversion of their waters for irrigation purposes. In the Great Lakes separating the United States and Canada, airborne toxic matter, artificial fertilizers, pesticides and animal waste are the main sources of pollution. The lakes most threatened in industrialized countries are the ones in areas of extensive agricultural activity.

The same World Council on Water repeated, **in March 2002, that 1.2 billion persons in the world are suffering because of the lack in drinking water supply. By 2025 the Council foresees <u>that over than half people on the planet will suffer a lack of drinking water</u> if this tendency is maintained.** Even certain regions of Canada could face shortages in drinking water, confirms scientist David Schindler from Alberta, Canada (*La Presse de Montréal, March 7th, 2002*).

Drinking water

In the twenty-third edition of the summer campaign on the Economy of Drinking Water, published at the beginning of August, 1999 by the Ministry of the Environment of Quebec, we learned that 5% of the drinking water is consumed in the mining sectors, 46 % in manufacturing and 40 % by municipalities to meet the needs of people, institutions and businesses.

According to this document, the residents of the Province of Québec consume about 800 liters of water a day, compared to 522 in nearby Ontario. The campaign president said in a radio interview that in Germany, one uses only 150 liters of water a day per individual. It seems that we distinguish ourselves as champions for wasting drinking water. There is in Québec an abundance of water: lakes, brooks, small rivers, a majestic river and finally the ocean. Some years ago, no one would have imagined that Québec would one day lack of drinking water. But we have already exceeded the reasonable acceptable and are still acting unconsciously by managing to contaminate it, to pollute it and to consider it as an inexhaustible good.

The figures of these annual reports show the distribution of our excessive and daily consumption of treated drinking water. Each day we use 25 liters of water a person to flush the toilet, 95 to 189 liters to shower, 8 liters to wash our hands, 4 liters to brush our teeth by leaving the faucet open, and 38 liters in the dishwasher compared to 76 liters when dishes are washed in the sink. We use 136 liters for a bath, 25 liters if a faucet is leaking and 260 liters if the toilet leaks.

Filling the swimming pool requires 54,000 liters of water on average and we must figure another 22,730 liters for maintenance. Lawns require 1,000 liters of water every day of the summer, which accounts for 50 to 70 % of all our needs. Washing the car in an automatic car-wash requires 30 liters of water. If washed by hand, it amounts to 300 liters...

On the wasting of water

The water in our homes has been treated, filtered and disinfected in expensive and gigantic installations before being forwarded to us. Water is a tangible, daily and important necessity in everyone's life on the planet. Drinking water enabled us to become physically healthier. Still, we are the champions in wasting drinking water...

This data of our shameless use of drinking water in Québec is confirmed year after year. Wasting drinking water is a crime against nature and we waste a resource which is not ours. It is reasonable to use water is for gardens, leisure, meals or for hygienic reasons. But must we take two ten minute showers every day or use 20 liters of filtered water each time we use the toilet? Must we water our lawns with treated water? One day, authorities will have to install systems to control drinking water, because of our lack of public-spiritedness towards this natural resource given to us by nature.

Every day, we waste millions of tons of water treated for consumption in completely unnecessary ways. The article of July, 1999 adds that Quebeckers spend over 20 % of their drinking water for reasons as stupid as washing their parking lots, watering the lawn or for other unreasonable purposes. We have all noticed that even when the medias warn us over the radio and on television that watering the lawn is forbidden or restricted on very dry summer days due to a lack of water, a considerable number of our fellow countrymen, with no civic sense, continue to water their lawn anyhow, some going as far as hiding at night to do so! What irresponsibility!

The pollutants mainly of water

To sterilize the drinking water we have in our homes at great costs, technicians must add chlorine to the water. **Chlorine disinfects the water to eliminate pathogenic microorganism that could cause diseases.** But these treatments do not mean that the water called "drinkable" is exempt of pollutants. This "drinkable" and treated water can contain loads of more or less toxic substances which should not normally be there. Depending on the time of year, the proximity to factories and in spite of much precaution, drinking water is filled with large quantities of soluble, insoluble or toxic substances from manufactured goods, agriculture, drainage water, sewers, pharmaceutical and industrial products, etc.

Technocrats tell us what they want on the subject of drinking water. When speaking about water on television or in newspapers, they only give the results of the concentration of the E coli bacillus in the water. These analyses consist in looking for and counting the fecal coli Bacillus per each 100 ml of water. They only speak of the bacteria from people or animals feces that can cause diseases or death. In certain larger cities, the concentration of witness viruses is measured, such as the three known viruses of poliomyelitis, as a means of controlling other viruses that are generally killed by the chlorine or the fluoride added to "disinfect" the water. Rarely, if not ever, will technicians tell us about the other polluting or toxic substances contained in our drinking water, simply because they do not measure them. As a rule, results are communicated by bacteriologists or virologists who know very little about the other types of pollutants that can also be found in water.

Industrial pollutants and others

Faucet water can contain, depending on the season, a variety of substances as surprising as they are unusual. Very few studies have been done in this field. However, they can demonstrate that, apart from the controlled microbial and viral micro-organisms, there can be loads of other soluble or insoluble, biodegradable or non biodegradable substances, resulting from people, consumption waste, industry, agriculture, etc.

Water, as well as earth and air, can contain thousands of toxic or non toxic polluting substances which appear periodically or are forwarded to our faucets. We are totally ignorant about the harmful effects on our health caused by most of these substances taken individually or of the possible synergistic effect of one on the other, because they have not been studied enough. But the facts are there. Thousands of different polluting and toxic substances can be found in the water coming out of our faucets.

Acid rains

Researchers had discovered that sulfur was the main pollutant of acid rains and responsible for the acidification of lakes and rivers. The flora of acidified lakes is modified when sulfur is transformed into acid sulfate (H_2SO_4). The United States managed to control their emissions of sulfur and decreased by 60 % the acidification of their lakes after several years of control. Years after the decrease of sulfur emanations, the nitrates (H_2NO_4) are now significantly increasing in the formerly acidified waters. Scientists thought at first that the use of nitrates to counterbalance the acidity would cause no problems, because these substances can help cultures grow. But now it is the earth itself that is soaked and even saturated with nitrates in certain areas. Transported on the surface waters of brook and rivers, nitrates now act as contaminants causing other problems, some of which we only suspect. And this is how these pollutants contaminate the water in villages and in wells.

Herbicides

Atrazine, a weed-killer used to control weeds for over forty years in corn fields, in a proportion of one to four kilos per acre of culture, year after year, eventually ends up, mostly in the summertime, in water streams surrounding these plantations. This weed-killer can be traced all along the food chain on the maritime coasts, because it is not particularly biodegradable. The paraquat, another weed-killer highly used in potato culture was found in large concentrations in the waters of the maritime coasts. Methyl bromide is a pesticide largely used in agriculture, and it has been scientifically demonstrated that it destroys the ozone layer.

Dozens of other herbicides and pesticides abundantly used in agriculture end up, year after year, with the chemical fertilizers in the drainage water of agricultural lands. Researchers even identified several pesticides in the vegetables and fruit that we eat. Even though no particular diseases have been formally associated to weed-killers, we should not bask in our glory. Studies are beginning to incriminate these substances in certain pathological diseases. To reassure the public, some very obliging scientists have already reported that, in experimental conditions and at concentrations of 8 milligrams per glass of water, atrazine has no carcinogenic effect. Very well, Mr Scientist, but what exactly do you know about the potential synergic effect of all the other toxic matter gathered in the flora stream or in even in a glass of water? Not a single thing.

Certain weed-killers and pesticides can be toxic and cause intoxication or diseases. Once again, very few studies have been done on the subject. However, it has been observed that individuals working in banana plantations, for instance, could become sterile because of their exposure to the weed-killers used in that culture... Let us mention here that in *The Medical Post (Canada), October 23, 2003*, authorities deplore the fact that over 40% of the pesticides on the market cannot be detected, for lack of tests to measure them. Would it not be preferable to simply withdraw these substances from the market until we know their potential toxicity?

Medication

However surprising it may seem, the water from our faucets can also contain, besides infectious agents, herbicides, pesticides and artificial fertilizers used in agriculture or for the maintenance of our lawns, a large quantity of industrial waste, as well as drugs used in treating people and animals. All we must do to find them is to look for them.

The European researchers were the first to demonstrate that water, so-called drinkable, could contain many drugs made by pharmaceutical companies and used by patients. Antibiotics, birth-control pills, hormones, painkillers, aspirin, blocking betas (to lower blood pressure), so far up to 32 such drugs, more or less biodegradable, have already been found in tap water. Different levels of these substances can be found in our drinking water as well as in the food we eat. They were found at different times of the year and some can be there permanently. Generally, one can assert that if these substances are measurable in our drinking water, it proves that there are important in quantity and that they are not biodegradable or very little.

Over thirty different drugs were thus found in the water of lakes, rivers and aqueducts in Europe. I would not have been surprised to hear researchers say, under their leaders' pressure, that these assertions are false, if Health Canada had not confirmed it. One thing is certain: competent scientists have already observed and published these facts in scientific and medical annals.

No one would dare say today that these drugs are not found in North America's drinking water. An article published in *Actualité (Current Events),* a Québec magazine, mentioned that doctor Stuart Levy from Boston's Tufts University, declared that medication found in water could cause, among others undesirable effects, a resistance to antibiotics. Claude Soli, a chemist at the Charles-Débaillets factory in Montreal which supplies Montreal and over twenty other cities with drinking water said, in the summer of 1999, that he had already found in certain Québec sewers sufficient quantities of estrogen to affect the health of fish.

In September of 2001, researchers from the Canadian Federal Government's Ministry of Health said they were worried because they too have found drugs in drinking water. Health Canada confirmed on a news bulletin on the Internet on September 5th 2001, that estrogens were indeed present in sufficient quantities to affect the health of fish, and that the drinking water of Canadians can also contain antidepressants, medication against high blood pressure, antibiotics and birth-control pills.

Health Canada researchers are urging the people not to throw their pills in the toilet, thinking it was the source of the problem. It is one hypothesis which explains the presence of medication in drinking water, but others must be considered. Nobody has yet demonstrated, for example, that these medicines are not a result of the rejection of complete molecules or their partially biodegraded metabolites by the body, and then found in the liquids or feces of the individuals themselves. This hypothesis is valid until proven otherwise. It seems that the researchers from Health Canada were not interested in looking for drugs in drinking water before the events of the contaminated water in Walkertown, Ontario. We certainly recall that the water of this small village was contaminated with coli Bacillus by feces from surrounding farms and caused at least five deaths and numerous very ill patients in the spring of 2000.

Genetically modified organisms

Traces of insecticides, tobacco and BT corn plants developed in biotechnological laboratories (see GMO further on) have also been found in the earth and in the waters of rivers.

These could be other sources of water pollution to be added to the pathogenic micro-organisms, the toxins, the pesticides, the persistent organic pollutants, mercury, nitrogen, phosphoric acid agents, factory water waste, industrials dumping, city surface waters, dumps, agricultural and forest exploitations, medications, natural earth contamination, hydroelectric dams, etc.

Sewage water

What must we think of sewers, developed and set up by scientists and governmental engineers, as elements of pollution of our drinking water? Studies have showed that these gigantic systems often stop working for months on end, probably and regrettably much more often than we really know. What scientific logic allowed us to decree that accumulating in one spot only, billions of tons of human waste, was the ideal solution to sewage problems? Would it not have been preferable for each district or each house to partially purify their sewers before returning waste water to the main sewer without treatment? In so doing, the gigantic polluting load now transported for kilometers would have been disposed of in several smaller places! Sometimes, the distance is so great that the pumping systems are unable to take care of all the garbage. Such was the case in Lanaudière, a Quebec region, in 1999 and in 2000, for months if not more.

In other countries, each house or district must transform and deactivate their sludge and spread it on their lands, flowerbeds or gardens in their neighborhood, as was formerly done. Would this not have been a way to partly replace numerous polluting artificial and toxic fertilizers presently used? Was it really justifiable to build these very expensive installations and to gobble up billions and billions of dollars to dump in one single place, all the waste thrown in the sewer of a city or of several cities? Can we logically think that all these systems to purge cities and villages will still work in ten or twenty years? The analyses of certain installations are disappointing. We have witnessed that purging fields, which all end up in the water such as in streams, etc., are often overloaded with bacteria and non functional after only a few years in operation.

At the end of November, 2001, the Ministers and the authorities responsible for the environment of 101 countries, members of the United Nations Program for the Environment, met in Montreal to assess a plan to protect drinking water. They wanted to reassert their political will to control the pollution of ocean waters. We learned during that congress that about 80 % of the pollution of oceans originates from the earth and that the main sources of pollution come from waste water, streams, agricultural and industrial zones and the destruction of habitats. The Canadian Minister of Fisheries and Oceans, Herb Dhaliwal, indicated that sewage water, which represents about 20 % of this pollution, is one of the most important causes of water pollution and should be considered a priority in all countries of the world.

Without wanting to seem too alarmist, we cannot reasonably claim that even the best scientists, doctors or biologists, will be able to speak with competence before decades about all the pollutants in our environment and the negative impacts from thousands of polluting substances thrown in the environment. These problems are often beyond imagination.

We will certainly be long gone when world authorities start taking this matter seriously and deal with all the toxic polluting substances from chemical, industrial, agricultural, pharmaceutical, medical or other sources. Will we ever know how much these products contribute to the development of new diseases in our civilization, or to the new illnesses observed in people, animals and plants or even to our very extinction?

The pollutants found in human tissue

On March 21 2001, the CDCP (*The Center for Disease Control and Prevention*) of Atlanta, U.S.A., published the results of a study showing for the first time the ratio of 27 polluting substances found in the blood and the urine of 5,000 individuals from 12 different areas in America.

These 27 chemical substances belong to four subgroups: metals, (lead, mercury and cadmium), cotinines (smoke indicators in tobacco), organic phosphates (chemicals highly used in pesticide manufacturing) and phthalates (highly used for manufacturing soap, shampoo, hair spray, nail polish and flexible plastic products).

It is the diethyl phthalate (DEP) and the dibutyl phthalate (DBP), two of the seven most used metabolite phthalates in manufacturing products that were found in greater quantities in the blood and urine of the individuals in the study. Lead, measured since 1976, has decreased by 2.7 micrograms per deciliter in children aged 1 to 5, indicating that the measures taken to reduce them were effective. It is interesting to note that the rates of the cotinines have also decreased by 75 % since 1991.

The phthalates are profusely found in our every day consumption. Their presence does not inevitably mean that they cause disease. Special consideration will be taken in further studies regarding women of child-bearing age who have been exposed to DBP. These substances have already been associated to possible effects in animal development. The meaning of these bio-monitoring studies does not give any indication of their toxicity or of their pathogenicity (how they induce a disease). Further research will be necessary to learn of their possible toxicity and pathogenicity.

The CDCP plans to measure, as of now, over 100 substances in various populations, during subsequent studies. The group of experts on the esters of phthalates of the Association of American Chemists said in a press release that the high rate of these substances remain within the safety limits established by the Agency of Environmental Protection. This is what we always hear in the beginning...

In the case of ecological accidents which can leave substantial pollution on large territories, the polluting substances can be extremely varied. Hiroshima, the accidents of Tchernobyl, Three Miles Island, Exxon Valdez are examples of it. Over 12 years after the drainage of 39,000 tons of petroleum from the tanker Exxon Valdez in Prince William Bay, Alaska, in March of 1989, residues still cause problems. Eight years after cleaning the banks and after the deaths of thousands of maritime mammals and birds, the maritime coasts are still filled with the stench of this slick oil. One can never account for the number of deaths and the contamination due to the explosion of a nuclear power station, even decades after the event.

The effects of certain pollutants on life

For years, researchers have observed a feminization of the male fish in the Saint-Laurence River, congenital malformation in several species of fish or living organisms in the river's flora. All these contaminating agents come from factories, agriculture, domestic waste or medication and can often found in our drinking water. This should lead us to think that the toxicity, the teratogenicity and the pathogenicity of some of these products should at once be taken into consideration. A teratogenic component is a substance which can induce congenital malformations in cells and a pathogenic substance is one which can induce disease.

Scientists have already observed that the native flora changes with the currents. Studies demonstrate that the inhabitants of this flora are often chronically modified, no one being able to intervene or to find the cause of this. It could be that more or less biodegradable estrogens, as those used in female contraceptives (anti-ovulating pills) since the beginning of the sixties, are partially responsible for this, and many studies will need to be conducted to demonstrate it. Hormones similar to human estrogens and used in abundance as weed-killers in our flowerbeds could also explain the modifications of the flora in streams. Herbicides, such as mercoprol and dicambra are hormones chemically similar to human estrogens. They have been largely used as herbicides since several decades to control the bad weeds in lawns (dandelions, etc.) and could hypothetically combine themselves to the same receptors as the estrogens on human or animal cell tissue. As these substances are more or less biodegradable and have little molecular weight, they could very well be transported by streams right up to the source of our drinking water. Yes, vegetable hormones have similar chemical components as human estrogens. It would not be surprising that some of these hormonal weed-killers could get linked chemically, by covalent links or otherwise, with the estrogen receptors on human or animal cells. This hypothesis is as plausible as any other. It would explain the large percentage of male fish with their testicles being transformed into ovarian tissue, or the decrease of sperm production by animal and man since about forty years! It would not be surprising to find that under the influence of this hormone-like substance, male fish and people begin to feminize… Did you ever think of the social and human impacts if this is one day confirmed? We will see later on in this work, that many other polluting substances can affect brain tissue and modify reactions leading to violence, drinking habits, people's moods…

Today's systems for cleaning up water, which already cost billions of dollars, cannot filter all the substances liable to be found in drinking water. Most of these products, if not all, are the result of incredible technological applications which have enabled the material evolution of man and society. **These problems come from what we call evolution and progress!** It is in no way excluded that one or several combinations of these substances can be linked to the increase in cancer, regardless of the age of the individuals in any given population (breast, ovarian or prostate cancer, for example). Committees of inquiries will conclude, as it is often the case, that "nobody has yet scientifically demonstrated that any one or any combination of these substances is carcinogenic, pathogenic or teratogenic". Nobody has demonstrated the contrary and only studies will answer our questions. Some will plead the absence of convincing data, which is "politically and scientifically correct", and which is profusely used by our corporate authorities and the governmental agencies of control. What if a group of pathogenic products or a particular combination of some of these products causes these diseases? One thing remains certain: the day we will have all the proof to satisfy scientists, technocrats and the forever slow governments, damage to human beings, animals and the environment could indeed be so great that certain living organisms will already have disappeared from the surface of the earth.

Let us mention that reports are beginning to show, for instance, a much closer relation between the simple uses of trichlorohydrex in antiperspirants and the incidence of the breast cancer... This chemical has been measured in the cancerous cells...

Let us add to these considerations the astronomical quantity of polluting substances, more or less biodegradable, already observed to in the food we eat, in the seafood and in the fish at the end of the food chain.

A televised broadcast on water pollution of the Saint-Lawrence gulf-stream, prepared for *Zone Libre* (French-Canadian weekly show on topical subjects) diffused in December of 2001, **reported that overseas commercial vessels pour into the oceans 13 tons of waste per hour, and this throughout the year.** Draining the oil from ships in the sea is catastrophic to water birds. According to a scientist from *Zone Libre*, over 300,000 birds die every year because of oil waste. **This waste is also <u>excessively harmful to the maritime plankton, which synthesizes a great quantity of the oxygen in the air</u> and, added to the industrial waste of the planet, finds itself ending up in the oceans.**

When waste was more biodegradable, sea turtles could still manage to eliminate it. But today, these animals die suffocating in plastic waste thrown in the sea. Numerous fish become prisoners of fishing nets lost at sea as do many birds caught or tangled in fishing lines.

As we write this manuscript, a tanker bearing the flag of the Bahamas and belonging to the Mare Shipping Company and registered in Liberia, breaks in half at sea near the west coast of Spain, after having dropped a large part of its 70,000 ton shipment of Russian petroleum coming from Saint-Petersburg. It sank with the greater part of its load into the Atlantic Ocean at the north of Gibraltar.

Before speaking about chemical substances already linked to the development of particular diseases, let us mention two other examples of contamination: the water in Bangladesh and contaminated fish oil.

Who would have thought that among others "acceptable" polluting substances, the water of the wells dug by UNICEF in several regions of Bangladesh could be contaminated with arsenic?

An important scientific article on the prevention of cardiovascular disease, published in the Lancet (354, 447-455, in 1999) confirmed, to the greater satisfaction of supporters of disease prevention by natural substances, that the consumption of fish oil rich in omega 3 oil **<u>decreased by 20 % the mortality rate in patients recovering from coronary</u>**

thrombosis, by 30 % the deaths following a cardiovascular accident and sudden deaths by 45 %. Fish oil rich in omega 3 also strengthens the immune system, protects against cancer, increases the metabolism, improves the skin and prevents aging effects. The PCBs (polychlorinated biphenyls) are recognized by the World Health Organization (WHO) as carcinogenic and immunosuppressive agents capable of depressing the immune system. *Medscape News* reported, on January 20th of 2000, that researchers from the Association of Hong-Kong Consumers had identified 5 kinds of shark oils with high contents of PCBs, which were 1, 3 times higher than the 240 picograms recommended by the WHO. The toxicity of these substances is known. These fish oils are sold everywhere in pharmacies in Canada, in the USA as well as in other countries of the world. People use them precisely because omega 3 oils are beneficial to the health of arterial cells as well as several other cells of their body.

Recent studies (*Nature, September 19, 2003*) have just demonstrated that the North Pacific salmon that die after reproduction contaminate the water of those lakes 7 times more than natural pollution, through the PCBs accumulated in their carcass at sea.

Let us remember that it is not difficult to accuse scientists of being accommodating by having deliberately hidden for decades the fact that tobacco was carcinogenic. In spite of the overwhelming proof, scientists kept maintaining that tobacco was not responsible for lung cancer. More important is the fact that these **dishonest scientists without any moral standards went as far as to genetically modify tobacco to increase consumer dependence!** These scientists hid these facts in their scientific reports and let thousands of people die before acknowledging them. Rather tragic, isn't it? We recognize that these scientists are sorry individuals rotten to the core and that they prostituted themselves for their employers: the tobacco companies. Similarly, several scientists deliberately omit to consider, in their publications, the effects of pollution and toxicity of some of these substances taken together. No one mentioned the possibility of the synergic effects of certain substances with one another.

We try to control pathogenic micro-organisms, because we know that some can trigger diseases, sometimes pandemics or epidemics such as those which formerly decimated entire populations. We learned at our expense that epidemics are caused by pathogenic micro-organisms. We are just beginning to study the pathogenic effects of other pollutants. Had a project been presented by Canadian researchers before the deaths by drinking contaminated water in Walkertown, Ontario at the end of the last century, it would have been rejected as being "non-pertinent". Before this tragedy in Ontario, such research was not a part of government priorities, which are established by scientists!

Governmental authorities will one day authorize the necessary funds to study such contaminants, but only when they have no other choice. Even though such priorities are here now, the true results will be known only decades from now…

Ground water

In March of 2000, a meeting took place in The Hague in the Netherlands, reuniting members of the World Council on Water. We learned that over 1, 5 billion individuals depend on subterranean waters for their domestic use. More and more cities and countries draw from these waters to irrigate their lands.

The waters of rivers and lakes are usually renewed by the rain and by the cycles of evaporation / precipitation. Groundwater recharges very slowly and sometimes not at all. In several regions of Europe, the United States, Mexico, India, China, Africa and the Middle East, stocks of water called "aquifers" buried 2,000 meters deep are also threatened by impoverishment. The report of the congress, partially presented in *La Presse* (Montreal) on March 25th of 2000, specified that, because the global consumption of water increased a sixfold over the last century, people now over-exploit this blue gold. **According to the World Council on Water, the aquifers could be empty in 50 to 100 years from now.** These reserves, looking more like sponges than underground lakes, have collected the rainwater over 10,000 years ago in formerly lush pastures, and are now little more than deserts on the surface.

Oxygen isotopes enabled us to analyze these fossil waters and to determine their age. The Sahara hides a huge underground aquifer 20,000 year old, called the "Nubie Stoneware". It is situated under Libya, Chad, Egypt and Sudan. Libya will use a large part of it within the framework of Colonel Kadhafi's Big River project whose construction began in 2001. It will cost over 25 billion dollars and its objective will be to pump the groundwater and forward it northward through a pipeline 4,000 kilometers long.

Researchers found a great aquifer under Brazil, Argentina, Uruguay and Paraguay. There would be, according to scientists, other numerous unknown aquifers. One of the conclusions in this congress was that we still hardly know anything about these reserves, but damage is already done.

The aftermath of drilling is catastrophic, says Robert Becht, professor of hydrogeology in Einschede, Holland. One of the problems is that water is drawn without collecting data on level measurement, which would allow us to estimate and establish reliable forecasts.

Another even greater problem comes from the fact that man is not only content to use water in a disorderly manner, but he also pollutes the higher ground water. Exhaustive agriculture, large urban agglomeration, pesticides, herbicides, industrial waste and waste water, of which we spoke earlier, could be the most likely pollutants. No has yet studied the presence of industrial waste in these water supplies.

A third problem is foreseen in the horizon. By always digging deeper, man disrupts the natural balance of the underground. In maritime coasts, drilling lower than sea level draws salt water by capillarity. For example, **in 1955, the people of Thailand pumped 8,000 cubic meters of water a day. Today, they use 200 times more water and the level of the ground is declining. Bangkok, the Asian capital, has declined to this day by two meters for the same reason.**

Underground waters buried less than 1,200 meters deep would be, according to these reports, 5,000 times as large as the water of all the rivers on the surface of the earth. We do not know exactly where this water is and in what condition. Spatial missions will soon have for objective to find the beginning of an answer. Buried in the depths of the earth, this water shows abnormalities in gravitation due to the fact that its density is of 1 compared to 2, 5 for the earth's crust. Satellites will show a perceptible slowdown as they fly over water pockets. If two satellites 1,000 kilometers apart from each other circulate on the same orbit and their speed decreases at the same place, it becomes possible to detect these pockets and define their dimensions, by treating the data in communication computer systems.

In a German-American project called Gravity Comparative Experiment, the first phase of which began in 2002, an electromagnetic wave was used to constitute a link between two satellites. Computers can produce from the cumulated data more and more precise hydro-cartographic maps of the underground. Similar procedures are used to localize oil spills or regions where certain minerals are concentrated.

B. Deterioration of the air and of the earth

Thousands of tons of smoke loaded with multiple and varied toxic products from factories producing the goods we use daily emanate in the air every day. Carried with the winds and the rains, tons of industrial matter travel in the air and sow their distressing polluting substances over hundreds and even on thousands of kilometers of land. **They destroy plantations, acidify and poison the air we breathe, farmlands, lakes as well as our drinking water and the food we eat.**

Gases from these fumes contribute at creating greenhouse effects. They mainly result from volatile gases used in the industry, such as sprays and cooling gases. Many also result from factory chimneys and a strong percentage from motorized vehicles. Carbon monoxide (CO), carbon dioxide (CO2), nitrogen oxide (NO), ozone (O3), anhydride sulfur, organic volatile compounds and other numerous particles in suspension in the air are the most harmful to our health. The combustion of fossil fuels such as firewood, coal and petroleum also affect the ozone layer. Among gases other than the CO2 creating a greenhouse effect but to a lesser degree, there is methane (CH4) produced by animals during digestion or through decomposition of products in dumps and swamps, sulfur hexafluoride (SF6), polyfluorocarbide, hydrofluorocarbide and nitrous oxide (N2O), which are all synthetic gases caused by industrial petrochemical production. Their warming potential is markedly superior to the CO2's. The SCI-FI possesses 23,900 times the warming capacity of the CO2. These synthetic gases also persist much longer in the atmosphere that the CO2.

Every year, <u>an excess of 13 million tons of CO2</u> is produced over and above what the planet can biodegrade (*cf. Radio-Canada, "Découvertes" with Charles Tisseyre, on January 6, 2002*), mainly because they are fossil fuels. **We pollute our environment to a point where we exceed the capacity of cleaning up the earth biomass. <u>It is not exaggerated say that we would need a second planet to help to clean up the planet earth we live on: it is a scientific truth.</u>** These polluting substances, of which the CO2 is one of the greatest, have two major side effects. Firstly, they destroy the ozone layer that protects the earth from the harmful rays of the sun through chemical links. Secondly, some of these substances have been found to be responsible for numerous diseases. **Thousands of individuals die every year in every country of the planet because of these air pollutants. They are responsible for shortness of breath and for decreased respiratory functions. They aggravate respiratory diseases such as asthma and bronchitis as well as several other diseases. Several thousand if not several million persons die each year in the world because of these air pollutants.** Produced permanently through technological applications, these pollutants are also responsible for an important **<u>increase in hospitalization time and for health care costs</u>** which increase at the same rate as the factory productions. According to certain studies, an increase in hospitalizations of 2 to 10% is related to smog.

Gases causing greenhouse effect (GHG) were regulated during the meeting on the environment in Japan in 1997. In October 1999 and in 2002, scientists still find deplorable that the GHGs continue to accumulate in the ozone layer, as if nothing had been done. Due to their emanations and probably to their concentration, and because they are produced in urban factories, the great urban centers are the worse source of pollution. At such a rhythm, there will be 18 % more harmful gases creating greenhouse effects in 2010 than there are today. Are we on the right track, since the World Meteorological Organization (GENEVA'S OMM) reported in September 2003 that the hole in the ozone layer had now reached a new record in size?

Adopted in 1997 by 159 countries in Kyoto, industrialized countries and those in the process of transition towards a market economy decided to reduce by 5, 2% the gas emanations between 2008 and 2012 based on the 1990 level, these gases being responsible for the hole in the ozone layer protecting the earth from the sun rays. During their meeting in Japan in April of 2000, the Ministers of the Environment of major governments confirmed that they were going to force their government, in spite of their reservations, to fight against global warming. Because of the urgency of the matter, countries such as Germany, Italy, Japan, France, Great Britain and Russia would have preferred a 2002 term, but opposition came from Canada and the United States! This hole in the ozone layer is as large as the surface of the United States, Mexico and Canada together!

The thinning of the ozone layer, the pollution of groundwater, drinking water, water waste and overexploitation of biotic and non biotic resources are not the only effects caused by numerous technological applications and science on our environment. Let us reproduce here in its completeness an appropriate article on "smog", its composition and its effects on health, published on the Internet by "Service Vie Inc." and *Health Canada* on June 21st, 2003.

"Smog" is a word invented over thirty years ago to describe the combined presence of smoke and fog in the environment. Used originally to indicate a combination of industrial smoke and fog in the atmosphere, the word "smog" now refers to the chemical "soup" often observed over urban regions forming of a brownish yellow mist. Smog is in fact a mixture of atmospheric pollutants consisting especially of ozone particles, acid sprays and sulfates, nitrogen oxide, organic volatile components and carbon monoxide. Although smog is often produced in large cities, its concentration can also be very high in suburban and rural communities because of the wind and other factors. Whatever the place, visible or not, smog can be harmful to health. One of the main preoccupations is its potential harmfulness to the heart and lungs.

The sources of atmospheric pollution

Atmospheric pollution can result from natural or from anthropogenic sources and can also be caused by chemical reactions between pollutants in the atmosphere. The main sources of pollution consist of industrial productions and the combustion of fossil fuels used to produce and transport electricity.

Particularized materials (PM) are constituted of microscopic particles of variable chemical size and composition. For surveillance purposes, PM levels are classified for total particles in suspension (TSP), particles of 10 micrometers in diameter (μm) or less (PM10), particles with a diameter between 10 and 2,5 μm (PM10-2,5) and particles of a diameter of 2,5 μm or less (PM2,5). PM's main sources in the atmosphere include industrial emissions, exhaust systems from vehicles, dust on roads, agriculture, construction work, wood combustion, forest fires, pollen, spores, bacteria and volcanoes. The PM can be also be formed by chemical reactions in the atmosphere.

The tropospheric ozone (from the troposphere) comes from nitrogen oxide (NOx) and organic volatile compounds (OVC) that combine with solar light. The combustion of fossil fuels is the main source of NOx and OVC. The OVC is also produced by the evaporation of liquid fuels and solvents. The most important episodes of ozone are often produced when stagnant or slow air masses retain pollutants during long periods of time.

Other pollutants include sulfur dioxide (SO_2) and carbon monoxide (CO). The SO_2's main sources come from power plants using coal and ore foundries other than iron. The SO_2 interacts with water, solar light and the ions in the atmosphere to form different acid particles (sulfates) which are important constituents of the PM and the acid rains. The CO is a gas mostly produced by the combustion of materials containing some carbon (ex.: the fuels used in vehicles).

The tendency of atmospheric pollutants

Although the quality of the air in Canada is rather good compared to that of other countries, it remains worrisome to a certain extent. The levels of SO_2, dioxide of nitrogen (NO_2) and of CO dropped between the mid seventies and the mid nineties; however, the SO_2 levels have slightly increased since 1995. PM levels also dropped, but PM10 and PM2,5 levels remain high enough in certain regions to continue presenting health risks. Furthermore, in several regions of Canada, the ozone concentration in the troposphere is superior to the current quality standards of the surrounding air. The regions particularly at risk concerning the ozone layer are the Windsor-Québec corridor (Ontario and Québec), the valley of the lower Fraser River in British Columbia and the Southern part of the Atlantic Ocean. Considering that the forming of ozone in the troposphere is related to the rays of the sun, the levels of ozone are higher in summer, with peaks being observed between 12H00 and 18H00.

What are the potential effects on health?

Being a mixture of pollutants, smog can have variable effects on our health. These depend on the concentration of pollutants, the duration of exposure, the potentiality of interaction between the various pollutants; the effects of a pollutant can be amplified when combined to another, to the weather and to atmospheric conditions.

Potential effects on health during a short exposure to smog can vary eye from irritation of eyes, nose or throat, to a decline in respiratory functions, an increase in respiratory and/or cardiac diseases or even premature death. Certain groups of persons are more sensitive to smog, such as the elderly, the young children, cardiac patients and persons either allergic, asthmatic or affected by other chronic lung diseases. Research continues on the effects of prolonged exposure to atmospheric pollution.

Some particles are small enough to be inhaled and remain in the lungs. Particles can carry other toxic substances such as acid and metal, which could be a key factor on the effects on health. Studies have demonstrated that the increase in concentrations of particles and ozone can cause light reversible reductions of lung function in persons affected by lung diseases, particularly asthma, and also children not necessarily suffering from lung diseases. Other studies indicate that a regular and prolonged exposure to these particles increases the risk of lung cancer.

The role of other pollutants such as the CO, the SO2, the NO2 and acid sprays becomes clear and helps to understand the effects of smog on health. The effects of the acid sprays, the SO2 and the particles are difficult to distinguish because they often form a mixture in the atmosphere.

What is Health Canada doing?

By basing themselves partially on the results of the research made by Health Canada, the federal and provincial governments have collectively elaborated national qualitative objectives regarding several pollutants in the pervading air. Sufficient scientific proof having been established to indicate that exposure to trotospheric ozone and to particles can have important harmful effects on health, the Canadian council of Ministers of the Environment (CCME) recently established cross Canada standards for the PM2, 5 and the ozone. These standards will contribute in an important way to the reduction of the risks these pollutants represent to health and the environment.

Health Canada will continue to evaluate the effects of short and long-term exposure to the pollutants causing smog. The results of these studies will allow the development of standards and effective ways to better protect the health of all Canadians.

What *you* can do

Here is how you can reduce your exposure to smog and its potential effects on your health:

- Check the quality of the air in your area, particularly during smog season (April to September) and choose your activities accordingly (see below).
- When the smog level is high, avoid exhausting outdoor activities, particularly after noon when the level of ozone is at its highest, and choose indoor activities.
- Avoid doing exercise near heavy traffic zones, at least during rush hour, to minimize your exposure to smog.
- If you suffer from heart or lung disease, follow your doctor's advice regarding what you should do when the smog level is high.
  In the long term, you can also help to reduce the emission of atmospheric pollutants as follows:
- Whenever possible, use public transportation instead of your car, walk to work or use your bicycle when the smog level is not too high.
- Replace by something else vehicles and other motor driven machines that require gas (motorcycles, boats, lawn mowers, etc.).
- Take into consideration the energy efficiency during the purchase of a vehicle and maintain it properly.
- Reduce the use of energy at home. Inquire on sources of energy replacement.
- If you must use solvent based products such as fuel, mineral oil and oil paints, manipulate these products with care and put them in the garbage by respecting municipal laws.
- Burning wood, grass and leaves increases atmospheric pollution in the neighborhood. Think of recycling or fertilizing organic matters.

- Think about joining citizen committees campaigning for clean air in your community.
- Teach your children the importance of adopting ecological ways of life and help them become aware of the importance of the rational use of energy and community participation.
  Source:

Health Canada

The Sword of Damocles

We constantly live and will continue for a long time to live in the pollution of factories and toxic products, developed by technological applications and intended for our physical well-being. Each of these substances and pollutants represents a Sword of Damocles hanging over our heads, and we are incapable of understanding their real danger. If science has considerably increased our life expectancy in the last century and enabled us to enjoy an easier life than our ancestors, it also subjected various forms of life on earth to harsh tests and greatly damaged, as we shall see in the following chapters, the humanity in man.

Global warming

Because of the GHG and the toxic pollutants from factories, from our cars and probably others not yet identified, the temperature of the planet increases and consequences are more than disastrous. Polar ice is melting. Seas are warming. Scientists estimate that cities and entire regions will disappear during the twenty-first century because of this climatic warming. Boreal flora will be modified. The animals, and especially birds such as the wild geese and waders, will no longer find their secular reproduction sites. Analyses on all these repercussions are being done.

This phenomenon occurs because the melting of ice does not allow light from the sun to reflect on large areas of ice as it formerly did. Heat from solar energy is then absorbed by the water of the polar seas which are no longer protected by ice. The deep maritime currents circulating in the seas and whose function is to balance the temperatures of ocean water, transfers this heat to the water of other seas. Evaporation is modified and the formation of clouds changes the climate. Because the seas are warming, the whole surface of the earth is also warming.

Robert Watson, President of the Intergovernmental Sample Group on Climatic Changes, made the alarming prediction in his report presented at the WHO conference in Shanghai in January of 2001 that average temperatures during this century would increase by 5, 8 degrees. This report is, to this day, the most complete on global warming. It brings new proof demonstrating even more clearly that the increase in temperature is an important cause of pollution and not at all a natural factor. The rhythm of climatic changes during the present century should be higher than it has been in the past 10,000 years, estimates Sir John Houghton, co-president of the Shanghai conference. Drought could strike agricultural zones. The rising of oceans could flood populated coastal zones in China, in Egypt and elsewhere. The poor countries or those developing would be most affected. Pessimistic forecasts even mention that the melting of the Antarctica ice could provoke a rise in sea level high of three meters during this century.

Warming of the air and climatic changes

Where is the improvement, after over 25 years of hard work, in trying to control gases creating greenhouse effects, acid rains, exhaust fumes in nature and all the numerous pollutants? When scientists speak of the gases responsible for creating greenhouse effects, they simply mean that the ones already identified are the main sources of the tragic effects observed in climatic changes. It is very likely that among the estimated 100,000 pollutants, several could contribute to this effect, some of which could be just as disastrous, but have yet to be identified. There are thousands of other toxic products which we should one day control in our environment. So far, no procedures have been undertaken to identify them, much less to eliminate them. **When we see the United States still refusing, in 2001, 2002 and 2003, to conform to the Kyoto agreement of 1997 in order to continue making their economy turn, we are shocked by such selfishness!** Considering that over 25 years of work was needed only to reach the Kyoto agreements of 1997 only try to control the harmful effects of the GHGs and the emanations of fossil fuels from factories and vehicles, we can be nothing but pessimistic for the future!

Let us remind ourselves that polar ice can thin down by 3 % certain years! Since 50 years, climatic changes have caused a thinning of approximately 40 % of the polar ices.

We can not yet be certain that the murderous disasters of 1999 only as well as those in the following years are not connected to global warming or to the greenhouse effect. They were particularly more numerous in recent years than what would have been normally expected. Several disasters connected to this pollution happened in 1999.

1- Floods and landslides in Venezuela in December: 50,000 deaths; 2- An earthquake in Izmit, Turkey in August: 19,100 deaths; 3- A cyclone in Orissa, India in October: 15,000 deaths; 4- An earthquake in Nantou, Taiwan in September: 3,400 deaths; 5- Floods and landslides in Mexico in October: 1,300 deaths; 6- An earthquake in Colombia in January: 1,185 deaths; 7- An earthquake in Duzce, Turkey: 834 deaths; 8- A tropical storm in Pakistan: 751 deaths; 9- The flooding of the Yangtze river in China in June: 725 deaths; 10- Floods in Vietnam: 662 deaths. We do not know the number of deaths caused by the Occidental storm that swept Europe in December of 1999.

The costs of these 1999 disasters were compiled by a Swiss insurance firm, the Swiss Munich Re Company, and published in a Montreal newspaper on March 8 of 2000. The Occidental storm in Europe in December cost 4, 5 billion $ US. Cyclone Bart in the South of Japan in September: 2, 98 billion $. Hurricane Floyd on the United States East Coast: 2, 36 billion $. The storm in southwestern France in December: 2, 2 billion $. The earthquake in Izmit, Turkey in August: 2 billion $. Tornados in the American Midwest: 1,485 billion $. The earthquake of Nantou, Taiwan in September: 1 billion $. The hail storm in Sydney, Australia in April: 982 million $. The snowstorm in the American Midwest: 755 million$. The explosion of the Ford factory in Dearborn, Michigan in February: 650 million$. The fact that these disasters are apparently caused by the waste of our so-called prodigious civilization gives us much to think about!

If the year 2000 gave us some respite, with 10,000 deaths caused by natural disasters, the year 2001, on the other hand, accounted for 25,000 deaths on the planet, according to the same insurance agency, Munich Re. Earthquakes, floods, landslides and cold spells are responsible for those damages. The state of Gujarat, in India, was shaken on January 26 with an earthquake leaving over 14,000 persons dead. Central America, with its earthquake and its landslides, totaled 850 deaths. A cold spell in Afghanistan, in February, killed 750 persons while floods in Vietnam and in Cambodia accounted for 440 deaths. Between August and October, floods caused the deaths of 750 persons in Algeria. Financially, the tropical storm Allison, which struck the coast of the United States in June, caused over 6 billion dollars US in damages, including 100,000 vehicles either damaged or destroyed. The earthquake in India cost 4, 5 billion $ US.

A news broadcast on *Radio-Canada's "Technologie"* in May, 2002 reports that our world, as described in the latest report of the WHO on world environment, is nothing short of a gloomy prospect. Natural disasters are more and more frequent and environmental problems endanger the lives of human beings as well as those of animals and plants, according to a document made public in May, 2002 by the Environmental Program of the United Nations. By studying the last three decades, this document tried to foresee the future of the environment in time for the next world Summit on durable development which was to be held in Johannesburg in August / September, 2002.

We also learned that the biodiversity of the earth is seriously threatened. Although not immediately threatened by extinction, some 1,130 species of birds and over 4,000 species of mammals could disappear within the next 30 years. Among the species in the most precarious situation, are the black rhinoceros from Africa, the tiger from Siberia and the leopard from Asia, according to the WHO.

Fish are no better than the other members of animal kingdom. At the world level, one third of the sea reserves are very low because of excessive fishing. Fortunately or regrettably, the European community has just made its effort in cutting by 10 % its fishing quota for the next year. Canada has already made the same exercise some years ago. These animals are mostly threatened by the loss of their environment caused by industrialization, mining activities and agriculture. Some of these problems may get worse. The report predicts that the infrastructures of industrialization could cover more than 70 % of the surface of the globe within the next 30 years.

Fortunately, the report of the Environmental Program of United Nations is not the bearer of only bad news. The document mentions notable progress since the 70s regarding the quality of the air and water and an increase in parks and natural reserves. "We have hundreds of statements of agreements on the treaties to control and solve environmental problems, and we must now find the political courage and the necessary funding to take care of these ideals", declared Klaus Topfer, Director of the Environmental Program of the United Nations.

We take very lightly the commitments against pollution

Climatic changes caused by the destruction of the ozone layer are progressing. In April 2000, we learned that our own country, Canada, was not respecting the 1997 Kyoto agreements concerning the respect of control periods for its emissions of greenhouse gases. As I write these lines in June 2002, these agreements are not yet signed, but they finally will be at the end of 2003.

We have just celebrated, at the turn of the year 2000, the 20th anniversary of "the United Nations Convention on Long Range Transboundary Air Pollution (LRTAP)". The protocol to decrease acidification, the eutrophisation of lakes (choking) and the ozone at earth level is just beginning in most countries in 2003. Hundreds of tons of chemical substances, such as chloro-fluoro-carbons for coolers or the halones, were still being used in Canada and in the United States, in 2002. These gases are still used in shops and factories all over the country and continue polluting the environment. At the opening of the Summit of the Earth in Johannesburg, South Africa in August, 2002, the South African president Thabo Mbeki blamed rich countries for not having respected the recommendations made during the 1992 Summit of Rio de Janeiro on the Durable Development. He exhorts them to put an end to this "world apartheid". According to Hubert Revees, out of the 2,500 Rio recommendations accepted by the international society, less than 20% were carried out.

We have mentioned that the United States, the greatest polluter on earth, still refused in 2001 to respect their deadline on controlling their GHGs - same refusal in 2002. They evoke their lack in electricity for their internal immediate necessities. It is true that California and certain eastern American states lack in electricity, because of poor economic planning by its deregulation. In the coming years they will need to build several power plants using coal, petroleum and maybe nuclear power to compensate for their lack of good economic planning. All these sources of polluting fossil energy will only increase the pollution problem... Reassuring, isn't it?

In July of 2001, we learned in an article in *La Presse* that the United States discharged 24,8 % of its GHG against 10,9 % for China, 6,5 % for Russia, 5,0 % for Japan, 4,0 % for India, 3,7 % for Germany, 2,5 % for the United Kingdom, 2,5 % for Canada, 2,0 % for Italy and 1,8 % for France. As for the emissions of carbon dioxide, the United States are always first in line, with 5, 6 % of its emissions. Australia comes next with 5, 0 %, Canada with 4, 9 %, the Netherlands with 4, 1 %, Belgium with 3, 7 %, Germany and the Czech Republic with 2, 8 % each, Russia with 2, 7 %, the United Kingdom with 2, 6 % and Japan with 2, 4 %.

In Canada, Quebec is best quoted because of its hydroelectric power plants. With 24 % of the Canadian population and 22 % of the gross domestic product (GDP), it is responsible for only 13 % of the emissions of GHGs. Alberta, with 10 % of the Canadian population emits 30 % of

GHGs, because of the production of petroleum and the procedures used to extract petroleum from the bituminous sands. Ontario, with 37, 5 % of the population and 43 % of the GDP emits 29 % of GHGs. In November, 2000, the National Program on climatic change indicated that Quebec should reduce its emissions from 10% to 6 %, Ontario from 43% to 21 % and Alberta from 41% to 27 %.

It is sad and inconceivable to observe that countries as rich as Canada and the United States are the worse pollutants on the planet regarding the emissions of greenhouse gases. These countries remain the greatest fossil energy-consumers of all the countries in the world, even though in Japan in 1997, Canada had committed to reducing by 6 % its greenhouse gases. In an irresponsible and scandalous way, we have managed to increase them by 3 % between 1990 and 2000! Nevertheless, every year, as mentioned above, over 13 million tons of CO2 is produced beyond the earth's capacity to biodegrade. This is irresponsibility at its best!

The consequences of these delays will be disastrous for several countries which will see their climate changing and new diseases appearing. One can foresee, for example, that Florida, a favorite spot for many Canadians to spend the winter, could be infested with anopheles, a kind of mosquito responsible for malaria in warmer countries.

Our governments had the obligation to react accordingly and warn us that we should quickly modify our lifestyle. It is not what happened. Contrary to the democratic spirit of a society which respects itself, certain unscrupulous leaders, either in our governments or in private companies, carry their cynicism and irresponsibility to the utmost by buying rights of pollution for their companies. Can we possibly imagine a greater materialistic spirit and a greater lack of concern from our leaders? We must no longer wait for the government to act. Why not use the Greenpeace slogan and start war against our irresponsible leaders: "Think globally, act locally". Let us act individually and collectively.

If rich countries such as Canada and the United States can not respect their commitments, we can imagine how poorer countries will ever be able to conform! (*Béatrice Olivestri's report, presented at the LRTAP Congress held in Switzerland at the beginning of 2000*).

The extent of the current destruction processes

The World Fund for Nature (WFN) presented in October, 2000 a report entitled "Living Planet 2000". **It indicated that the biological richness of forests, rivers and maritime ecosystems have decreased by one third between 1970 and 2000, and at the same time, man's**

ecological pressure increased by approximately 50 %, exceeding the rate of regeneration of the ground biosphere.

The report brought undeniable proof that the production of natural resources for consumption, necessary to the absorption of the CO_2 emitted since 1961, had doubled. **Since 1996, it exceeds by 30 % the total available de-polluting surface of the planet, and consequently "begins to seriously affect the nature's wealth"**. Jonathan Loh, author of the report, wrote that if every human being used as many natural resources and emitted as much CO_2 as Americans or Europeans, two other planets would be needed to insure de-pollution. "Consumers of industrialized nations are the first responsible for the decline of the biodiversity in tropical regions" also wrote the author.

Rating the ecosystem pollution

By using for the first time as a marker an "ecological imprint" to measure the pressure of human beings on ecosystems, Ruud Lubbers, President of the organization of the WFN, demonstrates that the surface of productive lands necessary to meet the world needs in food, wood, infrastructures, as well as the necessary land to store the CO_2 emissions, exceeds nature's limits. It is now clear, according to the data, that drastic measures must be imposed, by all the countries in the world, to reduce at once the pressures of pollution on nature (*cf. La Presse, (Montreal) October 21, 2000*).

The decline of the ecosystems

Christopher Flavin, president and co-author of the annual report of the Worldwatch Institute, predicted an important decline of the ecosystems. He based his assumptions on the failure of climatic changes, confirmed at the Hague Conference in November, 2000, by the indifference of political leaders, the uncertain political and economic climate throughout the world, the relaxing of environmental laws and the refusal of several countries to respect international agreements.

Maritime biologists give credit to this forecast by estimating that one quarter of the coral reefs, very sensitive to climatic changes, are already ill or dying, and that in certain areas of the Pacific, this proportion reaches nearly 90 %, besides compromising fishing and tourism.

The melting of arctic ices, caused by the combustion of fossil fuel, accounts for the disappearance of several kinds of amphibians and coral reefs. The observation of frogs is important here. These tiny creatures act as a barometer to the health of the earth. Because of their greater sensitivity to environmental stress, to deforestation and to the thinning ozone layer, frogs act as biologic markers of health on earth.

A research team at the Institute of geophysics at Alaska University in Fairbanks confirms that 85 % of the glaciers were melting at an alarming rhythm. The study of 67 glaciers from the fifties until the mid nineties shows that the glacier Columbia, in Prince William's bay, decreases by about eight meters in length every year, and that the glacier of Bering, in the Mountains of St-Elias by about three meters. **The rhythm of the melting of ice of the polar ices has doubled since 1990.**

The Program of the United Nations for the Environment (PUNE) reports that the Munich Re Insurance Company, which studies since the sixties the cost of natural disasters, estimates that in 50 years the annual cost of global warming will exceed 300 billion dollars. These figures were obtained by calculating the cost of tropical cyclones, the loss and the damage to the land caused by the increasing sea levels, the excessive fishing, agriculture and water supply, all provoked by global warming. Damages to the ecology, in lagoons or in coral reefs could exceed 70 billion dollars a year. In Europe, the most important losses are linked to the increase in mortality rate and to the expenses in health care costs which reach 21, 9 billion dollars a year. They are presently estimated at 30 billion dollars just for health related costs and water management. No country, however prosperous, can assume the astronomical price of pollution …

C. Expansion and thoughts on the processes of environmental destruction

Thousands of regrettable effects other than those already presented are the result of technological applications. Some of them affect people in a more or less important manner or have different targets from the ones already mentioned. Here are some of them.

Shameless and unprecedented urban spreading progresses on farmlands, even in our beautiful "civilized" country…Let us look at the constructions of houses built without discrimination, on the farmlands of our neighborhoods and around several large cities of the world… We will better understand that municipal leaders can no longer continue this shameless appropriation of the agricultural lands that feed them, without diminishing the culture potential! The side effects of this urban spreading are already visible on lands formerly cultivated around cities. Nevertheless, we have continued since several decades to spread by using these productive lands! What carelessness!

The continuous burying of domestic waste, sewer water or their diversions in brooks or rivers is so natural that we do it almost automatically. With the droughts, as we underwent in the summer of 2001, our rivers and our "majestic Saint-Laurence River" will be lacking the necessary water to carry the rubbish we discard in it and to wash it down towards larger maritime reservoirs!

Globalization will only delay the measures which should already have been taken. The problem seen in the horizon stems from the fact that more and more production is made in countries with no rules or regulations concerning pollution and welcoming industries with open arms, unaware or uncaring of the danger.

Let us recall the diseases and expenses to the owners of houses isolated with MIUF. Approved by scientists of the Canadian Federal Government in 1977, the sale of this insulating material was forbidden in 1981. This now infamous first-rate insulating moss, regrettably used by Canadian owners to insulate their homes on the recommendation of governmental scientists, was even subsidized by this same government! These scientists had chosen to subsidize the MIUF as the insulating product for Canadian homes following the petroleum crisis of 1973. Shortly after applying this product in their walls, people noticed that the product released volatile substances which were toxic to them. Toxic vapors are released during the whole degradation period of the product, which takes several years. Furthermore, the product is not stable and its composition varies from one company to the next. It even prevents air circulation in the walls, aggravating the problems of humidity in the houses. The product becomes ineffective after a certain number of years and cannot be removed without demolishing the walls because of its great adhesion. Never had a product been so badly studied by scientists before its promotion. Never before had scientists made such a huge error by choosing and promoting such a product. Nevertheless, even though the government removed the product from the market a few years after its release, the errors made by the scientists were never repaired. The thousands of citizens who had considered themselves protected by the recommendations of their government were granted a mere 600,00$ per home !

Other dreadful processes of environmental degradation have just begun to strike the minds of the men in power and are in full progression. Let us see what some of them might be.

The gigantic machines "to destroy" the environment continue to grow. More and more powerful, they know no limits in the destruction of our planet. These mechanical monsters open up the ground and increase the deterioration of the planet more quickly than all the cracks on earth made by past generations.

Man continues building machines to rip forests apart. Giant machines continue to devastate the earth. Others destroy mountains, but who cares!

The earth is not only destroyed through mining: the exploited sites are left to disfigure the landscape. The erosion of chemical mining waste and the proximity to these mines are often poisonous to streams. Sometimes toxic liquids originating from excessively toxic metal extraction pollute rivers, as was case with the Danube at the beginning of 2000.

Mountains are holed like cheese or simply destroyed to give way to desolate looking landscape.

In Amazonia, earth is recuperated from rain forests and transported to culture fields in order to feed people. This new earth cleared from rain forests is not rich enough in humus to really become agricultural land in an effective way. Quickly abandoned because of its rapid impoverishment and insufficient production, it leaves immense deserted areas. This massacre of rain forests also contributes to the desertification of the earth.

In other countries, forests are also constantly subject to destruction by man. There is no more natural development and the desert continues to grow. It is case in Amazonia, in the Sahara desert and the Goby desert. Desertification increases everywhere on earth because of the inconsiderate interventions of man. The inhabitants of these regions must leave the forest where they formerly found their food, and in so doing, inflate the shantytowns of large cities in the hope of finding work to survive. Poverty increases concurrently in these ill exploited regions.

Gigantic fishing boats equipped with floating factories, the most modern equipment and, with large nets extending several kilometers long, travel across the seas of the world and literally empty the oceans of their fish and seafood. <u>Since 1950, oceans have lost over two thirds (66 %) of their consumption fish.</u> Scientists estimate that the decrease presently exceeds 2 % a year. Several species have disappeared or are in insufficient quantities to be identified. At such a rhythm, the oceans will be totally emptied of their fish in a predictable future. We can fish anywhere we want and the maritime animals have no protected areas of reproduction. To think that man is the greatest predator!

Numerous maritime species are mutilated by these murderous nets which are in no way ecological. Over 80 % of dolphins found dead on the beach presented wounds caused by these nets and the World Fund for Nature blamed the nets for the deaths of 300,000 whales every year. The ultrasounds emitted by the fishing boats could be responsible for beached whales…

By loosing their fruit and their fish, the seas become impoverished, fishermen lose their jobs and humanity loses a product considered essential to the survival of people and marine animals. Soon, the price of fish and seafood will become prohibitive.

Considerations on the physical effects of technological applications

Irresponsible and inhuman, scientists and technicians applying such technologies go on destroying the environment, believing that their scientific education produces only material benefits, because in their eyes, the matter is eternal and the matter is god (see Hegel and Karl Marx, Chapter 7), this material god dominates his creatures: the people, the animals and the environment. The damage to man and the environment increases as quickly as scientific knowledge progresses, because it is only material. As new technological applications appear because of the new knowledge on the matter, they go on developing without any ecological discrimination and without humanism by other scientists who believe only through reason and scientific, linear and rigid methods, and who are incapable of studying anything but the matter itself.

Early on in this twenty-first century, the damages caused by man on the environment are so important and generalized that no one knows how future generations will manage to control them. **Many believe, wrongly, that scientists will soon find the solutions to their past mistakes. Most of these damages on the environment are visible, tangible, measurable and often caused by irresponsible and often well educated technicians and scientists.** The events of September 11, 2001 and the outrageous demonstrations by the opponents of the shameless exploitation of wealth in each country should make us become aware that there is an overexploitation of the resources of the planet for the only benefit of a few individuals, countries or industries, and this is unacceptable.

The cries of alarm of environmentalists to save the planet are noble, but insufficient in themselves to stop the evil. If the polluting countries have such difficulty in respecting their commitments to controlling just greenhouse gases, how will we be able one day to stop and control the myriads of other polluting products caused by technological applications, industries and man, and which destroy our environment?

These examples of the deplorable effects of technology on the environment demonstrate without a doubt that science and technological applications are intimately linked to the processes which deteriorate our environment and our life. Proof incriminates the thoughtless decisions of the uncaring and often incompetent people in power who decide what is convenient for us. It is criminal for a government to let unelected administrators who are not medical doctors take charge of the people's health, for instance. It is also criminal that only government scientists can control the health of the environment, the health of education, the forming of people, etc. **Right in the middle of this laxity and shameless damage, the planet is running out of resources, the environment suffers the horrors of technology, of science, of man, and life is dying of pollution.**

D. Diseases caused by the effects of technological applications on life

We have so far indisputable proof to confirm that we are witnessing a massive and growing deterioration of our environmental universe. It is seen through the accumulation in the air, the water and the ground of countless polluting and too often toxic substances which threaten all forms of life on earth, including ours. We are surrounded by so many different pollutants that we have become used to it. Nevertheless these pollutants modify our lives. They contaminate the air we breathe, the food we eat and the water we drink. Some of these substances cause diseases or insidiously affect our physical health, our activities, our work or they induce, more often than we know new diseases that biology and medicine are beginning to identify and that no one suspected until quite recently.

The air loaded with pollutants, traveling with the wind from one country to the other, affects people from countries that do not pollute. These pollutants are sometimes found hundreds of kilometers from the contamination source and by falling with the rain, contaminate the ground and the water, the food we and the animals eat as well as the fish and the seafood. Unsuspected forms of life go through the horrors caused by pollution. **The volatile toxic substances mentioned above fill the air, cause diseases and increase the health costs which progress in parallel with the implementation of the new technologies.**

Scientists have just started to notice that the 500 million tons of calcium salt that Canada has been dumping for decades on the frozen roads every winter also contaminate the ground and the streams. The fact that we still know nothing of the possible physiological disturbances on life caused by each of these pollutants does not change the effects they can have on the fragile biological mechanisms of the cells of a living body. When thinking about the potential diseases of 100,000 different toxic industrial substances that enter, every day, in the production of the commercial products we use, it is easy to understand that more than likely certain toxic elements of these substances will one day be found in the water, in the air or in the ground and disrupt living organisms.

So far, it was only by chance that one observed the relations between substances and certain specific symptoms in a person or in a population. **We still know nothing of the pathologies these agents can cause in living organisms and we can only suspect their participation in the health of people and animals.** The majority of the toxic pollutants have not yet been identified, much less studied in relation to health. It is only quite recently in the history of humanity that researchers have observed a relation between atmospheric pollutants, nevertheless known for decades in the Los Angeles area of the United States, and certain congenital malformations in the newborn.

Even the domestic waste incinerators can spread toxic matter, contaminate the air, the water and the earth for kilometers on end and cause diseases of which we never suspected the existence. Nothing better that an anecdote to understand how this can occur.

The people of Savoy (France) suffered for years from different symptoms that medicine could not identify, before it was scientifically demonstrated that the people were being poisoned by the toxic fumes from the smoke of the regional incinerator. These toxic products were found in forage, in pastures and in the grain animals consumed. They were found in eggs, milk, cheese, butter, cream and even in the meat of these animals of consumption. Even the vegetable and fruit from kitchen gardens were contaminated with these toxic pollutants and people were more or less ill, depending on the various agents emanating from the incinerator fumes. The problem is so serious that the pollutants from this particular incinerator created health problems in the people of this region of France for about twenty years, before scientists could demonstrate that the incinerator was responsible for these diseases. Savoy "Tomme" cheese, a famous French cheese, was contaminated by the milk used to manufacture it and farmers stopped exporting it. This "delicious" contaminated cheese was nevertheless exported for years throughout the world before scientists could demonstrate that it was the incinerator that spread the disease to the neighborhood inhabitants!

According to biologists there would be, in France alone, several other defective incinerators contaminating meadows, neighborhoods, gardens and, depending on the wind, spread disease for kilometers on end (cf. *Report on TV 5 (French TV station), January, 2002*). What scientist could be conceited enough to persuade people that local incinerators throughout the world do not produce such pollution in their environment?

Following these observations, scientists are slowly starting to establish links between pollutants and cellular genetic alterations of the animal organism, certain cancers observed in people and animals and even the aquatic flora. More or less soluble and toxic wastes, as beryllium or dioxins, contaminate houses and inflict upon us diseases without us knowing exactly the symptoms they have on our health.

Yes, the earth is sick with pollution. The air we breathe is unhealthy. The water, so-called drinkable, as shown above, is often unfit for consumption and human beings become sick with pollution. These words of yesterday's wise men come to mind: "If man contaminates all the water he consumes, all the air he breathes and the food he eats, he will die from the contamination he himself produced". Nobody has yet demonstrated that the constant growing toxic pollution that surrounds us will not implacably produce catastrophes on earth! The industrial revolution is quite recent and is only starting to show its side effects on health.

Nutrition scientists have demonstrated only recently the major impoverishments in nutrients in the today's food (*The Impact of Nutrition, Environment and Lifestyle on the Health of Americans, by Joseph D. Beasley, M.D. and Jerry J. Swift, M.A., in the "Kellogg" Report, 1989*). By continually using chemical fertilizers or cultivating for years fruits and vegetables in the same spot, the food not only lose their flavor, but also their contents in nutritional value. **These processes literally empty the food of their nutritional values. Protein deficiency, sugar, vitamins, trace-elements and minerals have been observed in several varieties of food. As beautiful as it is, the vitamin C deficient orange contains only a fraction of the nutriments it had fifty years ago.** By feeding animals "high performance" mixtures of foods labeled as "successful", the beef from stock breeding contains eight times less protein per hundred grams of meat than the red meat from wild animals. Hundreds of similar examples have been reported by research scientists in nutrition sciences worldwide (*cf. "The Kellogg Report" references).*

Scientists have recently confirmed that new diseases are intimately linked to the contamination of the environment by specific toxic pollutants. Some are caused or aggravated by pollutants of the inhaled air, while others are connected to chronic or acute poisoning.

Diseases directly linked to environmental pollution

We can distinguish two categories of diseases directly caused by or corollary to technologies applied without precaution for living organisms. In this chapter, we will see how some diseases are caused directly by the toxic products rashly disposed of by man in his environment or through the interventions of man on nature. In Chapter 4, **which treats of the side effects of technological applications on man's humanity,** we will discuss those diseases resulting from indirect and stressful effects of technological interventions on the psyche of the individuals. That is to say that these diseases are born through material technological applications with their inflexible rules and the stressful acts of certain "individuals" on their fellow men and their society.

Even though the control of pathogenic micro-organisms in drinking water is in place in all the developed countries since several decades, even though there exists in all these countries surveillance technicians, and even though they themselves are under the supervision of scientists from the Ministry of the Environment, **the water, contaminated with micro-organisms, is still responsible for one third of all the cases of infectious gastro-enteritis in Quebec.**

During the seminar on Water and Health, presented at the ACFAS congress (the French-Canadian Association for the Promotion of Science), which took place at Laval University in Québec on May 17, 2002, biologist Pierre Payment of the *INRS-Institut Armand Frappier* (Montreal), stated that patients hospitalized in Québec because of infections caused by "drinkable" water only represent the tip of the iceberg. Then comes the infectious gastro-enteritis discovered by doctors during offices visits and finally the whole mass of the iceberg in the cases of "infectious gastro-enteritis" caused by infected water that every citizen experiences one day and that the health system fails to acknowledge.

In spite of the control of micro-organisms, infections through faucet water still remain a major problem in our environment. According to Doctor Payment, these diseases cost about one billion dollars in absenteeism from work or school, in medications, in medical and hospital visits, without counting the 10 to 30 deaths that occur every year in Quebec because of infected water.

If water is responsible for one third of the infectious gastro-enteritis that affects every Quebecker about once a year, the other sources of infectious gastro-enteritis would result from food and from animal contact. Imagine now the origin of the numerous diarrheas commonly called "turista", contracted by travelers during their trips in several countries of the world!

Apart from this acute and fundamental problem of infections through "drinkable" water, the individuals themselves, rather than the doctors and the scientists, are the first to be struck by the toxic side effects of certain environmental pollutants on their health. They are the ones who experience it first.

The publication of Rachel Carson's book, *"Silent Spring"* published in 1966, is an eloquent testimony of the fatal impact of pesticides on people's health. That same year, the world learned that a drug called thalidomide prescribed as a tranquillizer during the first stages of pregnancy lead to congenital malformations. These two publications were to destroy a medical myth that went on with no scientific backup since the mists of times. Biologists and doctors had always claimed that the placenta acted as a barrier to protect the fetus against outside toxic matter. The detection of highly toxic chemical substances in Love Canal, U.S.A. in 1977 and the discovery in 1989 of Alar, a carcinogenic pesticide, would lead us to a way of thinking different from what was already taught until now in medicine regarding substances found in the environment.

Several publications reviewed the toxic chemicals capable of interfering in the normal growth and development of the fetus and the child. A certain number of chemical pollutants, potentially capable of interfering in child growth, have already been inventoried (cf. *D.T. Wigle, Child Health and the Environment: Patient Care, volume 11, no 2, February 2000)*. Much however remains to be done in this area and we will present only a few examples of the diseases caused by certain of these pollutants.

Diseases caused by metals such as lead, mercury and others are better known today. The absorption of lead causes attention deficit in young children, a decrease in the I.Q. (intelligence quotient), sickle cell anemia and directly affects the development of the nervous tissue. Aftereffects on the child are almost always permanent.

The first publication on environmental components other than infections seems to go back to the 1950's when people observed that their children could be poisoned by mercury salt used in dental surgery (*Am I Dis Child, 1951; 81, p. 335-373*). Medicine discovered the real toxicity of mercury through the disease *of Minimata*. Researchers observed that children, exposed to mercury *in utero*, can present a normal appearance at birth. A few months after birth however, the children, poisoned *in utero*, start presenting signs of mental retardation, cerebellar symptoms, dysarthry, hypersalivation, strabismus and pyramidal symptoms.

Researchers identified for the first time in 1959 the poisoning caused by mercury methyl in 1,422 individuals. Four years were needed to demonstrate scientifically, through courage and perseverance, and to convince the industry that these cases of intoxication were inescapably linked to the mercury methyl thrown in the water of a nearby bay by a local industry producing acetaldehyde. Researchers first demonstrated that the fish in the bay were contaminated with mercury. They further established that the consumption of these fish filled with mercury induced the disease in the inhabitants consuming the fish from this bay. In the face of evidence, chemists confirmed that they had indeed used mercury chloride as a catalyst in the manufacture of the acetaldehyde.

In 1971-1972 in Iraq, over 6,000 persons were poisoned with the mercury from the treatment of seeds used to produce bread, and caused 400 deaths. (The *American Journal of Internal Medicine, 1992, # 21, pp. 275 – 280).*

Over 95 % of the ingested mercury is absorbed and distributed throughout the body. The mercury easily goes through placental and meningeal barriers as other substances do. An important source of poisoning with mercury, today, comes from the consumption of sea fish. Fish become contaminated with mercury salts disposed of by industries using mercury in their fabricating processes and eventually discharging it in rivers that finally reach the sea and the ocean beds. An incalculable number of the other non biodegradable polluting toxic substances end up as well in this immense pool of waste in the oceans. Because of all these pollutants, health scientists recommend eating no more than 2 or 3 meals of sea fish a week. Mercury is thus added to other numerous polluting and non biodegradable substances reaching the sea, the majority of which are not yet identified, much less measured.

The other main sources of mercury come from fluorescent lamps, computers screens and from manufacturing products we can't seem to do without. Thousands of tons of mercury have been buried with domestic waste since many years. From there, the mercury and/or other insoluble or soluble products find their way into streams and finally to the sea. Another part infiltrates the earth contaminating groundwater. Another part is carried in the water and contaminates the food. Still another part is dissolved in the water and reaches brooks and rivers and then the sea, contaminating fish and seafood.

Among the products that succeeded in contaminating the sea fish and sea food we consume and that have been identified are the PCBs (polychlorinated biphenyls), the dioxins and their by-products. We find these in the flesh of fish, but they are less toxic than mercury. We also find them chicken fat, in beef, and in the mammary tissue of women and animals, even several years after contact. It was also found in the milk of breastfeeding women and in Taiwan's rice. These pollutants are only the tip of another iceberg. We can now suppose that some of these polluting substances could be involved in the increase of certain cancers, in intellectual deficiencies, in Alzheimer's disease and even possibly contribute in triggering certain cases of violence, insanity, madness and depression… When we know for a fact that 450,000 persons on earth or 1 person out of every 13,000 is afflicted with brain disease, it is food for thought.

There could very well be other links between pollutants and certain diseases. **We have found in the urine, in the hair and even in the brain of certain persons with chronic diseases, allergies, neurological problems and certain cancers, heavy metals, pesticides used in agriculture or spread on lawns to destroy dandelions.** These symptoms have been reduced in persons who were subjected to procedures of exclusion of these substances and had no more contact with food containing pesticides, additives, "purified" water, or when the chemical products used in domestic chores, dishwashing, laundry detergent or body care products had been banned… We may be on the right track.

The World Health Organization (WHO) declared, during the March 2002 meeting on "Environmental Threats for Children's Health", that at least three million children (3,000,000) of less than five years of age were victims of their environment every year. Even though children of that age group represent only 10 % of the world population, research demonstrates that over 40 % of the diseases linked to the environment affect children under five years of age. In 2000, 1, 3 million young children of developing countries died from intestinal diseases caused by polluted water and poor hygiene. Over 2, 2 million, or 60 % of these children, die each year because of respiratory infections associated to domestic air pollution, such as stoves and heaters. Over 400,000 deaths are caused every year by road accidents, drowning, burns or poisoning.

The statements on radioactive fallout and radiation, done county by county by the National Cancer Institute, a branch of the American Government's CDC, have been analyzed by the Institute for Energy and Research on the Environment (an American institute independent of the government). The researchers observed that the radioactive fallouts from nuclear attempts, made since the 1950's, provoked about 80,000 cases of cancer in the people living in the United States between 1951 and 2000. Over 15,000 persons suffering from these cancers would have died (*cf. News, "Le Journal de Montreal", March 4, 2002*).

Diseases associated to air pollution

In November, 1999, the magazine *Euréka* published an article which appeared a few weeks later in an excellent scientific review. **We learned that, every year in France 30,000 persons die because of the atmospheric pollution alone!** The air pollutants that destroy the ozone layer such as carbon monoxide, carbon dioxide, ozone, sulfuric anhydride, organic volatile compounds, particles in suspension and other polluting substances by industries or resulting from the exhaust systems of our cars are greatly responsible for these deaths. This is a dreadful scientifically calculated revelation obtained by researchers who finally dared measure the unavoidable. These results strongly demonstrate that the atmospheric pollutants directly affect people's lives and are indeed responsible for an important increase in health costs.

Nevertheless, it is the people that take advantage of the industrial benefits that have caused this pollution! One need not be very shrewd to come to the conclusion that these 30,000 deaths caused by air pollution, in France only, represent only a small portion of the deaths caused by the same pollutants on the whole planet. The global figures are not available, naturally, but projections can easily be made from these data... It would be very disturbing for industries and governments if the true figures were known. But, no one being afraid of what he does not know, the study is postponed as long as possible...

Similar results were published in the English medical magazine *"The Lancet"* on September 2nd, 2000. The number of persons having died because of air pollution in France and presented in the magazine *"Actualité"* (French Canadian monthly magazine on current events), is not far from the truth. Indeed, the study in *The Lancet* **shows that 40,000 persons also die, every year, in three other European countries, always because of air pollution.**

It is estimated that in Mexico City alone, over 100,000 individuals die every year because of the disastrous quantity of the polluted air. Several cities in other countries, as in Germany, Japan and elsewhere, are at grips with similar problems.

Do you think people will let go of their cars, get rid of their trucks to which they have become so accustomed to and that make us so happy? With the revolting quantity of industrial substances polluting the planet, it is not surprising that thousands of deaths are caused every year because of the technological applications that contribute to our wellbeing.

We are totally ignorant of the more chronic morbid effect of all these polluting toxic substances on the health of people and animals.

Let us think only about the thousands of persons who suffered for years because of the radioactive fallouts from a large nuclear power plant in Russia.

Canada has waited for the results of the European epidemiological studies demonstrating that about 8 % of all the non violent deaths in the country are linked to the atmospheric pollution, before starting such studies. Scientists of the Canadian government quickly calculated that about 16,000 Canadians would die prematurely every year, due to problems attributed to air pollution. However, no serious study has been made on the subject. In the United States, for the same reasons, **there would hypothetically be 160,000 deaths. In fact,** we know strictly nothing about it as we write these lines. This kind of research has never met government priorities who prefer not knowing these things and burying their heads in the sand.

An article dated December 31, 2002 in "Reuters Health Information" confirmed our pretensions on the noxious effects of pollution on people's health: "Until basic environmental issues are dealt with, improved personal behaviors and medical care will have limited impact on childhood asthma, according to researchers".

When we see the thickness of a toxic cloud over our head, we can only reflect at all the particles of death that this cloud can contain. Can we only visualize the social burden, the medication and the hospitalizations that will shatter hospital budgets and those of the Ministries of Health? Pollution increases breathing difficulty, bronchitis and asthma and people who already suffer from these diseases will die prematurely. Let us think of the economic impact in the face of these disasters. If the Canadian Government could not sign the Kyoto agreements before the end of the year 2002, it is because the main source of pollution comes from the extraction of petroleum in the bituminous sands of Alberta. There is a lot of money to be made there!

Atmospheric pollution does not produce only noxious effects on asthmatic people or the elderly. It can also cause birth defects innate from fetal life... The Los Angeles *"Times"* published, on December 16th, 2001, the results of a study done by Beate Ritz, of the University of California in Los Angeles (UCLA). This study demonstrates that smog (*cf. p. 72 - 75*), known since dozens of years in this area of California, can have harmful effects on the health of newborn babies. The epidemiological data of the researcher shows that pregnant women, exposed to high levels of ozone and carbon monoxide, are three times more susceptible of giving birth to children with harelips or of innate heart malformations (*The American Journal of Epidemiology, December, 2001)*.

Certain biologists are beginning to think that weed-killers, pesticides and toxic industrial pollutants and even antiperspirants could be involved in the increase of breast, prostate and testicle cancers in young adults of 20 to 40 years of age. The incidence of breast cancer has tripled in the last 50 years in the world. **It went from 1 woman out of 22 in 1950 to 1 out of 8 today.** There could possibly be a relation from cause to effect between the increase in breast cancer and other types of cancers, and the monstrous proliferation of synthetic chemical substances used by the industry since the last World War. Think about the putative scientific results that are starting to link the use of antiperspirant with the increase incidence of breast cancer! Barely 7 % of some 100,000 synthetic chemical substances identified and used by the American industry have been studied to this day for their toxicity. The researchers are now getting ready to study such relations in California because breast cancer in this American state reveals that 1 woman out of every 7 will develop this type of cancer during her life. Nobody has yet been able to prove that such polluting agents could be responsible for the rapid increase in diabetes in the last ten years, in the young as well as in the adults.

A report from the American Agency of Environmental Protection (EPA) published on the Internet on September 5th, 2002 reports that diesel fuel mostly used by heavy vehicles could probably be linked to lung cancer. This American agency recommends stricter rules for this type of fuel.

In spite of all these deplorable incidents, the Summit of the Earth on Durable Development, held in Johannesburg in August / September, 2002 ended on a pessimistic note. **The report ends with a reflection of the durable development "slow, unspectacular, sometimes fruitful and strong of promising initiatives, but submerged by problems affecting hundreds of millions of people on earth" (an article in "le Journal de Montréal").**

We know almost nothing of the diseases caused by food contaminated with toxic matters. I am not referring to diseases caused by pathogenic micro-organisms or viruses in drinking water which we more or less try to control by chlorinating or by fluoridating drinking water. I am thinking more about the new diseases that could be caused by the chemical substances polluting the water, the ground, the air and the food. After the Tchernobyl accident, scientists were able to measure within hundreds of kilometers and for years after the disaster, the radioactive contamination of food, kitchen gardens and prairies where the cattle fed. People ate the meat not to starve, and milk and eggs were also contaminated by the radioactivity.

Before the European studies mentioned above, no one could explain the significant increase in the mortality rate of asthmatic patients observed everywhere on the planet over the last twenty years. We do not know as of yet the cause of the increase in certain types of cancers in man and animal, or the real cause for the decrease in sperm cells in both man and animal since about 40 years. **Let us remember that it took twenty years to demonstrate that a simple incinerator in Savoy, France, was responsible for diseases nobody understood (see above)!**

The National American Institute on Health (NAIH) reported in December, 2002 **that estrogens used in hormone replacement treatment, contraceptive pills as well as ultraviolet light and fine sawdust are now considered to be carcinogenic.** These substances must be added to the 228 substances already considered dangerous and possibly carcinogenic. New substances as metallic nickel (used in certain metal alloys), vinyl bromide, vinyl fluoride, chloramphenicol (an antibiotic) or even quinoline, a mutagenic substance found in food cooked at high temperatures such as eggs and meat, or cigarette smoke are now suspected of causing cancer.

We must become aware that on a planetary scale, **it is millions of individuals that die every year, uniquely because of the toxic pollutants from technological applications or from the waste of our consumer societies.** What scientist dare contradict us only because all the studies have not yet been made?

Scientists have already observed severe pathologies in the organisms of the flora of streams and marine animals, the belugas of the Saint-Laurence River for instance. Researchers strongly blame the pollutants thrown in their environment, but ignore which are the most perverse. Nobody has yet identified which product or association of products is mostly responsible for the deformation or the genetic alterations observed in marine animals or the aquatic flora. They know, however, that some of these products accumulated in the river mud in permanent contact with the aquatic flora and found in the flesh and organs of marine animals, could very well be involved. Researchers are reluctant to study this problem because they can foresee the results. **Scientific studies in the evaluation of chronic diseases or deaths possibly linked to toxic pollutants are just beginning to interest scientists and people in power.** We can understand why!

What must we think of the dreadful disasters on the health of animals in regions where accidents such as the complete or partial discharge of petroleum cargos, fuel oil or other toxic products that travel the seas by the thousands every year? **Let us remember that at least one major** petroleum discharge a month has been registered in the seas of the world and that boats pour 13,000 tons of various products every hour, year long year in the oceans of the world! What will be the long-term impact of the petroleum discharge at beginning of 2001, on the fauna of the Galapagos Islands Galapagos which have been up to now exempt of fatal human interventions?

Who can assert that there are no chronic effects caused by pesticides, herbicides, or the multiple toxic, non biodegradable products found everywhere in our environment, simply because we have not yet demonstrated it? It would naïve to believe that the polluting charge in the air, the water and the ground is of no consequence. Each of us can buy, at the corner store, any herbicide and pesticide we want to preserve our magnificent lawn from dandelions without giving it a second thought...

Here is another unfortunate effect of a deplorable scientific intervention on nature. Egyptian scientists planned to eliminate schistosomiasis from their country: it is a serious disease caused by parasite worms living in the blood vessels of mammals. The disease is contracted when the worms or the larvae of polluted waters penetrate the skin or the blood vessels. By meaning to inoculate their fellow countrymen against this plague, sanitary authorities inadvertently transmitted to several million Egyptians a much more serious disease, hepatitis C, by using without knowing it, badly sterilized syringes!

Impacts of certain technological applications on the future of botanical and animal species

Apart from the direct intervention of polluting and toxic elements which cause poisoning, disease and death, there is a whole new field of possible diseases to consider. I want to speak about the consequences to come because of man's intervention on natural selection. Certain diseases could appear decades and even hundreds of years after the interventions of man. We still know nothing of the long-term implications of vaccination technologies, for example, which nevertheless allow people to protect themselves and to survive the epidemic plagues that formerly decimated every year by the thousands the inhabitants of cities and countries. What we know is that by using the scientific knowledge that we have, man has succeeded in eliminating smallpox from the surface of the earth. He has strongly decreased the mortal impacts of several other infectious diseases through immunization, measures of hygiene, antibiotics and certain treatments in genetic diseases. Nothing is known, however, about the long-term impacts of these human interventions on the descendants and on the appearing of new diseases in their offspring.

The considerable prolonging of life during the twentieth century, from 50 years of age to over 80 years, is one new benefit from science that hopes one day to bring immortality to man's physical body (cf. Chapter 7). This incredible increase of about thirty years of life, within a century, is mostly the consequence of three groups of main discoveries. New knowledge and techniques have been acquired by laboratory experiments and in animals: **1- the application of measures of hygiene** since Pasteur identified bacteria as agents responsible for certain diseases, 2- **inoculation against serious epidemic diseases** and 3- **the discovery of antibiotics.** Specific medication, the improvements in surgery, blood transfusions and all the progress in science have still just begun to influence the prolonging of life in both human and animal populations.

We still do not know if man, during all these technological applications on disease, unconsciously modified the physical resistance of human beings and animals. We can believe that by increasing the survival of people and animals with vaccines, antibiotics or medication, man can intervene in the laws of natural selection. We can also believe as well that these interventions, although very relevant and tangible on disease, were able to modify, in a much more important way than we would have thought, the rules of the nature and the life of affected animal species. It could be that following their interventions on the life of human and animal populations, scientists have unconsciously modified the genetic codes of the individuals who have survived and, consequently, their descendants might be less resistant to certain infections.

Let us look at this hypothesis by talking about genetically modified organisms (GMO) and then about the possible future manifestations caused by technology and science on the future of the species.

Medicine and GMOs

Scientists have always believed that their reasoning and their intelligence was superior to "the wisdom" dating back millions of years and to the raw forces of nature. They consider themselves superior and capable of correcting "what seems to them as flaws of nature". Scientists in molecular biology can, for example, add a gene of resistance to the cold or any other gene or part of the DNA (genetic material), to the genes of plant cells or animal cells not possessing this gene. The DNA of plants so genetically modified is called "genetically modified organism" or GMO.

The GMOs are organisms genetically modified by scientists in laboratories to compensate, or so they think, for nature's imperfections.

Controversies are numerous regarding GMOs used recklessly in nature. Those knowing a bit of history are pertinently aware that very numerous interventions of man on nature have not always been beneficial to the environment or to the people themselves, as these scientists had thought! **When scientists modify their environment apparently to improve it, they often create, paradoxically at the same time, disaster and catastrophe. The history of science is filled with misdeeds caused by the hand of man.**

People remember that multiple developments and technological applications ended in bitter failure for nature and life on earth. Can we blame them today for being afraid of the new technological applications that other scientists, as intelligent as they may be, want to implant in their neighborhoods or their factories, without knowing the real repercussions on their life, their environment, on man and animals? **Scientists are always unaware of the complications which could arise from the implementation of their technology.**

People will never forget that **it is the scientists in research who invented decades ago, these thousands of dreadful pollutants that we know today to be toxic to life on earth.** Several of them, as mercury salts, polychlorinated biphenyl, greenhouse gases, carbon monoxide from our cars, to name but a few of these pollutants, would be responsible for the numerous damages to the environment and for hundreds of thousands of deaths each year. These scientists have perhaps solved a certain specific, at a given time, **but had not at all foreseen that the combustion of petroleum would produce one day more CO_2 than earth can degrade, or that these gases would contribute to the destruction of 85 % of the ozone layer and modify the climate for decades, if not hundreds or thousands of years to come!**

None of the scientists who contributed to the development of combustion engines had foreseen that the combustion of petroleum would produce enough CO_2 to one day melt the polar ices, flood the maritime coasts of numerous countries, produce cataclysms, eliminate thousands of animal species or kill thousands of persons ... **No scientist had ever thought that over 2,000 conifers would one day be needed just to clean up the CO_2 produced by a single car, in a single year, or that another planet is now needed to help the earth overcome this CO_2 alone.**

By one day introducing foreign bees to increase the resistance of aboriginal bees against harmful insects, biologists created a new and dangerous species, much less productive, harmful to man and that nobody can get rid of.

Chemists who used mercury chloride as a catalyst to build up acetaldeide did not know that the water discharged would contaminate the rivers, the fish, people who would eat them and poison or kill thousands of individuals.

Today, the Southern part of Quebec is invaded by thick clouds of Asian beetles imported by biology scientists from South America to destroy harmful insects found in corn. Now, this Asian beetle has literally invaded all of America. It gradually migrated over the years to the North of the United States and then to Southern Canada. Today, houses, buildings, cultures and cities all over America are invaded by these thick clouds of beetles. They infiltrate everywhere, multiply, hibernate and wake up in your houses the minute the weather warms up. Scientists do not know how to get rid of this plague they caused…

In his book *"Virage global, l'effondrement de notre monde est-il inévitable?"* Éditions de l'Homme, 2002 (Is the Collapse of our World Inevitable), Laszlo referred to some of Robert Muller's figures on the degradation of our world. Muller spent over 40 years at the United Nations. **"Every minute", he says, "21 hectares (52 acres) of rain forests are destroyed, 50 tons of fertile arable soil is lost and 12,000 tons of carbon dioxide are scattered in the atmosphere every time 35,725 barrels of oil are burned for fuel in manufactures or industries. Every hour, 685 hectares (1,696 acres) of dry productive land becomes deserted and every day, 250,000 tons of sulfur acid falls on the northern hemisphere in the form of acid rains. This degradation of the water, the air and the earth is too rapid not to lead to disaster".**

It is necessary that scientists, whatever their milieu, whatever government they belong to, become less conceited, stand with both feet on the ground, take a deep breath, show remorse and stop thinking, once and for all, that the linear way of reasoning of their science that they elevated to the rank of god, does not make them gods!

More often than not, unaware of the adverse effects of their technological applications in the long run, it is understandable that people are reluctant when a group of scientists wants to implant new technologies in their neighborhood. In the case of GMOs, fears result from the fact that nobody knows how these hundreds of organisms modified in laboratories and then implanted in nature, can affect the reproduction of native species, modify the body of those consuming them or cause other unpredictable inconveniences.

Scientific literature reports so far only few possible disadvantages from GMOs, such as the possibility that genes induced could produce toxic substances or allergies to the living organisms that consume them. The implantation of a gene from a Brazilian nut in a Soya plant has already demonstrated the presence of the known allergen of the nut in these plants. So far, we do not believe that anyone reported the presence of toxic molecules strictly speaking, but tests are still rudimentary and incomplete to this day.

Scientists do not know if the GMOs they make can interact with the DNA of wild plants in nature. Nobody knows to what point these genetically modified plants can hybrid with wild plants, and in so doing transform the environment or one day induce unpredictable disasters similar to those reported above. No scientist knows how to react should such a situation arise. Nobody even knows the possible effects of these modified plants on the body of those who will consume them. In the face of so many fatal effects caused in the past by technological applications, people are more conscious and more reluctant.

Everyone has the right to protest when new technological applications, such as the installation of hog producing farms in their area, the bottling of water from the neighboring groundwater, the burying of waste, the passage of electricity pylons above their homes, the installation of a nuclear power station in their immediate environment, etc. They will be the first to suffer the consequences if their source of water is contaminated, if the fish in their rivers are not good for consumption or if their territory is contaminated by radioactivity, etc...

When scientists get involved in manipulating the DNA molecules of life itself, when they modify as they please the genes of numerous plants, animals or men, when they clone animals and human beings, not to save the species in distress but to improve them, **it is high time that enlightened people start asking serious questions about the future of our world and it is time for governments to stand on their own two feet**...

Let us imagine as an example, at what would happen if a researcher inserted a new gene coding for xylenase in the genes of a tree's reproductive cell. The xylenase is an enzyme which liquefies the lignin of the wood. And the lignin is this fibrous substance which gives trees their rigidity. It is the lignin which insures the tree to maintain a vertical position. If such a gene was inserted, let us say deliberately or maliciously, in the DNA of a tree cell, this one would synthesize the xylenase which would liquefy its own lignin. If this artificial laboratory mutation of the DNA were passed on to the other wild trees in the forest, scientists would have created another severe problem they could not solve. They would have started, because of their mindless intervention, a process of uncontrollable reproduction of trees bearers of the xylenase gene. Trees transformed by this modified DNA gene could not grow vertically any more, because the liquefaction of their lignin would make them grow horizontally. Terrorists would gladly sow these seeds on enemy land. Everything is possible, isn't it? Of course, these results would appear several years or decades after their sowing, perhaps even after the scientists who would have made them have disappeared.

It is completely legitimate for people to be able to preserve their individual right not to consume GMO plants if they wish. So, they required from their governments that the plants containing **GMOs be identified on the labels of products or mixtures of consumer foods containing it,** since the consequence of using these transformed genes in the long run is totally unknown.

Seeing this, how can we possibly understand that scientists in power in the Canadian Government **refused to include in the law on labeling, passed in October 2001, that the GMOs contained in a product be identified on the label? Is this not a beautiful demonstration of the violation of individual rights by scientists in power?** Scientists that were never elected by the people and that will never be sanctioned for this action! Is this not how scientists in power steal the individual rights from their citizens? No man, whether in power or not, should have the right to deprive individuals of their fundamental rights as if they were second class citizens.

It is always necessary to remind ourselves that a century or two in the history of the development of life on earth is very little. Let us remind ourselves that **the thousands of pollutants poisoning the air today, contaminating our drinking water, the earth and the food that we eat, were invented by scientists who one day lived...**

Researchers have just reported for the first time an undesirable effect of the GMOs. Corn is often infested by a small worm called the "pyrale of the corn". This insect makes a great deal of damage in corn culture and pesticides must be used in great quantities to get rid of it. To reduce the quantity of pesticides, bio geneticists (scientists) inserted in the grain of the corn a gene from a bacterium found in the earth, called Bacillus thurigiensis (BT). This gene codes a natural toxin that destroys the "pyrale of the corn". The new gene of the transformed BT corn now contains the coding gene for the natural bacteria protein capable of destroying the corn parasite. These transformed BT corn seeds are widely used in Quebec in about 40 % of the corn cultures.

Doctor Jean-François Narbonne, a toxicology professor from Bordeaux University in France demonstrated, during the International Colloquium on GMOs and Food held in Montreal in December, 2001, that the sediments of the Saint-Laurence River are now contaminated with the toxin of the BT transgenic corn. At the confluence of the rivers draining large cultures of BT corn, the sediments contain five times more BT toxins than the drainage water of the nearby sediments of agricultural lands. Researchers believe that the roots of the BT corn could pass on their genetic sequence to other bacteria in the surrounding earth and that these could in turn build up BT toxins; if this hypothesis is confirmed, it could represent a chain production of toxin transfer to other bacteria. If the gene of the toxin could be transferred to other bacteria, it would cause damages to other elements of the aquatic flora, by some sort of biologic chain reaction that no one had foreseen! Another hypothesis is that sediments retain the BT toxins in larger quantities.

Much worse is the discovery of the corn contaminated with BT toxins in Mexico. Let us remember that the Mayas living in this country gave humanity the biodiversity of corn. A study held in 22 rural regions of Mexico in September 2001, revealed that the wild corn is now contaminated with transgenic material coding for BT toxins. Indeed, from 3 to 10 % of the grains of wild corn now carry BT genes. All the scientists, naturally, have no clue on how this genetic material was able to reach such isolated locations, sometimes dozens of kilometers from the nearest road! Have these scientists forgotten that pollen travels thousands of kilometers carried by the wind? Whatever the reason, one thing remains certain, the Mexican wild corn, guardian of the biodiversity of corn for over 12,000 years, is now contaminated with this BT toxin! No one can yet estimate the impacts of this contamination on the biodiversity of wild corn...

In 2001, a field belonging to the Prodigene Company in Nebraska, USA, received genetically modified corn cultivated by a soya farmer for the production of medication. Moleculture is a new application of genetic agricultural engineering intended to fabricate medication. This experimental field of moleculture of genetically transformed corn, insufficiently cleaned, still produced medication in 2002 and contaminated silos of soya cereal. Greenpeace groups demanded that this kind of experience immediately end. At the beginning of November, 2002, the Minister of Agriculture of the United States ordered the Prodigene Company to destroy, at its own expense, the loads of soya in Nebraska where the corn was cultivated as well as in the corn fields in Iowa close to a Prodigene experimental farm. Charges leading to possible imprisonment will probably be brought by the Ministry of Agriculture against Prodigene and its leaders, as we finish this book.

If we must acknowledge that if GMOs are generally useful and bring profits of billions of dollars to the companies producing these modified grains, they are by no means necessary to life on earth that went on long before they arrived. The pursued objectives in the manufacture of GMOs are of a pecuniary order above all. Their use is not a question of life or death, although they aim at reducing pesticides or seem to improve certain cell functions. If someone could demonstrate that they intervene with the natural selection of wild plants, it would be necessary to forbid their use before a disaster occurred. To do so, we must have political courage and knowledge!

Let us repeat that it is inconceivable, in the actual uncertain conditions of the possible effects caused by GMOs on nature and life **that the Canadian scientific powers refused to register the GMO contents on their product labels.**

The fact that the cries of alarm of the people are not heard by scientists nationwide is very harmful. **It is also inconceivable that such a decision concerning every individual can be taken by the scientific powers who decide for each one of us.** These smug persons do not have the right to decide for people, any more than their colleagues in education have the right to decide of the education of our children or that their colleagues in the Health Department have the right to decide about my health and the medication I have to take. That is socialism at its best!

"Tell me, do you believe in a collective mind-destroying plot?" asked Foglia, a journalist from *La Presse de Montréal*, to his friend Renato? "There is no plot, it is an admitted objective. There is no plot, it is a system…" (*June 15, 2002*)

A short Radio Canada news bulletin seen on the Internet on February 25, 2002, reported that researchers from the National Council of Research for Canada, which belongs to the National Science Academy, estimate that the organisms regulating labeling (that is their colleague scientists) should worry more about the long-term impact of GMO cultures in the environment. "The introduction of a biological novelty", they write, "can have undesirable and unpredictable effects on consumers and on the ecosystem". This scientific report believes that although the American Ministry of Agriculture has improved the regulations on GMOs, that the labeling process should be reviewed by asking public participation, by encouraging contradictory scientific expertise and by presenting more clearly the data and the methods used to decide on a measure of regulation.

The Council did not mention any specific risk caused by transgenic plants, but indicated that studies are continuing even after the introduction of GMOs in commerce. The Coalition of Alert on GMOs happily greeted this report which confirms that the actual regulations on GMOs are weak and inappropriate.

The GMO is not the first intervention of man on nature

The GMOs presently creating a debate is not the first example of the intervention of man on nature. **The inoculation of people and animals to prevent epidemics is undoubtedly the first, or at least one of the first, direct interventions of man and science on the fragile balance between man and nature.** We still do not know to this day, after decades of vaccinations, if these interventions impoverished the genetic heritage of the survivors. It is important to know if these interventions of man on the natural selection of individuals have influenced the resistance of their descendants to infection. In other words, have animal species inoculated with all kinds of vaccines become genetically impoverished by such interventions?

Could scientists have unknowingly messed around with the genes?

No one can doubt that inoculation against smallpox allowed thousands of persons, who would otherwise have died, to survive. Inoculation against smallpox has succeeded in eliminating to this day this powerful natural selector from the surface of the earth.

Have the mass vaccines against smallpox, the implementation of hygiene measures, antibiotics, other subsequent inoculations against different micro-organisms or treatments used to prolong the life of patients, weakened the survivors? Have the survivors escaped from the process of natural selection because of these interventions? Very conceited would be the scientist who would dare assert today that individuals and their descendants, who survived because of these interventions, have not been ultimately affected as a result of their resistance to infections!

When trying to control or to check an epidemic with an inoculation or to cure an infection with antibiotics, scientists and doctors only cater to immediate and humanitarian needs. No one cares enough to find out in these painful moments, if these interventions are going to affect, one way or another, the future generations. Looking back, however, it is reasonable to wonder if these interventions by scientists did not stop natural selection from playing its role as it always had in the past! By using these therapeutic and preventive measures, medicine preceded the

GMO and acted on nature, without suspecting it. It was probably the first intervention of man on natural selection.

No scientist knows if these interventions allowed to "artificially select" the genetic characteristics that would weaken the descendants of those who underwent them. No one ever took any interest in finding out, to my knowledge, if the survivors passed on impoverished characteristics, meaning those that natural selection eliminated formerly by epidemics. It may be necessary to wait a very long time to get a clear answer to these questions. Interventions on natural selection can take years before appearing. In the long process taken by nature to perfect the life of the species, time is not an important factor.

Let us analyze some data suggesting that the inoculation of people and animals can effectively act on natural selection and produce a weakening of the species.

Let us keep in mind that the principle of natural selection "selects" the less resistant individuals more likely to die of natural epidemic or infectious selectors. Logic teaches us that infectious microbial agents capable of causing epidemics are natural selectors in all animal and plant species. By preventing these natural epidemic selectors from acting by vaccination, one can logically wonder if we did not allow the survival of individuals who would otherwise have been eliminated by nature. Are the survivors more apt to development other diseases? Do these interventions engender the multiplication of more fragile descendants less capable of defending themselves against other infections or the reproduction of defective genes of resistance in the survivors? **The matter deserves attention...**

Before trying to understand how the irresponsible interventions of man on natural selection have modified human and animal race, let us first consider the role of the infections and epidemics of yesterday as natural selectors for the less resistant individuals. We will then analyze the data that suggest that these interventions really affected the natural selection of these individuals and that of their descendants. Different levels of proof can be advanced to support this hypothesis.

Indirect proof that infections and epidemics are natural selectors

Darwin and Lamarque taught us that natural selection was a combination of mechanisms that allowed the strongest individuals to survive to insure the survival of their species. According to the principles of natural selection, it is not presumptuous to claim that infections and epidemics have always been nature's way of selecting less resistant or genetically weaker individuals to reproduce. Only the most resistant generally succeed in surviving for a long enough time to insure the survival of its species. Numerous proofs demonstrate that the weakest, less skillful and less resistant individuals are generally not selected as breeders for their species. Selection through combat, another mechanism which determines which will be the procreator, speaks for itself on these physical natural selectors.

Before modern preventive or therapeutic measures of mass intervention on epidemic diseases, individuals without sufficient immune baggage to defend themselves adequately during an epidemic did not survive long enough to assure the survival of their species. Such processes of selection are found in all human beings, animals, plants and other living organisms.

Numerous observations demonstrate that mortality at a young age, well before the age of reproduction, was a common phenomenon in nature. Those that survived were inevitably the most resistant and it is among them that we find those assuring the survival of the species. This process was rather powerful because the medical annals demonstrate that before these interventions, about 50 % of the population died very young, leaving one individual out of two to insure the survival of the species. Such a high mortality rate has been found in a recently discovered Amazonian population where preventive measures against infection have not yet been taken, as were the non inoculated African countries hardly fifty years ago, shortly after the Second World War.

According to this logic, the more effective the vaccination are against a micro-organism such as smallpox, the more the populations are left with individuals without natural resistance. Inoculation against smallpox was so effective and allowed so many individuals to survive that this natural selector is today eliminated from the surface of the earth officially since 1977. It is thus acceptable to suppose that several thousand individuals deficient against this virus have survived and reproduced until today, uniquely because of this vaccination. And so in the world, at the present time, there are thousands of "possibly deficient" individuals who were not eliminated by the natural selector that was smallpox.

If we add to the individuals who survived because of smallpox vaccination those that were immunized afterwards against other infectious diseases, preventive measures of hygiene or the more recent use of antibiotics, the current population would contain several deficient individuals who escaped natural selectors, and who would not be of this world without the artificial intervention of scientists. In other words, our world will contain large numbers of individuals who survive because of "artificial selective pressures". Must we conclude that, as beneficial as they were, these artificial interventions of man on nature could indeed have been realized to the detriment of the surviving populations?

Consequently, the more intense the natural selective pressures are on a given population, the more resistant the individuals become and the higher is infant mortality rate in these populations. Several other levels of observations suggest that this hypothesis can be verified. Let us see how things were before and after the interventions of man on natural selection.

Certain observations made before the intervention of man on nature are relevant to this subject. Just 75 years ago, except for smallpox, other inoculations and the interventions of people on natural selection were very rudimentary and antibiotics did not exist. In Africa, to quote only this example, inoculation was still far from being common at that time. **Statistics demonstrate that before the Second World War, infant mortality rate in Africa touched 40 to 50 % of the children.** They would die before the age of 5 mainly because of natural infectious selectors. So, because of the great natural selective pressure on these populations, only the best genetically equipped individuals succeeded in surviving beyond five years of age. It all happened as if the most fragile were eliminated before being able to reproduce.

It would be rather easy to demonstrate that the South - Saharan Africans, living before the massive interventions of man (say before 1945), were more resistant individuals to infection because their infant mortality rate was higher. It is almost obvious to conclude that the weakest were eliminated at a young age. These facts can be confirmed scientifically in various manners, even today. In the first place, by looking at past statistics, we could measure the average rates of total immunoglobulin or specific blood antibodies in a number of non inoculated individuals, and compare the results to those obtained in an equal number of individuals matched for sex, age, race and country of origin. We could also compare the rates of survival of populations using antibiotics since the end of the forties, with those populations which do not use them or very little. The rates of childhood mortality before the age of five and the average survival rate of the individuals of these populations would also help confirm these differences.

Let us compare the data with those acquired after the interventions of man on natural selection

We easily forget that before vaccination and hygiene measures, the situation in Africa after the Second World War was comparable to that of

Europe or America in Pasteur's time, more or less over a century ago. Child mortality rates before the age of five by infectious diseases resembled at the time those of Africa. The mortality rates in young individuals were enormous compared with those observed today. Individuals born a hundred years ago did not have an expectation of life of 78 or 82 years as we do today in the developed countries of the planet. The average age of survival in a group of isolated Indians, recently discovered in Amazonia, is barely between 25 and 30 years. It is not presumptuous then to say that these individuals have not at all benefited from the interventions of modern medicine on natural selection and that the average survival of these people is appreciably similar to that of the Romans who lived 2,000 years ago.

Nobody can deny that countries using since over a century the hygiene measures recommended after the discovery of microbial agents, the vaccinations of the very young individuals and the more recent use of antibiotics, infant mortality has dropped drastically. The average survival of these individuals increases in a very considerable way in parallel with these parameters, and reaches 78 years in men and 82 years in women in Canada.

Experiments on animals suggest even more strongly that the interventions of man on natural selection really influenced the survival rates and decreased the resistance of the descendants.

Biologists and animal breeders have repeatedly demonstrated that besides physical fights as a natural selection in possible breeders, there are always a percentage of more fragile subjects who do not reach the age of reproduction. Should they reach it, they do not possess the characteristics of good breeders. As an example, before the intervention of man on natural selection, the chickens that succeeded in reproducing were selected by nature as in all other animal species. Those that reached the age of the reproduction and that raised themselves to breeder rank were inevitably the most resistant ones. To the domesticated, cherished animal, inoculated and followed by a veterinarian, the control of undesirable subjects not wanted as breeders is essential. But breeders regrettably do not always take care to make sure of the quality of the breeders. That is why these animals do not live very long and are more subject to various diseases. For instance, is it a good thing that veterinarians treat domestic diabetic dogs with insulin? When the process of natural selection is no longer exercised, the populations weaken and are less resistant. When man keeps natural selection from happening, he breaks the rhythm of secular balance in the animal species on which he intervenes.

Studies on animals demonstrate without a doubt, that when a deficiency occurs in an animal because of a decrease in resistance to infections, this can very well be passed on to the descendants. Researchers even demonstrated that the descendants are sometimes more difficult to treat and protect if they receive the same vaccine as their parents. We observed that on the poultry inoculated against the New Castle virus, a virus that causes much damage in the non inoculated flocks, the descendants are much more resistant to the vaccine against this same virus. The inoculated parents do not only pass on their fragility to the virus, but their descendants have to receive more than double the dose of virus to defend themselves adequately. They seem to be doubly deficient. If the descendants are more difficult to immunize against the same virus, it may be because genetic characteristics were passed on.

To transgenic animals of which the sheep Dolly was the most popular representative, the genetic characteristics are passed on to the descendants in a rather strong percentage. In extensive breeding practices in several animal species, breeders often select their dairy cows for their dairy characteristics and cows for butchery by their length, their height and their muscular mass. In so doing, they do not necessarily select the resistance or the immunologic capacity of the individuals more apt to resist to infectious agents. It is thus frequent to notice that these highly selected individuals for one or several specific characteristics are less resistant in other characteristics and this fragility is transmitted to their descendants. So, when strong or weak characters are created, they are genetically transferred to the descendants. These data demonstrate that individuals selected for particular genetic characters can transfer just as much the selected characteristics as those not selected.

Veterinarians and racehorse breeders know pertinently that horses selected during generations for characteristics which favor fast racing are often more fragile in a range of other characteristics. These weaknesses are genetically transferred to their descendants and this is easily observable.

Genetic diseases observed following particular techniques of reproduction, used in the English bourgeoisie of a still recent era or following consanguine marriages in royalty, still demonstrate that transmission of pathological and unwanted genetic characteristics also appears in man.

Numerous studies made on human and animal populations confirm that less resistant or selected individuals for characteristics other than resistance to infection can very well be passed on genetically to their offspring.

The phenomenon of mass immunity

Researchers observed that non inoculated individuals can be protected, not by their own immune competence but because of what they call "the mass immunity". Researchers have demonstrated that when certain individuals in a group are immunized, the infection does not propagate as much on the non immunized subjects in case of an epidemic. So, those that were not immunized or that cannot be immunized are protected by the phenomenon of "mass immunity". For this to happen, it is necessary that a certain percentage of the subjects in the group be inoculated.

Other scientific data demonstrate that epidemics can be prevented in human and animal populations when about 60 to 65 % of the individuals of a group are protected against the disease. When this proportion decreases, the risks of recurrence of the infection are greater. Scientists in health services will have to re-immunize people against certain infections, because the rate of immune subjects or the rates of specific antibodies in the population became too low.

Individuals responding and non-responding to infection

Results from French researchers still reinforce these data, because they further confirm that the individuals selected by man interfere on natural selection. Biozzi's team of researchers identified in laboratory mice individuals "not able to respond normally" to some infectious antigens and presenting inadequate immune responses against a variety of selected micro-organisms: they call them "the non responding mice". Mice of another group responded well to the same micro-organisms and were showing good immune responses against these antigenic substances: they were called the "responding mice".

The researchers managed to develop for their studies a pure breed strain of the non responding mice and a pure breed strain of the responding mice to these antigens. Numerous studies in these two strains of mice allowed to identify and localize the immune defective mechanisms of the non responding mice and to compare the results with those obtained in the responding mice. Responding mice made more adequate immune responses against the antigens than did the non responding mice. Contrary to the mice which present good immune responses against micro-organisms and that acquired immune resistance to infection, the non responding mice were found to suffer of **an immune defect implying the presentation of antigens by macrophages** to the immune-competent lymphocytes. In other words, the non responding mice were bearing a defect of the antigen presenting cells. These cells, specialized in phagocyting (eating and digesting) micro-organisms, have a particular network of enzymes to digest matter and present particles of the digested antigenic matter of the micro-organisms to specialized lymphocytes for differentiating self from non self, one of the fist step to build up immune responses against the none self. The immune responses of the responding mice were different between these two groups of mice, because of a genetic defect in the non responding mice located at the level of the presenting cells. These results confirmed that the resistance and the immune response in one group, as well as the lack of resistance and the incapacity to acquire effective immune response in the other, were genetically transferred.

Proof by the common sense

Logic and common sense also suggest that the interventions of man on nature by any kind of therapeutic or preventive measure prevent the natural selection from manifesting itself normally. So, it is logical to believe that whenever man intervenes with vaccines, measures of hygiene or antibiotics, there are a greater number of less resistant or weakened individuals, resulting from the interventions of man trying to correct what he identified as a problem...

Who are these less resistant individuals?

Let us ask ourselves what happens to the less resistant individuals in nature, in other words to the humans artificially selected by the vaccinations of men. They are deficient and they survive, and these weaknesses are transferred to their descendants and so weaken their respective species. Who can assert that the individuals of our societies, more fragile in contacting recurring infections and requiring numerous doses of antibiotics every year, or those who suffer from chronic debilitating diseases, repeated nervous breakdowns, auto-immune diseases, cardiovascular diseases or other, are not descendants of these interventions of man on nature? We know absolutely nothing about this at the present time, but it does not mean that it is not so. If one day a new epidemic arose and thousands of individuals died, we would have an element of supplementary proof to that effect. This would strongly suggest that those who managed to escape natural selection because of this human intervention on nature are in fact those more at risk of being the first selected by a new infectious natural selector. Disasters could be just as dreadful if not more so than they were when the epidemics of smallpox, plague, diphtheria or scarlet fever formerly devastated human populations.

People who present hyper-reactions to inoculations, as those developing multiple sclerosis, those presenting hyper-reactions to vaccination itself or who contract poliomyelitis simply by changing the diapers of their recently vaccinated babies who excrete the virus in their feces, are not only more fragile individuals, but probably more "immunologically" incompetent. Medical literature often demonstrates that some individuals are unable to respond to immunization, even though they have been inoculated before. It is more than likely that these individuals will be the first to contract diseases in case of an epidemic infection.

"Life is finding its way" said Doctor Malcolm in the film "*Jurassic Park*". Nature has time for itself. It will take its revenge one day against all the injuries it endures from man. The quick resistance that micro-organisms develop toward antibiotics is an eloquent testimony of this. We must remember that during an epidemic, it was not rare to see 50% or more of a given population decimated in a few short months by a single infectious natural selector. We do not know yet if new diseases such as the insulin-dependant diabetes, auto-immune diseases, multiple sclerosis or other similar diseases could not be connected to these techniques of intervention of man on natural selection.

We mentioned earlier that before the intervention of man on natural selection, infant mortality touched almost 50 % of the individuals of a given population before the age of five. This important selective natural pressure could possibly explain the auto-immune absence of multiple sclerosis (MS) or other immune diseases in Africans populations. It could also explain why the first description of MS was not done before 1830 by Charcot, about thirty years after the first mass vaccination against smallpox. This terrible disease was endemic in Europe and in America at this time. No one has yet reported cases of multiple sclerosis in native Africans, except in Ethiopia. The cases of multiple sclerosis in Ethiopia could be connected to the strong migration of Italians in this country at the beginning of the nineteenth century. At the time, the white Italian emigrants as well as Europeans had already undergone smallpox vaccination before migrating to Africa. Who can assert that the natural selectors eliminated in early childhood only the weaker individuals who would have contacted this disease, had they survived? It is tempting to assume that among the less resistant individuals who migrated, there were still those that had not been eliminated by natural selection or by vaccination. How many diseases of this kind are present in our populations today because of the artificial selection by vaccination?

To the data demonstrating that genetic deficiencies multiply in humans and animals after vaccination, it is possible to add other data suggesting that natural selection could very well have eliminated certain genetic immune deficiencies. To support this hypothesis, let us add the scientific data suggesting that certain groups of genes or haplotypes, recognized to be associated to fragility to infection and auto-immune diseases, are nearly absent or have strongly decreased in African populations. The incidence of the HLA DR3 and particularly of the haplotype HLA A3-B8-DR3 - a haplotype of the HLA system (human lymphocyte antigens), is well known in Caucasians to be associated to auto-immune diseases and possibly to an increased susceptibility to infections. This haplotype is present in only 3 to 5 % of African populations, compared to about 35 % in Caucasian populations. The decrease of this haplotype in Africans could also suggest that the populations subjected to strong natural selective pressures at a young age could have lost this haplotype of "susceptibility to infection".

If the haplotype should increases in native African populations, now that man has intervened in the processes of natural selection for some decades, it would be necessary to review our data on the subject. If more diseases were to appear in the first vaccinated countries, it would also be necessary to revise the data. **They could indicate that the haplotype HLA A3, B8, DR3 is really increasing, and that it can be used as a marker of susceptibility to infections or auto-immune diseases and perhaps a "witness marker" of artificial selection by vaccination.** Are we entitled to expect one day the appearance of multiple sclerosis or auto-immunizing diseases known so far in the eastern countries but not yet in African populations? Only time will tell if we were right.

The history of smallpox in the world teaches us that the Arabs and the Asians were selected earlier than the Europeans by smallpox. They used a special kind of inoculation at the beginning of an epidemic, which consisted in inoculating people in the nostrils with the pus of the first pustule of smallpox. This proof could be delayed however because Africa and Asia are today ravaged by another natural selector even more powerful than artificial selection of the people, HIV (human immunodeficiency virus) of AIDS.

Man had already intervened on natural selection and modified human and animal genetics well before manufacture of the GMOs without knowing it. These interventions have influenced natural selection, disrupted nature's balance and weakened many species, and so today's populations are a mixture of more or less resistant individuals. These genetic defects and immune weaknesses were passed on to their descendants who have been artificially selected. It will soon be possible to identify these individuals as it is already possible to select, by adequate laboratory analyses, the profile of the animals most apt at reproducing to insure a better selection of breeders. So, if an epidemic arose following the appearance of a new virus, the rates of survival of these populations could be even worse than those observed 500 years ago in Europe, because of the presence of the immune deficient individuals in today's populations. Simple logic prevents us from thinking that it could be it otherwise.

Internationalization of danger

On June 29, 2000 in Washington, Mr. David Gordon, member of the National Council of Information (NCI), gave a lecture to the members of the Commission of International Relations of the House of Representatives.

The scientist presented data indicating that infectious diseases will represent in the next 20 years "an increased risk to the health and to the security in the world. New infectious diseases, added to those not yet controlled such as tuberculosis, will put an increased risk which could be catastrophic to health in the world".

Results show that this increased fragility to infections could have a negative impact on the armed forces in the future. "These diseases will put in danger the citizens of the United States in their own country and abroad. They will threaten to spread in the armed forces abroad and will aggravate social and political instability in countries and key regions, where the United States has important interests". Threats will be of various natures. There is a growing eventuality that must be looked at, and that in case of war or of biological terrorist attacks against the United States or its overseas interests, these problems could be aggravated. "The negative impact, in a large number of infectious diseases within the national army will also be felt in the operations of regional and international peace prevention. They will limit the efficiency of the army and create new vectors of disease spreading within the peace-keeping forces as well as in the population". If the Amerindians of the fifteenth and sixteenth centuries had received such warnings, maybe they would have been less compassionate towards the European invaders who decimated them deliberately with smallpox within a few decades!

Regarding the GMOs

All these data of man's interventions on natural selection have implications on the GMOs. Scientists of molecular biology fabricate GMOs today like it was candy. Their objectives are to compensate, as we have mentioned above, for the "weaknesses" of nature. Their objective, so far, is not aimed at saving species threatened with extinction. Under these circumstances, there is no doubt that the GMOs are interventions on nature unnecessary to man.

The side effects of GMOs are just starting to be observed after less than twenty years of culture. The fact that the BT gene of transformed corn cultures can be hypothetically transferred to bacteria or to other botanical species or animals of the aquatic flora is now starting to worry researchers. When the BT genes transmit themselves to other plants and are harmful to the conservation of the genetic patrimony of wild plants, there is a problem. Even though these dangers have not yet been completely and scientifically proven and that no one has yet demonstrated their harmless character in the long run, common sense tells us to be cautions... When researchers show that biologic agricultural lands are contaminated with BT genes in areas surrounding corn fields and that dozens of plants are already modified through this intervention of laboratory men, we must ask ourselves if it will still be possible tomorrow to make real biological farms. Can we hope that the collective action taken against the producers of BT corn by the farmers whose farms were contaminated by the BT genes will change the course of things? Only time will tell.

It is indeed necessary that scientists in high power acknowledge and confess their mistakes. They have already listened too much to the blackmail of their colleague scientists of private multinational companies making these GMOs. They must modify their laws and at least warn the population and the elected members that they committed other scientific errors: 1- To have authorized, without valid scientific proof, the use of the corn's GMOs; 2- To have authorized companies to not indicate the presence of GMOs on the labels products containing them.

It is abnormal and criminal for the Biotechnoloy Industry Organisation of the United States and the North American Association to even dare propose that the tests for detecting the genetically modified corn be forbidden in the Corn Belt of the American Mid-West and on the soya culture in the provinces of the Canadian prairies. Just thinking of acting so is simply criminal.

A few useful but unnecessary modifications made by the scientists making GMOs, aim once again only at satisfying the researchers' ego but mainly the commerce's ego. Nature took thousands of years perfecting what researchers undo sometimes in a few years through their "reasoning", but without the secular knowledge of nature.

Risky medical technologies

The interventions of medicine on naturally occurring genetic diseases are a part of the duties of medicine. The mission of the doctor is to treat patients, to the best of his scientific knowledge, by using the appropriate technology for each disease. The objectives are to assure that patients have the best possible life conditions. But there are problems in the horizon. Some years ago, for instance, few patients suffering a neonatal genetic disease could reach the age to reproduce and consequently, to reproduce the disease. It is not always the case today, and tomorrow could be worse, because of the present efficiency of treatments for certain genetic diseases. Science helps some of these patients and noticeably prolongs their life expectancy, due to the knowledge acquired on these diseases, to new medications and to the implementation of new genetic therapies. Some of these patients live much longer than before, in better condition and can hope to live beyond the age of reproduction.

Some years ago, an 18 year old girl affected with cystic-fibrosis decided on her own to have a child. She gave birth without even knowing or inquiring if her child would also suffer from this disease... **Patients suffering from cystic-fibrosis or other genetic diseases should at least know that it is possible to select and receive transplants of their own embryos exempt of genetic defects, and detectable through modern techniques.** All embryos do not carry the defective genes. The ideal means of prevention consist in the transplantation in mothers of embryos exempt of the genetic flaw. They deliberately break the laws of nature by reproducing themselves without taking that precaution.

The treatments that prolong the life of a patient affected by a genetic disease do not give him the right to multiply his disease neither does it give him the right to have society care for the sick children that were deliberately reproduced. It is not eugenic to believe that it is the duty of every conscious human being to remember these things. If such problems exist today, it is because the efficiency of medical treatments enabled patients to reach the age of reproduction. We should not have to wait for legislators to intervene with their big legal paws for doctors to teach their patients to discipline themselves.

From the ethical point of view

It is unfortunate that scientists have not yet begun to study the long-term effects of their technological applications before making decisions that could be disastrous. I do not remember having read many articles treating of this possibility. As usual, scientific papers rather speak of the immediate advantages of knowledge and technology for humanity, as it is the case for all those scientists making GMOs. Few publications signal their disadvantages, although there are many.

In front of all these uncertainties and past errors, natural morality at least imposes to scientists the rules of strict logic. **It is not eugenic to say that all patients should select healthy embryos to reproduce. It is not eugenic to state that medical and legal measures must be taken to prevent the reproduction of these diseases when patients do not discipline themselves.** Facing these eventualities is not bad medicine. Not to face them is a lack of simple common sense because society possesses today enough knowledge to prevent the transmission of these diseases. In certain cases and to all those in age to reproduce and who refuse to act responsibly, medicine should think of permanent contraceptive methods. Otherwise, why do research and have expensive treatments paid by society? Why make all this research on diseases, if it is to keep on allowing the patients to continue reproducing their diseases? Medicine would turn around in circles as it is done today in several areas of science and technology. Interventions on human and animal cells, human cloning, fertilizing an ovule with a single sperm cell is bad scientific application. Scientists have now demonstrated that in children conceived with a single sperm cell injected into the ovule, the frequency of health problems is greater than in children naturally conceived (*R.D. Lambert, Human Reproduction, 2003; 18(10): 1987-91).*

The popular press often creates false hopes for patients when making important statement of research work consisting in preparing, let us say "humanoid pigs" by transferring human genes in the ovules of hogs for instance, with the aim of palliating one day to the possible lack of organs to transplant. Xeno-grafts (grafts between species) will still represent for decades too many challenges to be used in organ transplant. To think that Health authorities will authorize the transplantation of live hog organs to a human receiver is still pure scientific fantasy at the present time. The risks of passing or reactivating an animal virus in a transplanted organ into the human being are of major importance. Is it necessary to remember that the Spanish flu, following the First World War and killing millions of people throughout the world, was caused by the activation and the transformation of the flu's virus passing through pigs. There are indications suggesting that the AIDS virus also results from the passage of a simian virus to man. The transmission of the mad cow disease in cattle possibly comes from feeding of cattle with food containing carcasses of sheep infected with the prions (virus) having Visna or Scrapie. There are indications that the Creuzfeild-Jacob disease in man comes from eating beef infected with mad cow disease. On the basis of current scientific data, it will be necessary to wait for decades before such transplants can decently be made. For all sorts of similar reasons not presently justifiable, science is turning around in circles in too many areas.

I observed during my years of medical practice that the majority of patients who knew they were bearing a genetic disease, did not want to transfer it to their descendants. It is however essential that these patients be treated as well as possible and with all our compassion and diligence. All human beings have the right to live in the best possible health conditions and as long as possible. The elementary natural rules of ethics impose however on each of us to take all the possible measures to not stupidly multiply a disease that science has allowed to control.

In spite of the ideologies of certain good people, it is simply inhuman to let reproduce severe intellectually handicapped persons who will never be able to care for their offspring. The society does not have to take care of all the intellectually handicapped persons who never asked to be born and who will never have true parents or true families... That also, is not eugenic, it is common sense.

Man's interventions to animals allow controlling mass productions of human food. Before extensive breeding for consumption, the less resistant animals were eliminated by natural selection or entered the food chain for consumption. Problems created today by extensive breeding are of a quite different order. To meet the immediate needs of the market (a lot of meat, low fat, longer, higher, etc.), breeders select special varieties of animals responding to the desired market criteria. These special demands on the market promote the elimination from farms of dozens of varieties of animals. Only hogs, beef, poultry possessing the requested physical genetic and reproductive characteristics are permitted to reproduce; hundred of others not meeting market requirement are eliminated. We are suddenly becoming aware that thousands of genes and thousands of different genetic properties, skillfully selected by nature and by man for centuries, were simply eliminated from the humanitarian genetic patrimony only to meet the immediate need of the market.

It is the same at many other levels. For example, dozens of kinds of graminaceae other than wheat, corn, oat, barley, rice and the rye cultivated in large spaces, were left wild or have disappeared because one day man decided to cultivate only certain varieties of these plants on a larger scale... Such losses for the genetic patrimony! Can we talk about ethics for plants? I do not know, but nobody knows if in terms of evolution, the animals or plants selected to meet the needs of the current markets will be still beneficial tomorrow. In terms of evolution, these human adventures are too recent to allow making conclusions. But, in the face of so many risky interventions by scientists and of many non essential technological applications in life, the greatest caution is necessary. I am not certain at all that scientists have really learned from their bad interventions on the environment, on life on earth and on people themselves.

One thing is certain: the human or animal populations that underwent the interventions of man over natural selection are now well marked by a decline of their respective species. Regrettably or thankfully, none of us will still be here to observe this. Maybe, in the face of all these only rational and inhuman situations, we will pay more attention to the disadvantages of technical applications than we did in the past.

E. Discussion and reflections

Simultaneously to the scientific and technological revolution, we are living today a deep revolution of the humanity of man and society. All things have a good and a bad side, even technological progress and science. The data presented indeed demonstrated that from the progress of science and technology appear numerous unwholesome aftereffects which destroy the environment and the elements absolutely necessary to life on earth. It also appears as a consequence of this progress that people think that knowledge and science gives them the power to destroy their universe and to think that they own the truth. All levels of human activity have undergone the negative impacts of this progress as they undergo the advantages. The negative aftereffects of this progress continue to be so numerous and disastrous for the environment and the people that one can foresee the beginning of the end of life on earth.

Technological applications also affect man by making him a slave to the technology he uses and by making him more and more materialistic. Worse still are the technologies which tend to deprive him of his rights and liberties, to dull him, to materialize him and to dehumanize him. Numerous proofs demonstrate that science is much less powerful than people believed at the beginning of industrial times. People are slowly becoming aware that scientists, technocrats, intellectuals, etc. have imposed all their thinking on humanity as if it were the truth, but it is deficient and incapable of solving the large number of problems they created themselves by applying their technology everywhere without proper judgment.

So far, science has been only an accumulation of trials and errors. The research method of scientists is based only on "reason" (logos). So, it is rigid, linear and mechanical. It can be applied only to the study of the material things of the universe. If science is limited to answering only to things relating to matter, where are the certainties and the truths? The method of "scientific reasoning" has spellbound man so greatly that it is accepted as the perfect "mistress and that her truths were impossible to circumvent". If appreciative errors in technological applications stemming from this methodology and this way of thinking of scientists are so numerous and harmful, it is that the method has its faults and that the scientists are far from being gods.

Today, the world knows that the matter on which scientists work is neither eternal nor a god, as had thought Hegel and his friend Karl Marx (cf. Chapter 7). But scientists and intellectuals continue as if all this is still true and as if their legitimacy rests still on the fact that matter is a god and that it can still subject man to its domination.

People are becoming more and more conscious that what scientists call a "technological benefit" is far from always being that. Scientists speak rarely about the inconveniences caused by their technological realizations. The scientific literature is also rather quiet on this burning subject. Nevertheless, the negative aftereffects of science and technology are so numerous that it would be advisable to speak now, not about the scientific revolution, but about the revolution of the side effects of technological applications on the environment, on life on earth and especially on human being and societies.

Edmund Burke, the politician, wrote one day in "*Attribution*" that "in order to triumph, evil only needed the inactivity of the good people". They are trapped in the science and the "reason" that continues to deny the sacred and the existence of the non material fundamental human necessities of man. People are still blinded by the light of this incredible progress and lose courage in the face their helplessness to change things. Too many individuals continue to delude themselves in the material "velvet" of technological applications without further pushing their reasoning. Too many individuals have not yet opened their eyes to consider the damage left around them simultaneously by science and technology. Too many individuals are still rooted to the spot in front of all this science of "reason" which allowed man to evolve quickly only on the material plan.

To the numerous examples of destruction of our physical environment by technology, indisputable proofs add their weight to demonstrate that science and technological applications profoundly dehumanize man, his leaders and his societies. The leaders in power even succeed in making him more and more materialistic: they dehumanize him, make him regress and dull him. **Science and technological applications has also greatly helped the materialized leaders to further extend their domination on man and society.**

CHAPTER 4

The dehumanising effects of technological applications on man and society

Introduction

The side effects of technological applications are not limited to the terrible physical damages on the environment they also lead to disastrous side effects on the psyche of people and society. Intimately linked to their materiality, these side effects maintain the eternal conflict opposing the materiality of things against the non material human side of man. Society's problems and chaos often take root in this perpetual conflict between the human and the material.

Nevertheless, the inconveniences affecting the human part of man are more subtle than those affecting the environment. As in the case of the environment, these damages only appear after a while. We must often remind ourselves that technologies are material, mechanical, automatic, inflexible and devoid of all human meaning. What is most harmful is the way they impose and dictate their way of life to people and society. Those who are not careful soon become their unconditional slaves and lose all their freedom.

These problems will increase in the future because man will keep on being confronted to the almost infinite multiplication of new technologies. He will soon be confronted to intelligent humanoid robots capable of thinking and of showing human feelings. These computers are developed in laboratories to imitate human behaviour in all possible ways and even help people in some of their tasks. First developed to compile and memorize knowledge, these humanoid machines not only compile knowledge, but also learn how to react to sensations, to emotions, to the ways of living and thinking of man. Just imagine, if man can easily become a slave to the material objects he uses, how will he react when confronted in his every day life with these thinking humanoids robots, these machines that can learn to master all the knowledge of humanity, that do not forget, and that are capable of integrating this knowledge to their feelings and who will soon be able to distinguish what man calls good and what he calls evil? The slavery of man in the face of these high technology products will be all the more corrupt.

By using these phenomenal and enticing technologies, man will also come up against the paradoxical and deplorable after effects these machines will have on him. He will literally continue to drug himself with their benefits, and like frogs in a heated pool that become accustomed to the water's heat that slowly warms them and eventually kills them, **people are becoming dependant of these objects that control them, and when they finally recognize the symptoms, it is too late to get out of the water…**

Whether they like it or not, each individual using a material good inevitably accepts the rules and regulations attached to these material objects. Repeated contacts with all these objects can only materialize man more in spite of himself. Modern occidental men are so imprisoned in this materiality that it is not surprising that in the long run, they lead their lives as zombies unable to extract themselves from it, whether they want to or not. So many technologies of all sorts surround us and eventually they literally drug us. **We strangely resemble the great majority of individuals who stupidly agree to become slaves of the logic and the rules imposed by technology.** We become accustomed to all material things and accept them without a word….

All techniques answer to a logical machine, non human, stemming from a series of material data. They have the power to regress those using them to the level of machines without a soul. These inflexible logical machines are making man forget his humanity. They force all the people using them, including civil servants, governments and societies to become their slaves.

We will soon need to create, next to private clinics for alcohol drug and depression caused by work abuse, new centres to treat the abuses by technology, gambling and the compulsive games our good governments offer their fellow citizens. These materialistic authorities presenting themselves publicly as "virtuous" persons are in fact maliciously exploiting the weaknesses of the most vulnerable individuals. And why should we not also create private detoxification centers against the abuse of answering machines, television, the Internet, cars and the heartless governmental machine, etc.? Thousands of other mind-destroying material technologies have disastrous effects on people's behaviour, their way of working, the way they are and the way they act.

Plan

Let us first illustrate how this logical scientific method, combined to technological applications, not only degrade men but reduce them to the state of irresponsible animals. We will review hereafter the materializing effects of certain phenomenal technologies that reduce men to slavery, and clearly illustrate that education through science and technology without the humanities (see further on) is transmitting its materialistic values to all the people.

A. Technologies dictate their own laws and materialize

Over fifty years ago, in his novel *La Vingt-cinquième heure*, (The Twenty-fifth Hour), C. Gheorghiu, (Éditions Pigeon, Montréal, 1949), already described with much detail how technical slaves transform men into slaves to the technologies he uses. After awhile, technology dictates to man how he should live instead of being at his service. It even manages to run people, institutions, societies, civil servants and all the authorities by reducing them to the level of things and of machine parts.

In this novel, Traian, Minister Koruga's son, describes admirably well the horror his heroes and their societies live, flooded with "technical slaves". He perceives and demonstrates in minute detail the influence of the "technical slaves" lighting our fireplace, warming our bath water, opening windows, airing rooms, instantly bringing a love letter to a loved one as well as hearing the loved one's voice from a distance, ploughing, painting, singing, dancing, flying in the air, diving under water, executing the death penalty, curing patients in hospitals, assisting priests as they celebrate mass, etc. He also demonstrates to what will lead "this tragic exploitation of man and society, trapped in the torment of all these technical means by the authorities, the police, the administrations, the governments and all their machinations".

Traian observed, over a half a century ago, that this multitude "of technical slaves" already represents a crushing numeric majority on human beings. "The technical slaves act according to their own laws on automation, uniformity and anonymity. They hold the power of communication, supply and industry. They form the proletariat and force people to learn their laws and their language. And so, bit by bit, without realizing it, they induce individuals into giving up their human attributes and their laws in order to adapt them to their own lifestyles and to materialize them".

The first symptom of this dehumanisation was the contempt of the authorities for the human being. Traian had become well aware that he and his fellow men have been transformed, by being in contact with technology, into elements that authorities could replace at will. **"The technical slaves create their own society according to mechanical and inhuman necessities"**...

This is a tragedy. **"Human beings must live and behave according to technical laws, foreign to human laws. Those not respecting the laws of the machine, now promoted to the rank of social law, are punished.** And the human being becomes a minority in the midst of the technical slaves he himself created. He is excluded from the society to which he belongs and which he cannot integrate without giving up his human condition".

"This deterioration of the human condition by technology transforms human beings by making them sacrifice their feelings, their social relationships, and by reducing them to something precise and automatic, similar to the relations between one piece of machinery and another."

"The rhythms and the language of the technical slaves are then imitated in social gatherings, in administration as well as in art, in literature, in dancing, etc. The human beings are transformed into the parrots of the technical slaves". That is how novelists and intuitive poets spoke decades ago, when the first repercussions of materialism were just beginning to materialize and dehumanize men, well before the important wave of technological applications invading our modern world.

The lives of people and societies have become saturated with this odour of technological materiality. Fifty years after "*La Vingt-cinqième heure*", the dices are thrown much further than C. Gheorghiu had imagined. Our world is on the edge of the abyss, in a disastrous state of materialism, of advanced dehumanisation and of decadence. Is it not surprising that people still wonder how it all came about?

B. With the help of technology, the authorities rob the citizens from their individual rights and liberties

Technologies have increased their influence on individuals and their behaviour. One of their most important materializing effects is to psychologically destroy the human being. Through technology, authorities have succeeded in literally depriving individuals of their rights. Technology gives those people more powers. They consciously confiscate citizens' rights, their personal freedom and their right to confidentiality and violate in an unforgivable way their fundamental rights. Even democratic countries take away the rights and liberties of their citizens, as formerly did the Marxist Leninist socialist countries. It is always done in the same manner: they deprive the people of their liberties to make them their own, using the same socialistic tools filled with scientific and social materialism.

Democracies that respect themselves have the duty to assure individual freedom of conscience, of association, of expression and of human equality to all citizens. A democratic state also has the duty to insure to every citizen the right to political equality, equality in the face of the law, equality in opportunities and in personal freedom. **It is also the duty of democratic states to make sure that political authorities are legitimate and do not use their powers to interfere in the private lives of their citizens.** It is then up to the citizens, not the powers alone, to make sure that their political institutions remain healthy and that each individual's right is respected.

But, as we know, the authorities use technology without discrimination to take hold of their citizens' rights and to violate their personal freedoms. It is mind boggling to see how these people can materialise everything on their path and this at the speed of light. Their scientific culture renders them uncaring about the fundamental principles that should guide democracies. They deliberately impose their material laws, values and scientific principles to legally justify their actions. They then use these values and material laws, always based only on "scientific reasoning" (logos) and force them on people, societies and states. They even oblige citizens to follow their way of thinking and their values, and have the arrogance to claim they are democrats, simply because they give the people the right to periodically change governments and have freedom of speech. We nevertheless know that the elected people are not the decision-makers: the thousands of scientists and non elected civil servants present in all areas of power are.

In order to satisfy their desire to dominate men, these non elected scientists and intellectuals are shamelessly using all the new technologies at their disposal to subject the citizens to their dictate. They impose unjustified obedience sanctioned by inequitable laws, in spite of the frequent reservations of the elected members. These people show no limits to their power to intervene in their citizens' private lives. They go as far as treating them as slaves who must absolutely obey and even convince the population to vote for laws to justify their actions. They constantly lie to citizens, proudly call themselves democrats, play with words and conceal their true comments behind what they consider "politically correct" vocabulary, which is in fact deceitful and absolutely similar to the one held by the totalitarian regime leaders of yesterday. Lies, disrespect, intransigence and absence of human feeling characterize their gestures as did the Marxist Leninist materialistic cultures before them. Equipped with these laws without humanity, voted by elected puppets, they can, in all legality, violate the citizens' rights as freely as they want. They never hesitate to violate the basic principles of democracy and handle individuals and societies as they wish, with their "reason" and their mechanical and inhuman logic.

Overused every which way, the term "democracy" has lost the meaning it once had. These men change as they wish the meanings and the names of their political parties so that no one really knows who they are and what their ideologies actually are. And so, on the ideological plan, "social democrats" are much closer to the totalitarian Marxist Leninist societies than yesterday's liberal or capitalist societies. These social democracies have in no way "purged" the ideologies of the scientific materialism from their vocabulary, their actions always totally lacking in human sense and their citizens treated like slaves.

Numerous proofs demonstrate that in countries claiming to be democratic or governed by a social democracy, individual rights and freedoms were blown away like leaves in the fall, transported by the winds of inhuman laws. These laws are always written by the lawyers in power who follow precisely the "scientific reasoning" of this linear and Cartesian materialistic logic, a logic stemming directly from the scientific methods of reasoning of their truths, not of "the" truth. They were all educated with the principles of the "Marxist scientific and socialist materialism". These ideologies betray themselves by making inhuman laws which glorify the values and the virtues of materialism, without taking into consideration the human side of man. They draw their morality from collective principles denying all individual rights. **We call them inhuman, because they are materialistic and only obey the scientific morality, never taking into account the human being** (*cf. "Is Science Inhuman?" by Henri Atlan*). Morality without responsibility is not morality: it is the absence of morality.

If you think I am exaggerating, let us take a further look. It is no secret today that our telephone conversations, our fax messages, our correspondence, our long-distance phone calls, our mortgages, our loans, our bank accounts, our purchases, our diseases, our shortcomings, our mistakes, our special features, our habits, our telephone numbers, our social insurance number, our Medicare card, our tastes, the way we dress, our language, our weapons, etc., are all spied upon and known by our governments and by their institutions. All our actions, good or bad, are noted, registered and used by all the authorities, governmental agencies and private companies.

The cooperation and the participation of governmental authorities in the stealong of our rights and personal freedom do not make any doubt and this, in all the governments of Occidental countries. **The individuals**

in power exchange this personal information without our permission and without remorse. This lack of respect and this robbing of our rights are against the human nature of man. To protect themselves from these intrusions, democracies had included them in their "Charter of Rights and Liberties". But the authorities don't care about these charters and no one can drag them to justice! They, on the other hand, use as they wish the billions of dollars we give them in taxes to protect themselves! They justify themselves with the lies they hide in their "politically correct" revolting vocabulary. They steal the individual rights and freedoms of their citizens, "to help them" or "to protect them" or still because "they love them so very much" ...

When we observe how casually these people treat their people, consider them as second class citizens and make them their slaves, we have the right to doubt their good faith! The answers they give quickly make us understand where they stand, to what philosophical system and to what code of ethics and deontology they refer to!

Proofs of violation of our rights and liberties

What is it about these "depraved and satanic" non elected authorities that make us want to vomit? They use technology as weapons to degrade their fellow men, just as Stalin and Mao used their powers to exterminate those having different opinions than theirs. How can all citizens lose their rights because of technology, rights that only the criminals lost not so very long ago? Do these dehumanized individuals in power want to persuade us that all citizens are wrong? They refuse to understand that true power in democracy belongs to the people. If these good socialist hearts with lax and inhuman laws allowed biological weapons to enter the country through terrorists, they can only blame themselves. They are responsible for this, not the citizens.

Governments supercomputers

Why do you think these people are equipped with supercomputers? In a praiseworthy effort to prevent computer intrusion, the American administration created, during the last decade, a federal agency called FIDNET ("Federal Intrusion Detection Network"). This organization is responsible for the control of all governmental computer systems. Up to this point, everything is just fine. But there is another side, not revealed and underhanded, that comes from the fact that FIDNET also controls all the computer systems of banks, communications, world transport, transport industries **and possibly several others we know nothing about.** FIDNET has complete access to the multitude of information you must give the bank in order to obtain a loan or a credit margin, as well as to all your financial and personal transactions. Did we ever give our governments such power over our actions and our private lives? Never! They stole them from us, and that was the end of that.

Are we stupid enough to believe that it is for our own security that these technocrats sacrifice our private lives, our rights and our actions? Why then do they consider us all as criminals? How can our countries boast of being democracies? How do they justify the fact that every individual is treated like a criminal? Did we unknowingly give such power to our hypocritical governments? Does a vote every four or five years give the elected or non elected members so much power? If, in the mind of the authorities, the conception of democracy is so distorted, we must at once restore what is democracy! These thefts are unforgivable and unjustified.

Canada, as other countries, is equipped with an even more powerful and corrupted system than the American FIDNET. The "Canadian FIDNET", or whatever its name, can verify over three billion (3,000,000,000) elements of information about Canadians every 24 hours. Can we imagine the power of these computers? Over 100 facts a day per person including children and the elderly can be legally verified by our good governments, the police and all the ministries in governments, the army and others... Furthermore, we are totally unaware of how many other agencies, institutions or countries have access to these data. The majority of the other governments on the planet possess similar systems able to communicate between each other. This means that all the data collected on every citizen could be accessible at once to all countries, even to Ben Laden!

A "unique government file" on each citizen

On May 17, 2000, the population of Canada learned by the Police Superintendent for the Protection of Private Life, Bruce Phillips, that the Ministry of the Development of Human Resources of Canada (DHRC) took upon itself to constitute a complete file on each citizen. This file contains up to 20,000 elements of information on every individual person in the country. This monster is the by-product of a completely dehumanized bureaucratic and antidemocratic culture. It is a monster that compiles, without our knowledge, thousands of data on each and every one of us. These data come from former ministries of Finance, Employment and Immigration, Health, Work, from the previous unemployment insurance system and from various departments of the federal government and even from private institutions.

This mega-bank of data, called **"The Longitudinal File on Manpower", contains confidential information on income statements, child benefit, provincial files on social welfare, training programs, files on unemployment insurance, work reports, immigration files and several other sources of information. The quantity and the details concerning information on every individual are as impressive as they are unlimited: schooling, civil status, language, ethnic origin, citizenship status, incapacities, income, professional background, requests for social security, etc.** Phillips, the police superintendent, revealed that the personal file of each citizen also contains information dating back two decades and that the DRHC is continuing its revolting task. The scientists of this ministry want even more information. They want to get personal information on old age security, the Canada Pension Plan, the Canadian Student Loan Program, the provincial fields of social welfare, and more! It is distressing to notice that these non elected scientists in power have reached such a dehumanized materialistic culture! It would have been justified had this information limited itself to criminals and to future immigrants, in order to be better protected against terrorists. This denotes a completely antidemocratic power culture, a trademark of the Marxist Leninist socialist power.

It is abnormal and distressing that hundreds of elements of information could have been sent by various ministries to the DHRC, in the last twenty years, without the citizens' knowledge and especially without their consent. This unique file is a perfect example of the violation and the appropriation of our personal rights and liberties in our private lives.

The authors of these indecent acts against the Charter of Rights and Liberties are the non elected scientists in power, the programmers and their sorry accomplices from the other federal and provincial ministries. These people should immediately be judged for their acts and fired. These scientists and intellectuals must be brought before the courts at their own expense, for infringing on the private lives of the citizens of a country in which I was until then proud to live in. These dreadful actions against democracy are unforgivable.

The public denunciation of Mr. Phillips confirms that all previous police superintendents for the protection of private life lied through their teeth for years, claiming loud and clear that their government never kept such a file on their citizens! Even if they admitted today that they had lied, these high and mighty public servants are nevertheless liars or incompetent persons, superbly paid with public funds. They betrayed the citizens with the complicity of other high-ranked scientists and intellectuals of the State, other members of parliament, as well as the Ministers and the other despicable technicians who participated in establishing this dehumanising political culture.

"Read my lips not what I say", said ex-American president Georges Bush (father). He was talking about people such as these police superintendents and high-ranking civil servants who denied, lied and shamelessly didn't care about the citizens. Those are the real culprits: the people in power who hide behind their despicable laws to commit their crimes.

The British organization HYPERLINK http://www.statewatch.org/ Statewatch, wrote in MSN.CA on May 27, 2001, that Enfopool, a European organization whose name comes from Enforcement Police, **asked the European Union to vote for a bill which would give them permission to force Internet suppliers to keep for seven years all the information on your comings and goings, all the shops and the sites you visit, all that you have read, looked at, heard or done….**What a dreadful intrusion in the private lives of every citizen and this without their knowledge! However unthinkable and unacceptable it may appear, the European Union said no to the project, officially because it knows that "Statewatch" is already functioning and can get all the information it wants. It maintains its right to discuss the project. The worst it is that Great Britain, Holland and Belgium had already accepted the project for a two year period. **What difference is there between this and the materialistic laws that gave Marxist Leninist countries such power?** Yet, all this happened well before on September 11, 2001.

If only these data were kept to help in criminal pursuits, maybe the people could accept the idea. If these data were used in a very strict and well defined manner by law, not allowing organizations other than the police to consult them, maybe this principle would be partly acceptable to people. But, since these data exist, someone somewhere will always want to use them and not necessarily for the right reasons. I have a lot of difficulty understanding that the Europeans, who are much more socialist than the Americans, can accept such facts and continue boasting that they are the champions of human rights and liberties and democratic values! How many Jörg Haider in Austria, Le Pen in France and Pim Frotruyn successors in Holland will we need for the intellectuals in power to finally stop making people believe that they still live in true democracies? Are the people not confused, by any chance, by so many socialist parties?

More demonstrations of power abuse

In 2000, **Canada created a law on firearm control**, only to have a clear conscience, because this measure will never prevent the reason for which it was adopted. This law was adopted following the tragedy of the *École Polytechnique de Montréal* (Engineering School) 11 years before. We will never forget that during this massacre, 14 young ladies were brutally shot and killed by a mentally sick man. "Forcefully" encouraged by activist groups, the "compassionate" government adopted a totally useless law to serve a first objective, that of preventing such slaughters! This law, made in the name of compassion, is inhuman. It hides the true materialistic culture of the authorities, by using this golden opportunity to continue violating their citizens' rights.

This law was transformed into an ugly and inhuman law, allowing authorities and civil servants to inquire into the past of firearms owners and authorizes the police to enter their homes without warrants. Once again, pretending to answer the activists' desires, the authorities give the police the right to legally violate the past and private life of the individuals asking for a firearm permit. Without any other form of authorization, this law allows the powers to dig into the private lives of every applicant for firearms permits. Besides the interventions in the homes without a warrant, it allows the policemen to ask very delicate and unfair questions to the "compulsory" permit holders. This law obliges them to reveal "if they ever tried to commit suicide, if they had ever been treated for alcohol abuse, if they had ever taken drugs, if they had any emotional problems in the past five years", etc. besides having to give out information about their former spouses of the last two years and to give the identity of two guarantors that the police can question… The ex-spouses, who had nothing to do with these permits, were also inhumanly forced to sign the document containing the answers they had given the police against their will. I do not dare write the word that comes to mind to qualify the inhuman culture of these sick lawyers, creators of such laws….

Several hunters and farmers fiercely denounced this other "invasion of their privacy" by the State, but the State is dehumanized and chaotic and does not care. The Police superintendent for the protection of private life has publicly said that he would make inquiries, but do you believe that the laws or these "illegal and legalized" practices will be changed for all that? Not at all, there are too many. When a materialistic culture is embedded in the powers of a country, it is there to stay.

This is how a seemingly good law is legally transformed by the dishonest lawyers in power, into a vile, inhuman and inequitable law.

Is this another example of the materialistic and corrupt cultures that guide scientists and intellectuals in power? Such inhuman capacities seem to be endless in civil servants and in governments that only obey the laws of scientific reasoning without humanity. Citizens are only objects to these scientists and intellectuals who enslave them to their all wishes. It is only the rule of "reason" (logos) that guides the policy of these scientists and intellectuals in power. What a dehumanized culture!

At the end of the year on 2002, the Auditor General of Canada, Sheila Fraser came to the conclusion that the cost of this program, which was to be about twenty million dollars, will exceed one billion dollars, several dozen times more than originally estimated... In one of the harshest criticisms by the government, the Auditor General considers this cost dreadful and unforgivable. It is the cost for those people to have a clear conscience!

Encoding

There are still many more examples of this Machiavellian culture invading our occidental world. When citizens, for instance, try to protect themselves against the interventions of the powers in their private lives and on their "Internet journeys", they use a coded language called encoding. It is abominable to find out that when the encoding exceeds the capacity of their computer system (about 48 symbols in the year 2000), the governments of the United States and Great Britain forced their citizens, by law, to reveal them their encoding! When a law legalizes such manoeuvres, it is no longer a human law. Even worse, the antiterrorist law, passed after the events of September 11, 2001, **contains a clause which obliges Internet suppliers to keep during 7 years, without the user's knowledge, all the sites visited on the network.** And still more, these laws oblige Microsoft, which controls 95 % of the computers, to keep during several years all our activities on the Internet.

In November 2001, before these antiterrorist laws, *Femme Québec (a magazine for women)* published on the Internet a research demonstrating that their service supplier's site knew their computer much better than they did themselves. All you have to do is ask "Anonymat.org" and you will obtain the following information about your computer.

You are connected to Internet with the IP address:
24.202.246.8
By way of the server of your access provider (or the proxy):
modemcable008.246-202-24.mtl.mc.videotron.ca
Your navigator and your operating system are:
Mozilla / 4.0 (compatible; MSIE 6.0; Windows 98; Win 9x 4.90)

Clearly, your navigator is: Microsoft Internet Explorer
Your Java is activated
VBS is activated < SCRIPT language=JavaScript > <!--
Document.write (navigator) // - > < / script >
< SCRIPT language=JavaScript > <!--
Document.write (" JavaScript is activated ")
// - > < / script > < SCRIPT language=JavaScript > <!--
Document.write (navigjava)
// - > < / script > < SCRIPT language=VBScript > <!--
Document.write (" VBSCRIPT is activated ")
// - > < / script > Address of the web site visited before this page:
http://pages.infinit.net/femmesqc/2910.htm
Number of pages visited during this session by your navigator: 0
< SCRIPT language=JavaScript > <!--
Document.write (visits);
// - > < / script > In other words, your operating system is:
Windows 98
The clock on your computer indicates: in November 9, on 2001 08:
27 57
< SCRIPT language=JavaScript > <!--
If (now! = null) {
Document.write (now.toLocaleString)} else {document.write
("unknown" .italics ());}
// - > < / script > You are presently using the screen resolution: 800x
600 pixels
In 65, 000 , 536 colours
< SCRIPT language=JavaScript > <!--
Document.write (resol)
// - > < / script > < SCRIPT language=JavaScript > <!--
Document.write (colors)
// - > < / script >
In the name of what principle of justice can the international police systems and governments appropriate information contained in your personal computer? More obnoxious and antidemocratic than that, it is better to die for our freedom. In front of so much injustice all around them, is it not normal that the people aware of these multiple obstacles drop out of society, question social structures, are put under stress, become ill, use drugs, commit crimes, rebel or lose all trust in their politicians?

A law to locate individuals

We have not yet reached the limits of the intolerable. Until 2001, private communications by telephone could not be intercepted by the police without a mandate. **Today, a new generation of portable telephones communicating via GPS satellites enables to localize the caller within a few feet from where he actually is.** Well, believe or not, in August, 2000, MSN NEWS announced that "The Regulation of Investigatory Powers Act" (RIP ACT) of **England, ratified a new bill permitting that no distinction be made between private communications and public communications. This law allows authorities to localize, without a mandate, the owners of portable telephones.** This is another good technological tool that the authorities are carelessly using to further control man.

The Associations for the Defence of Civil Rights and Liberties have good reason to worry about the threat such laws represent in the citizens' private lives. The debate was raging in the year 2000 in Great-Britain, well before September 11, 2001, because this law authorized police to keep an eye on all owners of portable telephones to localize them and know their every move. The Foundation for Information Policy Research" (FIPR) is also justified in worrying about this abominable law because these new telephones become a "tag" to their users. Is it what powers call democracy?

In order to justify themselves, governmental civil servants use hollow and inhuman words taken out of the depths of the sick scientific logic that Marxist - Leninist leaders used. "This law is allowed, claim the technocrats in power, because new technologies are more and more precise and are meant to improve the surveillance of delinquents". Great thinking! What degrading morality! This is another lovely example of nonsense from our dehumanized scientific and intellectuals in power, justifying their actions, not by reasoning at this point, but under the command of material technology. What a beautiful example of technological application imposing its laws and subjecting them first to the people in power, by its mechanical, logical, material and completely inhuman and intransigent logic. "If you never did anything against the law, you have nothing to fear", we were told by these deranged and irresponsible powers. Once more, I invite the readers to read about the dreadful consequences of such decisions in *"La vingt- cinquième heure"*, published over fifty years ago, (see complete reference above) to understand the deeply irresponsible comments made by materialistic technocrats of neo-societies, derived from scientific and social materialism.

History teaches us that the collected data are always used to impede on people's freedom. Private lives are thrown out the window to give way to technological applications and materialistic laws with no soul.

Let us look more deeply to see to where such materialized cultures can lead us. The results of an inquiry led in the United States and in Europe by the Organization of Consumer Defence (OCD) and published on the Internet in February 2001, demonstrated that, in spite of repeated warnings and numerous appeals from Internet users asking that their private lives be respected, about two thirds of the commercial sites – out of the 751 studied – continue to this day "to collect very personal and unnecessary information about the consumers". Even though family names, first names and e-mail addresses are privileged information, several companies go as far as asking, to complete a transaction - or by simple curiosity – the postal address, the telephone number, the annual salary, the number of persons in the home, etc. Is it justifiable that, in this materialistic culture, ministries exchange data? Statistics Canada asked in its 2001 questionnaire: the salary, the address of individuals, if the house they live in needs repairs, if it is in good condition, etc. These are also, according to the OCD, personal information, demanded by law and registered by computers to be ultimately used out of context. These data can also be coupled to other data bases "to get a more precise picture of a person". **Many ministries exchange and sell this information to other countries or to companies specialized in advertising or to analyse trends, mentions with regret the OCD. We know little about these schemes.** Why not to Ben Laden or to other terrorists, if they are willing to pay the price? Is not the money an important value to these systems and these materialistic governments? The OCD wants a reinforcement of the existing laws on the subject and also invites Internet users to caution.

The White House and the secret of identity

In October, 2000, Privacilla.org announced that the "White House for the Children" ridiculed the rules on child protection and on respecting the secret of their identity. Signed in the United States in 1998 by President Bill Clinton and effective since 2000, the COPPA law aimed at protecting and respecting children's identities. However, the information collected from the children during visits enabled their identification without their parent's knowledge or consent. Far from denying that the White House violates its own law, spokesperson Nanda Chitre confirmed that this law was effective, but that the White House was unique with laws and practices different from those of the COPPA. This is a perfect example of "two weights, two measures".

Social repercussions of these amoral and inhuman measures

A Canadian citizen is arrested in the South of the United States and deported to Canada in February 2000. His crime was to have omitted to pay a traffic fine in Quebec, twenty years earlier. "It is unacceptable that information of this nature be available and transmitted to foreign powers" said Mr. John Philpot, a specialist in international law. This is another example proving that citizens' rights can be scoffed for a simple peccadillo which took place decades ago.

A major problem created by this culture

Another important problem ensues from the violation of personal freedoms: medical consultations on the Internet. More and more consultations and medical information are made via the Internet. One only has to look at the number of medical sites on the Net to be convinced of this. A broadcast issued on *Radio Canada*, by Stéphane Garneau in July 2000, mentioned that between 11 and 12 million files transit on the Net every day.

The confidentiality of these files is greatly questioned because of governmental interventions in your personal computer and in all your surfing on the Internet. ***"Family Practice"* of March 15, 2000 raised the problem of protecting personal and medical information through this media.** The interest of patients in using the Internet for medical consultations will immediately cease if the confidentiality of the information cannot be guaranteed, and if technology continues dictating to people in power their laws and their activities. The confidentiality of medical information is thus compromised because of this continuous surveillance by the governments.

Computerized health cards also contain privileged information that rightly belongs to each one of us. In spite of the usefulness of this information to doctors and pharmacists, the advantages of such medical records on computerized cards could very well lead to all sorts of intrusions in our private lives in spite of the precautions taken to maintain confidentiality.

Other technologies, other violations of our personal life

How do you think that information as personal as your genetic code, which contains over three billion genes, can easily enter governmental surveillance systems? A 91 year old woman was raped in the village of Wee Waa, in Sydney, Australia. In order to find the rapist, the police decided on its own to have all the male population of the village take a DNA test. Of course, the objective is good, as usual. It aims at reducing the list of potential suspects: a very noble motive. But then, what will happen to the collected data after the fact? In what computer file will they be compiled? Knowing the cavalier, depraved and careless ways of the police forces, the technocrats and the governments, it would not be surprising that a citizen of Wee Waa visiting France or Great Britain one day, could be identified because of his genetic mapping.

The Royal Canadian Mounted Police of Canada, our great national model of virtue, has begun computerizing the genetic codes of the Canadian prisoners. Having lost all of their rights because of the crime they committed, it seems understandable than a record of their genetic code be kept. Soon, authorities will be requiring the genetic codes of all citizens at birth, justifying this intrusion in their private lives by yet another illegitimate materialistic right dictated by technology.

Today's sophisticated technology allows authorities to identify individuals' faces, the iris of their eyes, their fingerprints, etc. We all know that when an individual goes through Customs, all the Customs officers have to do is to enter the passport into the computer system to know all the information on the person. Now, using a much more sophisticated technology, people can be localized and identified without their knowledge. **Hidden video cameras, connected to specialized computers, can now scan crowds, select faces and identify individuals. Connected to data processing systems, this allows the identification of the people in a crowd and can transmit the information.** Manhattan has over 2,800 such hidden cameras that police are using to track down in a crowd any individual having a criminal file. In Great Britain, while filming a demonstration, hidden cameras enabled the police to know where all the individuals precisely were in the crowd, at any given moment. Even in a crowd, privacy does not exist any more. Establishing an individual's citizenship, his country of origin, the places he visits, which restaurants he goes to, is now common practice. With the exchange of information between computers, your boss will soon know if you were at the stadium watching a ball game the day you missed work!

Private information and insurance companies

So far, insurance companies assumed that the healthy majority financially supported the less fortunate individuals. Insurance companies already practiced some discrimination, asking you to fill a questionnaire on your family's health and on hereditary diseases before they accept to insuring you. Since September 2000, the British insurers have the right to test for genetic defects responsible for Hungtington's disease, a deadly neurological disease which induces intellectual degeneration almost always ending in death. When the gene is traced, the insurer may insure the patient with an extra premium or simply refuse to insure him.

Even though until now the decision of the "Genetic and Insurance Committee" authorizes to look for only this specific gene, it becomes another wide open door into the confidentiality of a multitude of other genetic tests. Several other analyses for the detection of genetic diseases, defective genes, or infections by the virus of acquired immunodeficiency (VIH or AIDS) and others are already in place. These techniques are reliable and results can enter the data banks of insurance companies. **Research helping, with excessively sensitive and reliable technology, all sorts of genetic defects will soon enable insurance companies to adjust their rates according to the known risks.**

Today, one can measure a newborn's Br1 and\or Br2 genes (two genes coding for breast cancer 1 or 2) and so establish the risk of developing breast cancer in the female and possibly prostate cancer in the male. In the same manner, we could detect the presence of specific genes which predispose to a disease or increases its probability. The detection of genes related to colon cancer, prostate cancer, Alzheimer's disease, ulcerous colitis and many others are already being looked at by bio-geneticists working for insurance companies.

In Australia, in Austria, in France and in the Netherlands, genetic discrimination for insurance purposes is forbidden, but for how long? In Canada, since the Industrial Alliance Company won its case against a carrier of the Steinert gene of muscular dystrophy, this jurisprudence is used. If the insurance company manages to demonstrate, for instance, that an individual is the carrier of a defective gene or another benign pathology that he omitted to mention, the company does not have the obligation to indemnify him.

After September 11, 2001

September 11, 2001 marks a major turning point in an incalculable number of fields of human activity. What resurfaces here and clashes is the forever old quarrel between two different schools of thought: 1) - that of the materialistic and dehumanized world, created by scientific and socialistic materialism, abundantly present in all our Occidental modern societies. 2) - that of the world of spiritual faith whose very existence is denied by materialists. These two protagonists of both the material and spiritual worlds are quickly brought back to the battlefield by religious Islamic extremists. A group of experts from all over the world should quickly get to work to re-humanize the way we control others... If we could only understand that we should start minding our own business instead of that of others, we would have taken a big step forward!

For over a century scientists, intellectuals, authorities, institutions, societies and all the individuals mostly educated in practicing scientific materialism deny the very existence of spirituality in man. We mentioned earlier that this materialistic system considers as nonexistent the things he cannot see, smell and touch, in other words, that the entities that cannot be analyzed though his senses do not exist. This is where these two worlds must again face one another as they did before the creation of scientific materialism. Materialists cannot understand why terrorists can kill themselves with bombs and those that believe in spirituality cannot understand such a great absence of humanity on the part of the materialists.

During the counterattack on the "Islamic Holy War", the Ministry of Justice of Canada, following the example of several other countries, established new laws against terrorism. Regrettably, these laws once more hinder in an unthinkable way the few rights and liberties remaining in individuals.

We worry along with the International Federation of Journalists and many other organizations about the new laws against terrorism, adopted since these events in numerous countries such as in the United States, Canada, Russia, Great Britain, countries of the European Union, etc. These laws show a disturbing eagerness to allow, for example, a more frequent recourse to electronic listening, to controlling the Internet and to the detaining of immigrants.

John Reid, the Federal Police Information Superintendent, asserted in front of the Senate that this law is not necessary to assure the protection of the country's neuralgic information. It will however allow the general prosecutor of Canada to keep documents from the Committee of the Law to the Access to Information. "The drafting of this law is too vague. It will allow too large an interpretation and will escape an independent review", he added. The government replies that "Canada will be deprived of strategic information if it cannot supply to its allies a tried and tested assurance that this information will remain secret". (*New Internet and Media of October 24, 2001*)

Here is a very human reaction on the part of the Europeans in this case published on *Vnunet.fr* on June 19, 2002. "The recent intensification of the European rule on the conservation of electronic data tends to impose measures without offering a public debate. Worried that these interventions will strike a blow to the private lives and the rights of the citizens, about ten European associations, such as IRIS (let us imagine a united Internet network) in France, decided to unite their forces to create a union: the European Digital Rights" (EDRI or European Numeric and Digital Rights).

Created on the 8th of June, 2002, the objective of EDRI is "to defend the civil rights of the European citizens in this era of the technology of information and communication". To reach this goal, the associated members "will work collectively to heighten the awareness of authorities and citizens to the dangers threatening our private lives and liberties".

The Committee of Human Rights and Security of the International Federation of Journalists, regrouping 450,000 journalists in over 100 countries, has seized these laws because they eventually almost completely eliminate the rights and liberties of the journalists and the individuals of our countries.

Thoughts and comments about people disgusted by the technologies which allow people in power to abuse them

"There are many ways to kill a man" said Félix Leclerc in his song. To steal man's intimacy and to deprive him of his individual rights and liberties in a country we call free and democratic is one way of killing a man. The mistake of scientist and intellectuals in power, creators of inhuman laws is to consider citizens never having been judged for a crime as criminals, by depriving them of their rights and personal liberties. Because this materialism and dehumanisation measure keeps scientists and intellectuals incapable from making the difference between good and evil, between legality and immorality, these administrators of our individual rights and liberties, the scientists and intellectuals in powers, will never be condemned or even reprimanded for their slanders.

There are all kinds of scientists of different cultures, moralities, philosophies, religions, atheists or not, who are in charge of occidental governments. There are no more philosophical systems left to guide the political parties or to identify them, there is only this "scientific reasoning" that intellectuals have adopted to guide their lives and ours. After over a century of such scientific and social materialism "Karl Marx style", the "scientific reasoning" (logos) is the only way of thinking that guides the men of the Occidental world. Scientists in power see in this scientific methodology the ultimate way of thinking. Today in the Occidental countries, science and "scientific culture" run the governments. Scientific reasoning and material values are well anchored in people's heads through their education in science and technology, and no one seems capable of noticing that the powers went too far in the dehumanisation of their actions. We have been so contaminated by scientific and social materialism that we believe there is no other way of thinking than the one taught by this cursed scientific method. Extremely useful and indispensable for research in our material world, we must soon realize that this method is completely deficient in studying and understanding ideologies not coming from the material world.

Scientists do not know or refuse to admit that the human being is not only made of matter; man is body and spirit, matter and life. We have all been dominated and reduced to infantile levels with scientific education, material values and the completely unscientific judgments arising from this "reason". Scientists in power have led people to believe that "they" are the boss, that they should be obeyed and that their diktats, often deceitful, are a part of their "normality". **No, diktats and scientific laws are not normal because they are not human. They are not normal because they impede on individual rights and liberties and because they turn people into slaves that they treat like cattle.** A scientific truth cannot be the result of the reasoning of these "pseudo" scientists, often working in non tangible fields of science such as psychology, sociology or politics, as they want us to believe. As long as the method of thinking is not modified, a scientific truth will always be the result of a rigorous analysis, obtained and demonstrated by tangible facts, resulting from a hypothesis and a study scientifically done, but **only done on material things**.

People are confused by such abuse and the dreadful interventions of the governments in their private lives. This abuse of power is so great that citizens are lost in the inhuman laws they can't get away from and eventually consider them indecent and antidemocratic. Many refuse to accept the many administrative blunders of their governmental machines, and they consider them as abnormal and indecent acts. True democracy no longer exists. It had a much more human way of protecting the citizens' rights, without abusing them so shamelessly.

C. Examples of administrative technologies propelling materialism

Administrative technologies play a dominating role in the management of all areas of human activity. However, they have a great disadvantage: the same techniques are used to manage all areas without knowing too much about the subject. In administration, it does not seem necessary to know anything about the domain in order to administer it. These technologies soon became a conglomeration of "administrative techniques" applicable to all areas, after their passage through the scientific methodology of research, better known as the "truth". Let us analyze how and why these technologies were marked very early on by science and the rigidity of scientific methodology.

During the nineteenth century, intellectuals and scientists were spellbound by the scientific research method based on "pure reasoning". All the knowledge acquired on the "matter" was closely examined using this Cartesian method of research, and administrative techniques did not escape this. No domain, no conglomeration of knowledge aspiring to one day become a "science" could escape scientific analysis without using this method of research. But often, as in sociology, psychology, politics, economy or other "inexact" fields of science, they were all forced to accept this reasoning as the truth, whether or not it was confirmed by scientific experimental work.

The application of this method of research is creating havoc in education by literally invading it. No domain, no sphere of human activity aspiring to one day deserve being called "science", has escaped being put through the mill of this scientific methodology. The scientists decreed that any domains not accepting this method could never be qualified as "a science". So, all the faculties in universities and all areas of teaching eventually became one day a science, but only after their passage through this logical, Cartesian and linear scientific methodology. Because of their intrinsic materialism, the sciences of physics and chemistry were easily integrated, compared to the more ethereal pseudo sciences of the mind such as philosophy, psychology, theology, etc. **All these fields of knowledge however became "a science" through the Cartesian reasoning.** If this method was easily applicable to research on the material subjects of our universe, it was quite another thing for the study of less tangible domains or entities, and not at all in the non material domains, such as theology and human behaviour, for instance. **The problem is the method of research itself; in its actual form, it can only be applied to the study of material objects of research perceptible by man's**

primitive senses. If it has been extremely useful for studying our universe and the material things of our world, it is very different for the study of the non material things.

It is in this whirlwind of study and change that administrative techniques of management have become "the science of administrative techniques". The principles of administration lend themselves generally well to the exercise of the method. And so it became one of the first domains to structure itself into a logical, mathematical, linear and acceptable science according to the scientific model of the search for the truth.

The administrative techniques began invading schools at the end of the nineteenth century. Administrative technique sciences were found everywhere: in the production of goods, in manufacturing motor vehicles, as well as in garage administration, in emergency rooms, in hospitalized patients, in disease management, etc. Every area of human activity is administered by theorists or technicians specialized in administrative sciences. Schools and departments teaching these techniques are the ones mostly filling the universities. They have been drawing large numbers of students for decades. Those graduating from these administration schools become scientific administrators. Administrations are like beasts which reproduce themselves, creating the need to be "administered", and these rigid administrations start to infiltrate everywhere! There is no place or job today that is not administered by people having been trained with these techniques.

As for many such pseudo sciences, the "science of administrative techniques" answers only to the linear and mechanical logic of the scientific reason (logos). It is not surprising that the logic of these techniques is always inhuman and only responds to the commands of scientific reason, which is based on the perceptions by the senses. The problem of administrative techniques is also related to the inflexible logic of the method, devoid of all human sense. It is not surprising, either, that the administrative techniques always make decisions with no humanity, no compassion, no affection and no love. Whether applied in the administration of human beings, automobile parts or hospital management, they always use the same applications and never take into consideration the human aspect of the people, even though humans are often an integral part of the administered business.

Imagine what happens when these scientific techniques of the "reason" are applied to the management of individuals in a society, to patients in a hospital or in situations where human behaviour is implicated and much more complex. As administrative techniques cannot have other objectives than those dictated by the senses and the "reason", they often miss their objectives. They indefinitely multiply the facets of the subject to be administered and must hire more useless personnel to meet these "new demands", and so the administered "object" becomes chaotic. The administrative techniques possess an inherent incapacity to administer individuals and this deficiency is linked to their mechanical rigidity and the scientific method used.

Considered rightly as an ideal method of research to acquire knowledge on material subjects, this method is great in science. But it cannot be applied to the study of non tangible entities such as behaviour, qualities, human values, beliefs, human questionings, etc. It is thus totally inadequate to administer matters that are not tangible. These techniques cannot comprehend feelings, human behaviour and values and therefore must deny their very existence. Being able to study only what touches the senses is a flaw inherent to the scientific method.

And so, all things that cannot be seen through the administrator's microscope, touched with his fingers, smelled with his nose, or heard with his ears, cannot be administered by administrative sciences. This weakness is inherent to all pseudo sciences including administrative sciences. This weakness is also inherent to the methodology and the scientific way of reasoning that deny the very existence of non material entities, such as human questioning, etc.

The method of scientific research, as it stands today, takes into consideration only the material and tangible aspects of administered things and uses only material values. Man can only be disappointed when these techniques are applied to him. As these materialistic laws with no soul are inhuman, these rational administrative techniques cannot arouse compassion, kindness and love in their administration. The administrators applying these techniques must put aside their own humanity, if there is any left to put aside after their training.

Because of their mechanical application, these techniques contribute in strengthening the links already existing between the machine and the material. As all other technological applications and sciences, they only contribute to further dehumanize the people, their professions, their ways of being and of thinking and consequently, the services they render and administer. It is not surprising to observe that everywhere these techniques are applied, material values replace human values. With time, materialized cultures of thoughts and dehumanized behaviour will rub off on the people working under such administrations.

The fatal effects of administrative technologies on human behaviour and the promotion of materialistic values are against human nature itself.

As long as those administrative techniques refuse to take into account the human aspect of the individuals, how can they take into account the negative impacts they have on them?

The inexhaustible amount of technological applications destroys man's consciousness and people are unaware of the degree of materialism that strikes them. These data illustrate how scientific applications succeed in infiltrating all human actions and materialize everything along the way.

Invaded on all sides by material technologies, people in our modern societies are thus subjected to the slavery inherent to the technologies they use, to the materiality of these techniques, to the slavery of administrations, and what is more, to the slavery of the people in power that administer them like interchangeable machine parts to quench their thirst of power. When we think about the crazy rhythm of technological development, it is enough to make us shiver.

If a few enlightened individuals are starting to become aware that technological applications affect them as much as drugs and inculcate them their materialism, numerous are those who need to be shaken by events such as those of September 11, 2001 in order to wake up. Certain scientists are also beginning to realize that all this materialism is "putting them on a tree branch that they are sawing themselves". For how long will man be able to go like this before reacting?

Examples of materialization of administrative decisions

Numerous administrative decisions show an inhuman materialistic culture and a lack of human morality. Here are a few examples.

Individuals institutionalized for mental illnesses during years found themselves on the street when the Quebec Government health administrators approved the de-institutionalization of about 40 % of them, with the inhuman and unjustified approval of medical authorities. Irreparable damage was caused to many of these patients. The after-effects of these decisions, dictated by administrative management machines, echoed for years on the societies of large the cities where these patients gather to survive. Thousands of patients found themselves on the streets without a home, with no medication and without the care society owed them. To substitute for this, they resorted to begging, to drugs, to prostitution, etc. Such administrative decisions, while solving their material problems of lack of beds, health care costs in psychiatric hospitals at the expense of populations and patients, are tangible proof of the dehumanized degree of these administrations.

Evacuated from hospitals, these psychiatric patients increased itinerant ranks and were condemned to beg in order to survive. In 1998 in Montreal alone, 9,000 homeless people under the age of 30 appealed to soup kitchens in order to survive. Social services admit that many of these psychiatric patients of both sexes turn to prostitution and this creates another important social problem. Having no fixed address, these mentally ill individuals automatically lose their social security insurance, their free access to the medication they badly need and their diseases can only worsen.

During the 2000 federal elections, the people in power pushed their underhanded materialism to the utmost: they registered these homeless Canadians on electoral lists. Authorities were expecting that over 75,000 such homeless individuals would answer this call. We still do not know how many took advantage of their right to vote. Allow me to presume that this preoccupation is totally unimportant to these patients who do not even have the bare necessities to survive. To believe that these 75,000 persons could go to the voting polls tells us a lot about the astounding number of homeless individuals in Canada and on the absence of morality of the authorities!

Quebec political decision-makers with no conscience created in the 50's "the Duplessis orphans". Other decision-makers created Japanese internment camps in Canada during the 39-45 war. Others had no more moral sense when they deported their Jewish fellow men and participated to their extermination in Nazi concentration camps. It took over fifty years for their descendants to become aware of the atrocities committed by their predecessors' administrators and they want justice to be done. How many human beings have been dehumanized during this century of "reason" because the dehumanized powers that administer them automatically obey to perverse and dehumanizing administrative decisions?

The materialistic culture and the irresponsibility of the authorities

The decision-makers should be held responsible for their actions within the framework of their functions. They should at least be dismissed when the things they do are inconsequent. It is regrettably not so in countries governed by materialistic moralities and principles. Where are, for instance, these other scientists in power who decided to build powerful computer systems to spy on you everywhere, wherever you go on the planet, or on the Internet (see above)?

Where are these other scientists of the Ministry of Human Resources of the Canadian Government who thought, decided, developed, computerized, and implemented this huge unique file compiling 20, 000 different information on each citizen and for over 20 years?

Where are these civil servant lawyers who participated in the planning, the writing, the decision-making and finally the execution of this inhuman law on the compulsory statement of firearms? Let us look into other matters: this is too disgusting to even think about.

Can we return to the past?

Returning to the past seemed to me very illusive until I read an article in "*Affaires Plus*" of June, 2002, entitled "*La revanche de l'humain dans l'entreprise*" (The revenge of the human being in the enterprise). Usually, employers do not hesitate to fire their employees. Lets us look at the massive dismissals of "jobs for life" in the Canadian Health System and in numerous private companies such as Nortel (Canada), Enron, Worldcom, Anderson (USA), Vivendi (France), etc. during the first years of the new millennium, to understand that the workers are always the first to be fired in a difficult economic situation. It seems that according to the article mentioned above, this dynamics is changing in private companies, not for any "noble cause", but because "businesses are forcing companies to take a more human approach". Administrators suddenly became aware "of the cost that stress and maladministration can represent for a company", says Mrs Morin, professor at the *Hautes Études Commerciales* (Commerce College in Montreal). A new image would be on the verge of appearing in companies to attract and keep the best elements of a staff. A staff manager confirmed "to have never yet seen a similar situation during his long career". Everything was done "to maintain contact with the employees". In his company, he says "the employees are formed, assigned to special projects and paid even if there is a lack of work. People who lose their job are voluntary reinstated in the company as soon as production requires them to". Other companies are grateful to their good employees, offering global payment, training, flexibility of work planning, decrease of working hours and variable schedules. This tendency towards more human rapports among employers and employees is seen worldwide, thank God! It was always touch and go in this area! Let us hope that this phenomenon continues.

I often wonder, however, if individuals could do without their polluting motor vehicles, without the comfort of their homes warmed by fossil fuels, their "refrigerators", their freezers, their air conditioning systems, their cooling systems which increase greenhouse gases and the other multiple possessions that increase the production

of the polluting load? It seems to me very unlikely that one day people will deliberately give up the assets that they so greedily use…

The simple implementation of a new public transport in a city, planned to decrease the emission of polluting gases, raises general outcries and creates insurmountable problems. What are your reactions when a technology such as television, computers or cars does not adequately meet your expectations? Frustration and violence frequently overcome people in these situations. Just look at the people driving like fools on the road, firing in a crowd for no good reason, killing the referee in their son's hockey club, consuming alcohol or drugs to feel better, stressing out at work, having nervous breakdowns one after the other because of constraints, etc. The authorities have used and abused of the freedom that people had confided to them and crushed them like pieces of paper.

D. Technological applications mind-destroying to man

Let us illustrate some of the materializing and negative effects produced by certain intrinsically dehumanizing technologies. I mean those technologies that, instead of helping people, degrade them. **Telephony with multiple answering machines is a striking example of an advanced high technology causing more and more deplorable effects on the psyche of human beings.**

The constant use of these technologies in our daily lives is clearly related to these unhealthy side effects. The careless use of these technologies eventually makes people turn around in circles. It makes them waste their time and degrades them instead of helping them.

Information that should be easier to obtain and also be more complete through data processing is often usually not. Why? Because they work through intermediary machines. Not only are they dehumanized but, very often, they are incomplete and mind-boggling. They are reduced to questions asked by machines which more often than not, are badly interpreted, formulated and programmed by "little" individual powers. We were all happy at first but quickly became upset with these Machiavellian machines that make us waste such enormous amounts of time.

Calling one day the Local Centre of Community Services (CLSC) closest to my home to set up an appointment for a simple flu vaccine, I came face to face with personalized dehumanisation. First I had to 1) - wait for over 30 minutes to get through, 2) – then be subjected to loads of questions by several answering machines, 3) – after all of this, the machine tells to me to call another CLSC because the one I was calling and which is two blocks away from my home did not have the same postal code.

Can you imagine? What stupid administrator can make such decisions and give such absurd directives? I had to start all over again: 4) – wait on-line another 10 to 15 minutes at this other CLSC, 5) – be subjected to another bunch of questions, which 6) - finally ended when the silly answering machine, programmed by a secretary on staff, told me that I had to call back to set up an appointment... Well, I never called for another appointment nor did I receive the vaccine that year.

To my knowledge, I had never felt so much disrespect and dehumanisation from a more or less useless bureaucratic appendix of our health system that had confided to a bureaucratic structure the responsibility of inoculation against the flu. The worst is that these small useless administrators have not yet understood that what holds up the system are the numerous small useless administrative constraints of their own system.

The CLSC system is not standardized. It varies from one place to another according to the knowledge of the small powers running it. It cannot even provide the essential medical care that any doctor can give in a medical clinic. They are not opened after 17H00 or during weekends, they do not have doctors, nursing assistants or the necessary workers to adequately meet our needs, etc. However, they do know how to participate in long meetings, discussing the case of a single patient, and this during working hours...

The following winter, in spite of the thousands of more fortunate persons who were finally able to receive the flu vaccine, the emergency rooms of hospitals were invaded. Patients had to spend hours, even days, in corridors waiting to see doctors for the flu which could have been avoided, if telephone operators had not decided on their own to stupidly create these insignificant messages on their answering machines. Here is another example of maladministration and of misuse of technological applications by incompetent and non conscientious individuals protected by powerful syndicates. Such absurd administrative technologies multiply everywhere and carry the weight of their uselessness.

In November 2000, I phoned the CLSC located in the area of my postal code for a simple appointment for a blood test, a common medical act that should not take more than a few minutes in a well organized health centre. In certain hospitals, for instance, blood tests for simple analyses are made one hour before the visit to your doctor. In veterinarian clinics it is done as soon as the animal arrives at the clinic. I know of certain health centers which do blood tests without appointments. Analyses are done automatically and the results are sent to your doctor before your arrival.

Now, my CLSC refers me to a central operator, somewhere on the territory. For unknown reasons, such as "making useless statistics" on the number of requests for blood tests in a determined territory (figures which have been known for decades) or simply to meet another absurd administrative directive, I finally got an appointment for 8 o'clock in the morning, one month and a half later, for a simple blood test!

On the day of the appointment, I left my Medicare card with the requisition for the blood test at the reception desk of the hospital, as requested on the panel at the entrance, and I sit…for over three hours…

I was trapped in another cursed organization of blood tests, badly planned by other incompetent administrators having decided on their own of a complicated protocol so that all the blood tests in the hospital are done at the same place! What a sign of superior intelligence and what lack of competence! I had to wait for three hours and a half for my blood test. Administrative directives given to the employees (about ten, I think) were to say the least absurd, disconnected from reality and useless.

The staff had to:

1) - sort out the patients who had a blood test and a meeting with their doctor the same morning;
2) -sort out the patients who had to have their blood formula verified before receiving chemotherapy;
3) - sort out the hospitalized patients;
4) - sort out the patients going to the operating room for surgery (under general anaesthesia and released the same day for lack of beds).

There may have been other reasons not to respect the schedule of the appointments, but this is what I observed:

- Patients passing in front of everybody caused great dissatisfaction and the staff wasted precious time trying to awkwardly explain these transgressions.

- In the office where they took the blood samples, the samples had to: 1)- be labelled with a machine that badly worked and on which all the delays were blamed, 2) - be reclassified in different piles before finally being forwarded to the respective laboratories. It was clear that the employee that day (a volunteer maybe?) did not know what to do with these blood samples.

After 3 ½ hours of useless waiting for this badly planned administrative disorder to end, someone announced that it was finally time for the blood tests of the patients who had been scheduled "arbitrarily" by the useless and incompetent operators of another administrative power plant, outside of the hospital, six weeks previously, knowing nothing about the place, the schedules and the directives to follow in the institution where your blood will be taken... What kind of maliciously severe and pointless administrative organization has been created by scientists who thought, I am sure of it, they had invented the wheel!

As for the results of analyses, which had still not arrived 21 days after the blood test and more than three months after the prescription, I had to phone to find out (new waste of time) what was happening. The laboratory employee told me that they did not have the address of the doctor that had asked for the analyses! The address was nevertheless well stamped on the requisition form of the said analyses. I was told that the requisition had been lost, either in the administrative mess of the place where the blood was drawn or somewhere else in the laboratory!

It is abnormal, sad and absurd that administrations can manage to spend so much time and money pointlessly in an already bad health system and all this because of the incompetence of administrators!

If we stop and think about it, it is rather easy to understand why all these administrative systems are deficient. Why are there are so many delays in hospitals? Why are waiting the lists multiplying? Why is no one happy? It is that the planning is made by incompetent administrators in charge that know only how to run their machines with illogical, mathematical directives without humanity? All this administrative mess is caused by to much administration. Three competent persons could have done the same work much faster than the 8 to 10 persons who paraded like they were stunned between waiting rooms and the blood collecting areas. The administrators only think of collecting information and compiling them on useless forms in order to make statistics to justify their pay checks... The application of these administrative techniques is repeated *ad nauseam*, everywhere and in all administrations.

Health care objectives should be easy to settle. They involve **quality care for patients**, must be **given by competent and devoted multidisciplinary teams having and controlling technological materials.** But administrative powers, not medical staffs, have decided along the way that the patient, (the beneficiary, the client, etc.) will not be treated in hospitals any more, but in hospital centers, short term care centers, long term care centers, or neither short nor long term cares center, etc. Unfortunately patients and medical staff are not involved any more in these decisions. **The cares are only defined by incompetent individuals...**

The number one solution is simple, we must eliminate: 1) - the filling out and compiling of useless data, 2) - the useless amount of unnecessary paperwork, 3) - the useless middlemen, 4) - the useless centralization and 5) – the useless sub-contractors. The administrators are so greedy for useless information that they confuse efficiency and management.

While I waited for my blood test, I tried to imagine what could be the attitude: 1) - of surgeons who had to wait to start their day surgeries, 2) – of the haematologists who swore because they had to wait to decide if their patient was or was not going to receive his chemotherapy treatment, 3) – of the nurses in the "chemotherapy" department who were waiting for results to install their serum, 4) of the doctors in their office who had to put off appointments to a later date because they had not received the test results necessary to make a diagnosis or take a decision. No human consideration whatsoever for the patients or the individuals who had to wait for hours, like machine parts on an assembly line, for a simple blood test…

Same phenomenon for administrative techniques applied in private institutions

I had thought, when the first technological tools came out that when private clinics or free enterprises would use these dehumanized administrative technologies, institutions and public societies were going to collapse. I was wrong to think that governmental agencies were powerless to adequately apply these technologies. I was totally mistaken. The solution is not there. The problem is worse: it is in the teaching, the forming, the sciences and the dehumanisation by technologies. The administrators and the technicians managing private companies were formed at the same schools, with the same rigid and dehumanized administrative techniques without humanity and human values. Later on in this work, we will look at the flaws of our occidental education system.

Another day, trying to reach my doctor in a large Montreal hospital, a secretary, quite "powerful" in her position, had programmed her silly answering machine to say "that there was no possibility of making an appointment or to speak to a doctor, and to hang up". The answering machine simply went on repeating "please hang up"! If it is not insolence as well as dehumanization, I would love for someone to tell me what it is. Who can tell to me where the progress lies in all this technology that simply seems to worsen things?

People are fed up of speaking to answering machines and to press on 1, then 2, then 3, then 4, then listen and too often have to start again in order to choose the right one! They are fed up of having to call several times and waster precious time fooling around with these dehumanizing and mind-boggling machines. These totally

dehumanized technologies keep on repeating themselves over and over again...

How could have believed that the sophisticated techniques used in communication by administrations would decrease the number of civil servants or facilitate communications between human beings? The more we speak to machines, the more we will find badly formed and incompetent civil servants to program them, and the less human contact we will find among people. So far, nobody has demonstrated that it was less costly to speak to machines than to real persons! Nobody demonstrated that people are better informed through these communication techniques. Everything happens as if the authorities multiply these useless functions to complicate and justify their salaries. In all this dehumanized "over bureaucracy", people have long applied the Peter Principle!

I have in mind these individuals trafficking child pornography using their employers' computer systems, and to the others who use these computerized toys for their personal use, to chat, to send e-mails, etc. I wondered one day how many millions of dollars in lost work, how much our hundreds of thousands of civil servants cost us in tax money every day, all because of these communication technologies. This specific technology encourages laziness and ends up boring both the people working in institutions as well as those using them *ad nauseam* for their work. These ways of working are purely materialistic and unjustified. They also allow those that could give adequate answers to lounge around without the responsibilities of answering, working and accomplishing the work for which they are paid!

Implementing computer techniques in administration

I had first hand knowledge of the arrival of computers in administration at the beginning of the 70's. **The first computer technicians were idealists happy to assure us that computerized administration would save enormous amounts of money to companies and institutions. Do you think that is what happened in the past 30 years? Of course not!**

Before the computer era, accounting departments were quite simple to manage. Computerized and considerably more complicated today, they bring no monetary relief to companies. Instead of having 12 to 20 columns in the ledger, there are now as many columns as there are items. Never in 30 years has there been a decrease in costs because of computerization. If anything, we experienced new complications, computers breaking, repairs, new computers, new programs, new training of computer personnel and data complications. Never to my knowledge, has there has been a reduction in the costs in administration exploitation.

The weight of administrative techniques in the exercise of a profession

The dehumanisation of the administrative techniques is felt in human behaviour and echoes in the practice of professions. Let us say a word about the harmful effects of these techniques in the exercise of medicine. The cumulative effects of technological applications combined with the transformations left by science and dehumanized administrative technologies in a profession practiced since hundreds of years, are striking.

With the exception of technological applications in data processing or in communications, few branches of human activity have witnessed the acquisition of as much knowledge from different technological applications as sciences related to and applied to human and animal medicine. Changes in chemistry, in biochemistry, in immunology, in biology, in microbiology, in physics, in biotechnology, in genetics are so important that they deeply modified the practice of medicine.

Never before had so many technologies allowed a group of technicians, such as doctors, to so rapidly change their behaviour and to abandon so rashly what the doctors liked calling "the art" of practising medicine. To earn their keep, today's doctors apply in their practice the numerous techniques discovered by research scientists in various branches of science. Never before had management technocrats better applied their administrative technologies than in medicine and in hospital administration.

The socialization of medicine by governments allowed administrators to rise to the commands of health care in states, hospitals, health centers, insurance companies, and this to the detriment of the doctors themselves.

Hospitable medicine was transformed at the contact of these new technologies into a kind of industry without humanity. The administrative materialistic techniques were used to administer at the same time and in a similar way the hospital building, the workers and the patients, and to control the administration of hundreds of sophisticated techniques indispensable for the diagnosis and treatment of all sorts of diseases. Practiced together, these totally different groups of technologies strongly contributed in dehumanizing the practice of medicine, medical care, the medical personnel and all hospital workers.

Patients became numbers in hospitals, as did doctors, nurses and laboratory technicians. They became a number that administration technicians quickly placed on patients' wrists as soon as they were admitted. A number, to whom technicians specialized in surgery change the bones of the middle ear, restore the hearing, transplant a new heart, a new lung, a new hip, a new kidney, bone marrow, repair or correct all sorts of specific pathologies, etc.

The administrators, on the other hand, contribute in sabotaging the health system that citizens had given themselves, by endlessly creating new programs, making budget cuts, activating strikes, arousing dissatisfactions of all kinds on health care and then trying to settle the problems that they created themselves by using the same type of reasoning and stiff administrative and logic techniques devoid of humanity. It would be too long to review all the bad decisions made by the "low" administrators who control patient care from a distance, some of which have never set foot in a hospital. This blind management is a characteristic of this administrative technology. The administrative techniques cost billions of dollars every year to the health system. The medical administrators substitute themselves to the doctors by managing the patients, the workers, the doctors and the other competent technicians, believing that they know more about the medical field than doctors in front of theirs patients and their diseases. **They have been administering for years the health system without the help of doctors.** They think they can better control diseases and doctors by managing the money necessary to health care.

In spite of all the mistakes they made, hospital administrators will go on controlling my health, my medical care, my medication, etc, and this for as long as we let non-doctors be responsible for my health. These administrators have imposed their incompetent powers. They interfere in the scientific fields they manage. They keep on proposing scenarios accepted by other incompetent administrators totally disconnected from medicine. Some administrators should be dismissed for seizing the power of my health. I never gave them the right to take my money to take care of my own health!

Let us take a closer look at these governmental administrations to further convince ourselves further of the real dehumanizing effects of these administrative powers. These civil servants are formed at the same school of science and technology. Throughout the years and with the help of data processing, administrations have complicated themselves to a point where they need more and more specialists in administration to understand their own forms, reports, financial status, etc. It is the same in all administrative levels of government and their institutions. In all the positions in public services, the administrators are the only ones that constantly increase in number! Administrations grow like inflation. It is the only profession that has never had a decline in costs. Administrations do not reduce the burden of taxes: on the contrary they have amplified them considerably over the years. The science of administrative technologies that aimed at simplifying work has paradoxically become more expensive and more complicated. They have made their own administrations so complex that only certain initiated individuals can understand the language.

As a result, all government administrations in the past 30 years have undergone explosive raises in their costs and thousands of other administrators and technicians have invaded the powers. Governments became indebted for decades to come, in order to enable other scientists to fill the thousands of new positions they created. The citizens must now pay, over and above their heavy taxes, for a number of unwanted services. Can anybody tell us where the progress lies? How can we get rid of all these individuals? Until when will administrators abuse our patience and maintain these parasite jobs protected by beautiful collective agreements?

There are many other examples showing to what extent our materiality is decadent. Just imagine: the city of Montreal has just created "garbage inspectors"! The data they will collect will allow administrations to know if you had flu, if you have a baby, who sends you invoices, if you waste food, etc. The administrators are happy to send you a fine which can go up to 136, 00$, if your garbage is not put by the roadside at the right time, if it is badly classified or still, if it is not placed in bags of the right colour... Imagine the beauty of this, as elected members and administrations interfere to control even your waste! Are they not stupid to waste precious time and money to organize similar controls? Film-maker Stéphane Thibeault of the National Film Board of Canada has produced a film entitled "*La loi et l'ordure*" (Law and Garbage) because of the aberration of such municipal regulations.

We could tell hundreds of anecdotes of this kind illustrating administration abuse. Some bring out *ad nauseam* the nonsense of using technology. The facts are disturbing: they demonstrate a growing decadence in the administrative powers using science and the available technologies in a deceitful way.

I often wonder what part of responsibility this useless and awkward materialistic culture has on social and school dropouts, the "squeegees" of our societies, the adversities in hospitals, health care, failures in education, suicides in young people, illnesses, the chaos in the materialized societies and all this social degradation, provoked by the abuse of technologies? Today occidental governments have become immense impersonal machines running only with this linear and inhuman logic of the scientists' methodology. Is this preferable to the religious administrative powers of yesterday? These abuses of administrative materialistic powers exasperate everybody exactly as religious powers did yesterday. Have scientists only replaced one extreme power by another extreme power, their own?

Other characteristics of administrative technologies

The deplorable effects of all these technologies affect the psyche of the people and contribute to their materialization, their dehumanization and their degradation. Everybody knows nevertheless that all these technological applications arose from the logical ideologies and the reason of scientific and social materialism. These technologies are taught in all the schools and universities. We will never repeat enough that the problems they engender are bound to the mathematical rigidity of their logic devoid of human sense. The evil of science comes from the fact that it always denies and rejects, without proof, the existence of all non tangible entities.

A French philosophy professor published one day on the Internet an article on the necessity to quickly change the method of reasoning used by scientists by a thought of "meta-science". He wrote: **the cosmos will not indefinitely remain subjected to the Cartesian thought of "object to master and possess". It will indeed be necessary one day for humanity to cut this umbilical cord, by creating a new thought, in a face-to-face with the otherness of the Cartesian method. A new thought that would be at the center of the resources of our world, of its scenery and the possibility of expansion of the universe, a thought that would not always be reduced to utilitarian considerations and substances. "Next to the scientific predetermined aspect of the process of forming of nature, there also exists," he wrote, "an evidence of determinative additional freedom we could not dodge much longer."**

E. When techniques create disease

We have already spoken about the technological applications linking toxic polluting agents and the related diseases directly caused by the multiple pollutants discarded in the environment by factories, agriculture, motor vehicles, our ways of life, etc. And so we drew up a panoramic picture of the extent of the damages and underlined that dozens of thousands of persons die, every year in the world, poisoned with these polluting loads of toxic waste thrown in the environment by our technological civilization. In this chapter we will talk about different groups of diseases caused by technology and affecting the psyche of the individuals using it. Relatively new, these diseases are more and more recognized as other major inconveniences related to technology.

Certain technologies can directly induce dehumanising effects and constant conflicts on the people's psyche. **Let us think of the constant stress induced by professions with high responsibilities or those using frustrating and repetitive techniques. Have we ever demonstrated anywhere that these professions do not create behavioural confusion ranging from aggressiveness to violence, nervous breakdowns, alcohol abuse, gambling, overwork and even suicide?**

Science had promised prosperity, comfort, mobility, information and better communication between the people. This inevitably clashed with the logical, mechanical and inhuman technologies such as those of answering machines endlessly turning in circles by constantly asking us to go back to the "main menu", and then to wait for long and useless answers, to the mechanical linear administrative and inhuman decisions mentioned earlier, to the stress of traffic jams, the endless waiting in hospital emergency departments, to the shameless surveillances of people by governments, the monstrous governmental computer files on each individual, the insane and unthinkable dreams of their politicians, the gigantic defects of management techniques, the increasing number of tax forms, the complicated tax system, the shameless and useless shambles of powers always thinking they have found Peru or the answer to all the problems they themselves caused or of these other effects which have transformed and continue to transform all the actions of human activity into indescribable chaos.

The promises of comfort, prosperity and modernization of science have only managed to make us realize our own absurdity. The endless waiting for the promises of science are often at the origin of certain psychological diseases, aggressiveness and violence that strike human beings. Aggressiveness and violence are indeed linked to stress and to the multiple nervous breakdowns affecting today about 30 % of all employees. Stress can also be linked to certain forms of madness, to missing school or dropping out of school, to socially unacceptable behaviour in individuals demoralized by living in such materialistic societies with no soul or humanity.

Several modern diseases are directly associated to the wearing or repetitive effects of technologies, such as the repetition of noises which cause harmful reactions on the nervous system, the organs and / or on the behaviour.

The "stress diseases"

Stress-related diseases are other pathologies increasing in employees stuck in their materialistic societies. The phenomenon of stress is more generalized than we think. An Ipsos-Reid survey, done for the Aventis pharmaceutical company, revealed that the level of stress passed from 47 to 62 % during the year 2000 in Canada. The proportion of Québec employees enduring great stress at work passed from 39 % to 59 % from the year 2000 to 2001. The events of September 11, 2001 simply aggravated this stress.

Stressed by their employers, overpowered by mind-numbing repetitive technologies, assaulted by an incalculable number of telemarketing ads wanting to sell all sorts of products or who try, every which way, to make them spend the few dollars left over after having paid their taxes, people become irritable, anxious and violent. They suffer from insomnia and miss more and more work. The stresses of their daily lives are slowly killing them.

Doctor Hans Selye demonstrated that stress causes a hormone imbalance in gluco-corticoids and thyroid hormones. These hormones in turn affect the blood sugar levels. "The rise in diabetes since a decade appears to be a real epidemic", declares endocrinologist Dominic Garrel of the Hôtel-Dieu Hospital in Montreal. The role of stress on gluco-corticoid increases and the effects on the immune system and in the development of certain cancers are being demonstrated. More and more researchers indeed link stress to the development of cancers.

In the face of possible causes of certain cancers and different diseases, health professionals suggest preventive measures: stop smoking, do physical exercise, change your lifestyles, consume less alcohol, eat less fat, more fruit and vegetable and learn to relax: these are measures that could warn off, we believe, certain diseases.

The preventive approaches to disease have become more necessary and are not taken lightly by the most conscious people. They give priority to preventive measures for disease. People are afraid, and with good reason, of having paid all their lives enormous sums of money for health care from which they may not be able to benefit tomorrow. This is not without adding to their other stresses.

The Canadian Institute for Health Information published in May 2001, a sinister report on the state of Canada's health care system. The report confirmed the hard reality of the nurses' stressful profession. These employees are the ones that miss work the most in a year: an average of 15 days a year. Long working hours, workload and stress are responsible for at least 75 % of occupational accidents in their respective domains.

Doctors have also become a concern. In spite of denials by governmental "tenors", the report also mentions that Quebec continues to lose doctors, formed at great expense by society, because the health system is poorly administered.

The numerous setbacks of the health system are forcing doctors to leave the hospital network. More than 160 doctors left Quebec in 2000 and 634 since 1996, in a population of about 15,000 medical doctors. It is abnormal and paradoxical to leave to legions of non medical administrators and technicians, often working somewhere in towers away from the population and the care units, the power of dictating insignificant directives that delay medical acts and directly affect the people's health. Until quite recently, we could count on the fingers of one hand the medical doctors administrators in the health care network. If the people in power do not recognize these facts, they have severe problems of inconsequence. Less and less medical doctors are interested in practising in hospitals directed in a "go as I push you" manner by inhuman, elusive and incompetent administrators. How can these administrators be so indecent for instance as to blame medical doctors for medical acts performed by a paramedical staff they do not control and that are not even under their responsibility?

How can we understand that governmental authorities confided to non medical administrators the care of health and diseases? Have we given them these powers? Never: we simply asked them to administer the system: they seized it like thieves just because they control the money!

Stress and human behaviour

An American study that took over 15 years of work, made by David Snowdon, professor of neurology in the University of Kentucky, U.S.A., published in May 2001 that negative emotional states stimulate anxiety, resentment or anger. After a certain time and constant exposure to this, people become more subject to cardiovascular diseases and to heart attacks. The study that was done on a group of nuns demonstrated that those that are happy and live with non material values suck as joy, love, hope and have positive feelings, live an average of ten years longer than the others. Psychologist Louise Descôteaux from Sainte-Thérèse (Québec), a stress management specialist, said (*Journal de Montréal, May 9, 2001*), that "strong feelings activate adrenalin, exult anger and strain about forty muscles including the heart. The rage, the problems, the loss of control provoked by slow traffic, a slow driver or stress at work, all produce side effects and illnesses due to stress and reduces our longevity".

Researchers have hardly begun to study the side effects of the numerous new social, contemporary and daily realities that trigger new diseases. The negative effects of stress have not yet actually been connected, to our knowledge, to the materialism of technology, to the valuation of scientific materialism or to material values alone. The numerous proofs we have advanced here strongly suggest that the individuals of our societies can suffer from the absence of humanity and the dehumanized behaviours that unconsciously invade them since childhood.

The *"Food and Drug Administration"* **confirms that over than 25% of the children of the United States suffer from depression,** and rendered official, at the end of 2002, the prescription of Prozac to young individuals. Prozac is a drug initially prescribed by psychiatrists to treat hyperactive children. In the face of such abuse, one should wonder if scientists would not be well-advised to require from their co-workers in education, a return to more humanity, public-spiritedness and human values lost in the education system.

The Canadian Institute on Health Information made public, in May 2001, a "black" report on the state of Canada's Health System and presented its shortcomings. Quebec has one of the worst performances in the country in regards to cardiac patient management. The report shows that there is a 39 % risk of dying in the 30 days following surgery in hospitals making less than 500 such surgeries a year. Quebec makes over 47 % of the heart surgeries in peripheral hospitals, against 4 % only in Ontario, because medical administrators decided that more operations should be done each year in less experimented peripheral hospitals than those of big centres, simply because they do less surgery, their clientele being more restricted. The surgeon who makes one or two operations a week can not develop or preserve the dexterity of the surgeon performing six or seven a day. The same applies to the paramedical staff of these hospitals. Results also demonstrate that post-operative medical and paramedical care cannot be as good in peripheral hospitals compared to the larger centres, which possess teams of surgeons and paramedics better medically trained.

Toronto's Saint-Michael's Hospital has four cardiac surgeons from Québec out of the five doing cardiovascular operations in this department. During this time, our Québec patients must wait for months to be operated at home.

The magazine *"Diabetes Care"* published, in the beginning of 2001, alarming statistics on the increase of type 2 diabetes in the adult population. The incidence of this type of diabetes increased by 33 % in the United States between 1990 and 1998 and affects today 6, 5 % of the American population. It has also increased by 76 % in the young adult population during the same period. Administrators should be aware that this disease is at the origin of arterial diseases, high blood pressure, cholesterol in the blood vessels, renal disease, heart failure, amputations, blindness and vascular cerebral accidents, and that these diseases are so important that they require billions of dollars every year for the health care system. Have we not heard somewhere that disease prevention could save lots of money to the health system?

Although researchers think that the increase in diabetes could be related to bad eating habits, excess of sugar intake and to carbohydrates for pancreatic cells, obesity and lack of physical exercise, nothing keeps us from thinking that the daily amount of stress at work in the every day lifestyle of modern societies is not partially responsible. A decrease in sugar and refined carbohydrates, weight loss and increased physical exercise may help prevent this disease, but not cure it.

Nothing prevents us from believing that the numerous frustrations of life endured by people are better assumed by those believing in material values than by those sharing human and spiritual values! The loss of humanity and the lack of human contacts - in spite of the fact that we are surrounded, as never before, by communication systems -, the lack of compassion, the rareness of human values in the dehumanized societies, the non-practice of renunciation, etc., can be also important in the appearing of certain diseases. The facts presented so far also illustrate that when the frustrations of every day life are drowned in a pool of inhuman values and material laws, individuals become less tolerant, more aggressive and more likely to become ill.

Behavioural diseases

An article published on the Internet in May 2002, in *"Services Vie Santé"*, described the syndrome of sudden explosive violence as "intermittent behavioural problems which translate into outbursts of rage and fury leading to the destruction of material objects, assault and even manslaughter". Psychiatrists have classified these acts of violence among the problems of impulse control, that present the same characteristics as other disorders of this category which is "as uncontrollable need to attack objects or persons, an increased tension immediately preceding the act, and a feeling of pleasure, gratification and relief during and after committing the act". It is estimated that about 80 % of the persons suffering from this syndrome are males. Other data can also be found to characterize these acts:

- Frequent and often unpredictable episodes of rage and anger leading to the destruction of objects or to assaults;
- The extent of the aggression is always out of proportion to what triggered the anger;
- The crises disappear as abruptly as they appear;
- There is no threat or no sign of violence between episodes;
- The symptoms are not explained by mental illness, personality problems, a medical condition or an abuse of medical substances.

The article also states that certain individuals can feel regret and remorse after the act, but they rarely take responsibility for it. They rather tend to hold their victim responsible for their lack of control, evoke outside circumstances or the fact that a third party provoked their anger. This perception allows them to relieve the sense of guilt they can feel. This behaviour regrettably has the effect of strengthening the individual's behaviour and justifying his refusal to make any changes. This intermittent explosive disorder creates legal problems and the individuals who suffer from it are often accused of assault and domestic violence. Domestic, professional and social consequences are often devastating and regrettably, no cause for this phenomenon has yet been found, but progress is on the way.

Stress and human errors

"Health Freedom News", an American magazine, reported in 2000 **that errors in medical techniques killed more Americans every year than road accidents, breast cancer and AIDS put together!** The American Institute of Medicine, an independent institution in charge of making recommendations on medical subjects reported that **between 44,000 and 98,000 deaths are attributable to medical errors in 2000 only. The medical errors constitute the eighth cause of mortality in the United States.**

The Institution asked the American Congress to finance campaigns to identify the causes of these errors and to reduce their number within five years. It asked for the collaboration of the Food and Drug Administration so that special attention is taken in medicinal safety. Even though errors are human, they are unacceptable in a health system which refers to the fundamental principle *"primo, non nocere"*, (first, do not harm). It is inconceivable that a person should die because he/she is being treated! This absurd and paradoxical situation must be quickly corrected.

"The therapeutic arsenal used in conventional medicine and the numerous names given to a same drug do not help correct this situation" indicated the naturopath Brunet in a chronicle of a Montreal Newspaper on July 12, 2000. "It is wishful thinking to say that if conventional medicine could resort to more natural medication, rather than to the current toxic therapeutic arsenal they use, problems would be partially solved".

Linking diseases

There are other diseases that we could qualify as links. They are caused indirectly by the numerous ways chosen and used by people to counter their stress. Smoking that leads to cancer, medication, drugs, alcohol, gambling are all means of evasion which often cause diseases depending on the use we make of them. Stress can cause dramatic changes in behaviour. It is also necessary to remember that psychic and physical dependency diseases can be caused by technology, and that people's infantile attitude can be caused by their governments, their ways of life or by work itself. **It interesting to mention at this point that the World Health Organisation says that 20% (one person out of 5) of the North American population will suffer from a mental disease during his life - such as anxiety, depression, schizophrenia, bipolarity, obsessive-compulsive disorders, personality disorders, etc.**

Let us mention that the scientists working for tobacco companies are a good example of this degradation of our humanity. Already in the 30's, the administrators and scientists of these companies knew that tobacco was harmful to health. At the end of the 50's, they had in hand all the scientific proof to incriminate tobacco as the agent responsible for lung cancer. No scientist could completely eliminate the participation of tobacco in emphysema, in cardiovascular diseases and in other related diseases. How can we explain but by a blatant lack of dehumanisation that the leaders and scientists of these companies were transformed into criminals without ethics, taking advantage of the billions of dollars resulting from this toxic matter? How else do we explain that the administrators of these companies neglected to act like human beings, by inciting and by spreading death every day through their companies? One must be seriously dehumanized and even criminal to consciously agree to kill more individuals than the First and Second World Wars put together, by letting people smoke!

These people even forced their useless and dishonest researchers with no notion of deontology to deliberately and consciously add to the tobacco certain ingredients that induced dependency in the users... To be so Machiavellian, one must necessarily be mentally ill, suffer from deep materialism, be distorted or be in a state of advanced dehumanisation. When we think that these cultures come from accepting without proper judgment the material capitalist values resulting from a dehumanized formation, education through science and technology without the humanities, we are deeply saddened.

These leaders were not without knowing the statistics that demonstrated that every individual brought them 80 000, 00$ in 50 years. These people also knew that over 100,000,000 individuals died from the consequences of their addiction to smoking in the twentieth century alone.

And to think that this materialistic culture is passed on all levels of education at the same time as the knowledge! There seems to exist in certain university faculties of high knowledge, a sick, shameless and unhealthy ideological tendency to make the promotion of scientific and social materialistic ideologies, political socialism and politically socialistic vocabulary (*cf. Dr Satel's book, Chapter 5*). Certain administrative faculties even teach that people are not responsible for their acts and even for their diseases! Politicized and dehumanized activist professors try indeed to persuade students and people that society is responsible for their diseases and not the individuals themselves. Would Jean-Jacques Rousseau have so influenced education, with his *"Émile"*, that he depraved societies?

Why try to hold at arm's length such false, exhausted and incapable educational systems forming the essential part of man's mind, if once dehumanized, people are not held responsible for their actions?

F. When the materialistic methodology of the search for truth takes charge of education

Scientists, intellectuals, technocrats and governments have only succeeded in immortalizing yesterday's religious dogmatism the day they decided that the citizens would from then on be formed at the school of science and technology. They only confirmed that they now held such powers. When they decided to change the former ways of teaching with the humanities by science and technology without the humanities, they essentially imprisoned the people in profound materialistic ideologies. The implementation of such a system, before the young even had the chance to think or to make any kind of judgment, could only have disastrous effects on the transformation of man and society. Learning these scientific social ideologies has branded materialism in all the actions and in all the ways of the people and continues to echo throughout their lives. Technology and science without the participation of humanities in education and in the training of individuals have just contributed to changing their behaviour and materializing them.

What scientist or intellectual is not aware that the integration at a young age of such knowledge and materialistic values would not be embedded forever in their fragile minds? Do you think that the Nazi leaders of Germany, Italy or other Marxist socialist countries, in Russia or elsewhere, did not know the consequences of their teachings when they institutionalized the teaching of Nazi or Marxist principles in their schools? Do you think that the Mullahs who taught young Moslems to engage in atrocious terrorism, by blowing themselves up with explosives attached to their belts, did not know what they were doing? Do you think that the decision-makers of the ministries of Education in the occidental world did not play the same game as yesterday's decision-makers? If the repercussions of such an education were not obvious to those who imposed them, we must immediately start all over again, because this education only produces individuals and materialistic societies, totally deprived of all human sense. It produces individuals who think and work in a dehumanized way, individuals without humanity who have no respect for human values, no sense of real love, compassion or respect for others, no kindness, no human morality or sense of ethics other than the materialistic one. It produces individuals who justify their actions only on their own material values such as money, wealth, greed, personal success, contempt and condescendence towards others.

This deep uneasiness of our societies comes inevitably from the sciences which take into account only the knowledge acquired on the matter. Scientific methods of teaching can only build their hypotheses by using material and tangible data related to the senses, in other words data feeding only on material knowledge and on material values. If science and scientific methodology have produced such fantastic and phenomenal knowledge on material things, it is precisely because they recognize and study only the material things that are visible and tangible with the most primitive senses.

These sorcerer's apprentices managed to penetrate the secrets of the matter by using this method of training. But by doing so they cut themselves off from any access to the non material truths **and to all the traditional human entities that they cannot touch with their fingers, see through their eyes or smell with their nose, in order words with their material primitive senses.** In its current state, this method of research is completely powerless to study the existence of the data of the less tangible entities.

When intellectuals and Occidental scientists chose education through science to form their citizens, they forgot that the learning of the human thought is as necessary for the individual as the learning of science and technology. **They also forgot that scientific and social materialism was first conceived by intellectuals and scientists to promote research on the physical material world, but not at all on the learning of thought in young children.** Scientists and intellectuals never thought, at least in the beginning, that this method would be used one day for the education or the forming of the children and the people. Descartes, in his "*Speech on the Method*", had nevertheless warned those that would use his method for research that it is only useful in the study of material things. In their haste to take hold of their power over the people and to change the secular method of teaching, scientists and intellectuals privileged this methodology and the materialistic education to form individuals and consumer societies.

If in a majority of industrialized nations, education quickly followed this method of scientific forming which was considered, wrongly, as an ideal method of training, it is because the authorities thought that the method would facilitate the forming of engineers, technicians and scientists that their consumer societies badly needed. This was their first mistake.

If, in almost all the countries of the planet, science has been elevated to a status of "infallibility", it is because it uses a method of research which lends itself to very little criticism. It became implemented as "a new find" to serve consumer societies, because people were dazzled by the knowledge acquired on the matter through this rational and rigid method of research. Scientists and intellectuals wrongly believed, however, that by using this scientific method for training people, they would form superior individuals, as infallible as the method itself... This was their second mistake.

This incomplete method was used for the training of individuals and especially the young people of our occidental democratic societies. It produced devastating side effects such as dehumanisation and permitted to study only the existence of the matter, without any notion of humanity and the less tangible values. The people of Japan and China and elsewhere are very fortunate that human values are still taught while using this method of research and of thinking.

The first changes provoked by these education technologies

The desire to change human behaviour, acquired for centuries by people and societies, was a powerful element for scientists and intellectuals to justify this decision to choose science and technology to form individuals. The purpose was to make people forget their former training which was based on traditional humanistic and spiritual values. By removing the humanities and the secular formation from education, behavioural changes rapidly occurred in individuals formed at the school of materialistic sciences. The natural attraction of the people for material values largely facilitated the implementation of this new culture in consumer societies. The authorities were the first to hire these individuals newly formed in schools of science, technology and materialistic culture that invaded all fields and as well as all "corridors" leading to power. Little by little and from one generation to the next, more and more individuals became imprinted by these sole material values. After a few decades, materialistic values were eventually found in all spheres of human activity and society.

It is easy to demonstrate how education through science and technology quickly got the upper hand over traditional humanistic education, over humanities, religion and domestic or social values which were pushed aside bit by bit and finally completely abandoned by the education system. Religion, called the "people's opium" by occidental materialists (Lenine) was practically eliminated from education, to the benefit of science and technology. Everything sacred, as religion and traditional human values have never even been considered in science. Philosophy and history suffer more or less the same fate. They have been reduced to their simplest expression. This is how education through science became established and got rid of, by a simple stroke of the pen, all things that scientific methods could not explain. It is thus not accidentally that people having gone to such schools become atheists!

By using their scientific methods of education with science and technology without the humanities and armed of their imperturbable linear, rigid and senseless logic of "pure reasoning", scientists of the nineteenth century searched for the secrets of their god, the matter. They adopted the principles of Hegel's philosophy making "matter" the "supreme being", without ever demonstrating the veracity of this ideology, as in their other sociological, political and intellectual imaginings.

Scientists continue verifying the reality of their material hypotheses on their new tangible god, the eternal matter. This method of research produced such great resources of knowledge and techniques on the material world that people forgot that they were also human. They still refuse to recognize the less tangible facts that their method, limited to the study of material entities, does not permit to study. Science has thus deliberately refused to study the less tangible entities that the primitive senses of man do not permit him to handle: human values and spiritual entities, for instance, which are a fundamental part of the human being. Education through science thus insures only a part of the forming of the individuals: the material part. One must understand that this materialistic system of education is incomplete, because it is incapable to take into consideration man in his entirety and that it never should have been used as a system for training people.

The mass of the people is just beginning to realize that one of the end products of this materialistic education is a definitive secularization of public institutions by legalized inhuman laws, everywhere in occidental societies: religion was simply and legally removed from all institutions, schools, conference rooms and governmental courts. **The education through science and technology without the humanities obliges a professor to teach only what science can verify with primitive senses. It imposes science and the linear ways of reasoning (logos) used by scientists and their material values as the only tools for the forming the mind. In front of all these limits of the scientific method used in education, how can we not wonder if our educational system might be failing at least on the human plan?**

"La Guilde du Pain d'Épices" (The Gingerbread Guild), a volunteer organization which organizes literary workshops, publishes books, magazines and animates awareness campaigns to try to mitigate the fact that in Quebec only, 54 % of the population is illiterate, 10 % of those finishing high school do not know how to write and one child out of four is risking illiteracy. We forgive ourselves by attributing these educational failures to the people's poverty. We must nevertheless denounce this materialistic educational system as one of the main causes of this fact. By always subjecting children only to the material data of science, the method's logic, the cold scientific reasoning, the absence of humanism, the pressure to excel at all costs, can this system's responsibility be eliminated, especially when one considers the dropouts, the violence, the aggressions, the need to consume alcohol to forget, the drugs, the gambling, the suicides, etc.

Forming minds through science and technology to the detriment of a general forming of the mind and of the human being, have created individuals who can perhaps argue admirably well, but who are totally ignorant of their human counterpart. Indelible imprints of this dehumanized formation have been left on all the people of our modern societies.

Societies are saturated with too many so-called intellectual civil servants in all command posts of education and the people are fed up of seeing them spread their tentacles all the way from elementary to secondary levels of education. These authorities now want to spread their influence at university level! They want to subject universities to their materialistic judgments, their retarded culture and their influence, simply because they control the money! These administrators have not realized that they have no competence at all for this work, and it is not because they are paying the bills that they have the right to dictate their wills to these high levels of education.

The people in power at the Ministry of Education warp education

We must be aware that governmental directives force their ways of thinking and their directives on young children as early as in kindergarten, following the directives given by the Ministry of Education of Quebec in September 2000, obliging professors to teach children in elementary grades the rules of public-spiritedness of the scientific and social materialism. The children must respect the authorities in place, the governments' decisions and even denounce their parents' working under the table…because it is dishonest…

These same honest persons are absolutely not ashamed to advertise in the media and to try and educate their citizens with slogans such as "under the table work is stealing", "moderation has a much better taste", "gambling must remain a game", "persons on Social Welfare are abusers", etc. Have we gone back to the Nazis' youth education? If these administrators had learned history, they would know that every time a power, a monarchy or a dictatorship deceives its citizens by exaggerated and inequitable taxing, by stealing their personal freedom, by violating their rights, the individuals always react. They react by defying authorities or using self-defence mechanisms called the "service". The "service" is a normal reaction of mutual aid. It is a defence mechanism, a principle of natural and innate survival in human beings who still possess some notions of humanity. This mechanism "gets going" when situations become difficult for the species. Mutual aid is not "working under the table": it is a feeling that materialistic individuals no longer have.

To show they are good and compassionate, the sorcerer's apprentices in power try to impose their inhuman laws and their principles of materialistic morality by broadcasting advertisements that are televised and paid for by the citizens. Why are these powers trying to tell people that working under the table is a theft against the State? Do they not know that by doing so they hinder this notion of human service well anchored in people since centuries, a reaction that should have made them realize that they went too far by stealing their citizens' rights? Why do they want to persuade citizens that the services exchanged is robbing the government, while they legally steal by taxing over 70 % of the fruit of the labour of the people they administer? Who are the swindlers here? Who are the cheaters?

The atrocities of the powers

Powers dare give themselves very inequitable laws to legally monopolize a high percentage of the earnings of the people to justify themselves, added to an incalculable number of disguised taxes. These powers hide behind their inhuman laws when they try to educate their citizens in the principles of scientific and social materialism so that they can use them as slaves. A theft is a theft for everybody, the people in power included! Citizens have rights, but few possibilities of rebelling against this big institutionalized nonsense of their leaders, except during voting periods.

It would wiser for leaders to teach the young and the citizens of their countries what a misleading advertisement is... and what is really bad, depraved and unhealthy in their own administrative backyard. They are far from being good examples to their citizens with their obnoxious, dishonest acts and the betrayal of their own administrations. Let us mention a few examples of this in our state and it is putatively just about the same everywhere materialism is established as the "master".

The people in power exploit the credibility and the human weaknesses by promoting gambling and alcohols in all its forms:

By exploiting casinos, lotteries, and installing gambling machines on every street corner.

By acquiring thousands of new gambling machines to continue extorting still more money from their citizens.

By claiming not to be responsible for the suicides caused by the passion of pathological gambling which nevertheless touches 5 % of the players and knowing full well that 2, 5 % of this population will never be able to come out of it unhurt.

By quintupling gambling patients between 1998 and 1999 while they continue to increase, in an unforgivable way, the number of these infernal gambling machines.

The powers give themselves the right to administer all the alcohol outputs!, impose excessive taxes on products of first necessity, such as heating fuel, electricity, certain food, petroleum, etc. They limit to almost nothing the tax deductions for medical drugs or for patient home care.

The Defender of the Citizen of Quebec (the Ombudsman) denounced in her first report in November 2001 some dreadful flaws in several ministries and state societies. She publicly accused these authorities of always extorting the citizens in their favour.

The Department of Industry and Commerce is found guilty of mishandling transactions with citizens in 51% of the cases.

The Ministry of Family and Childhood is found guilty of having settled its businesses with the citizens in its favour in 50 % of the cases.

Workmen's Compensation is found guilty of settling their disputes in their favour in 45 % of the cases.

The Committee of Health and Social Services is found guilty of settling 41 % of their transactions in their favour.

The Public Security Department is found guilty of having extorted people in their favour in 39 % of the cases.

The Public Guardianship is found guilty for having negotiated in their favour in 37 % of the cases.

The Civil Servants of the Ministry of Finances is found guilty of settling their disputes with the citizens in their favour in 35 % of the cases.

The Administrative Committee of Pension Plans and Insurances is found guilty of having bad relationships with the citizens in 33 % of the cases. (*Cf. First Report of the Defender of the Citizen, November 2001, Quebec*).

When individuals of the Québec Ministry of the Revenue decide, from the height of their power, to forego fiscal privacy (January 2002) and to share, with no less than 17 different social and private organisms, the secret access to the confidential fiscal information that it possesses on each individual and this without their permission, are these people not expressing their deep materialism? This Machiavellian imposture is by itself an unthinkable lack of righteousness and human morality. It is against the Income Tax Law and even more against the democracy they claim to practice.

When dehumanisation reaches such alarming and gigantic proportions, it means that there is not one iota of difference between these ways of doing and those of the leaders of the Marxist - Leninist communist countries of yesterday. In the name of what retarded administrative principle can these people justify giving out this information to private survey firms in order to compile statistics for public administrations, to Quebec Pension Plan or to the police (*cf. demonstration of Mrs Diane Leblanc, at the Quebec Parliament in April 2002*)? Don't you think your colleagues at the Automobile Insurance Commission had enough problems regarding the theft of information on car owners made by their employers to the underworld? How will you control such thefts of information in these 17 organisms if you cannot control your own?

In front of such materialistic carelessness, I believe more and more that the mother of the young child accidentally killed 1995 by a car explosion provoked by the underworld people, was justified to **think that crime is organized and that the head is somewhere within the powers that steer us.** How can we think otherwise in front of such dishonourable stupidity?

Administrators take their authority from the power of money, their master. We could all name several private companies taken from their owners in the name of this corrupt materialistic and inhuman morality. These thefts are not different from those of the American materialistic companies in the face of their shareholders, such as Enron, Anderson and Adelphia in the United States, Parmalat in Italy, or the scandals of Elf, the frigates, etc. to finance the socialist party in France. It is characteristic of this materialistic culture to think that they own what they administer. The administrators seized our Health Care system, our treatments, our children's education, justice and law, road traffic control, the state of our roads, the environment, etc.

Let us think about the dehumanizing depth of technological applications

Materialism will never make any sense to the human side of man. Powers manage the people by using technologies without human sense and think like "Mr. Spock in the Cosmos Patrol" (U.S. television program). People are not just machines or robots with no human feeling.

Mary Midgley, an English professor and philosopher, wrote in "The New Scientist" on August, 1, 1992, an article entitled *"Can Science Save Its Soul?"* wondering about the future of science. "Current societies and particularly the occidental ones", she wrote, "have been so saturated by the only material values of science during their forming that they have come to think that there is no other mode of thinking".

This absence of human relations will continue as long as bureaucracy, civil servants and all implicated individuals continue to be formed by science disconnected of any humanity.

G. Summary

The irrefutable proofs demonstrated so far confirm that science, scientists and technological applications have been very beneficial to people in many aspects, **but at the same time they also produced catastrophic and disastrous side effects to the environment and have already killed numerous forms of life from earth.** Still more important, **they are producing excessively deplorable side effects on man's humanity by contributing largely to the materialization and dehumanization of man and his institutions.**

By exchanging bits of humanity for tangible assets and social measures which flatter their ego, the human dimension of people has rarefied and disappeared with this education. **Having damaged the environment and destroyed life on earth, we now observe that this same materialism is destroying man's soul and scattering his humanity.**

Occidentals were the first to lose their human, personal, religious, social and family traditional values because of this absence of humanity in education through science and technology deprived of any form of humanity. **The oxygen of love, of kindness, of respect and of feelings has become extremely rare in much of the lives of people at the beginning of this twenty-first century. People's hearts have dried out at the permanent contact with material values which today fill their lives and they are literally suffocating in the materialistic world they built. This absence of humanity shows a dehumanizing evil more or less evident in today's man and society.**

The most conscious individuals try to avoid these side effects by devoting themselves to sports, they work to the point of exhaustion, they play, go to restaurants, to ball games, on trips, take alcohol and drugs, etc. All these attempts are actions to try to escape from their dehumanized world, but they remain confronted to material solutions and cannot find peace in their soul. This is why some drop out of their schools without souls and from their dehumanized society in early adulthood. Other individuals believe they will find their missing humanity in spiritual ecstasy, in just wandering about or by disapproving and rejecting their materialistic powers and their inhuman laws. Others express aggressiveness, violence or become members of criminal gangs or by killing their fellow men for all or nothing... Some are transformed into compulsive gamblers, their behaviour changes and some even try to escape by committing suicide, etc. One can see in these human actions despair or a flight towards the "to each his own" which becomes a rule and a human reaction of survival.

Always hung on the technologies which dictate their rules of life, these human mutants educated through science and technology without the humanities try to counter for their missing humanity, but still enter the new millennium, dominated by inhuman materialistic laws, the same scientists and the same intellectuals and powers that enslave them. This state of dehumanisation transpires in all their actions, their ways of living, their social institutions, their work, their behaviour and their means of evasion, etc. All of them are confused because they have lost their humanity and their compassion, and they live in their fossilized social institutions.

A century of materialism has thus transformed the individuals of our world into dehumanized mutants, incapable of love, compassion or respect for others. And so this is what explains the changes incrusted in their soul by the materiality of science, technology and values deprived of humanity and traditional human values. These changes have come to stigmatize the people of occidental societies with "the dreadful evil of materialism" that strikes them today.

CHAPTER 5

Repercussions and impacts of materialism on the physical world, on man and on society

Introduction

Science and technological applications have not invaded our world without leaving disastrous and permanent impacts on the physical environment and the destructive materialization of human beings and societies. The numerous human and social metamorphoses that left their mark on the lives of Occidental people in the twentieth century are intimately linked to the explosion of science and technology, to the unbridled craze of individuals for all that is knowledge and technological application and mainly to the forming of people through these same sciences and technologies. As illustrated above, education through science took place all over the Occidental world to the detriment of education and teaching of the young with the humanities and values other than material. Such education precipitated people into a whirlwind of novelties, and in so doing, material values became the source of the deplorable side effects present in people and society today.

Once engulfed in this materialism, people have degraded their environment and their living conditions. Hypnotized by the charm of knowledge and technological novelties, as Ulysses was by the sirens' songs, people humanly changed without even realizing it. Some do not even feel the need to change their behaviour. If most people do not realize that the world is now in the dreadful grip of this materialism suffocating them, some are beginning to perceive the problems this materialism leaves on their behaviour, their institutions and society. The majority of the people still behave like the frogs placed in heated basins. They become more and more numbed until it is too late for them to get out. Scientific education and the attraction that people have for material values keep them from becoming aware that their behaviour is engulfed in this materiality. Some even prefer to ignore that the authorities managing them use this same science of technological applications and material values as tools to dull them more and more each day, while they keep on being dazzled by all this knowledge and the technologies which stream of everywhere.

Very few are conscious that all these changes are the result of their scientific education, the daily use of technology and that their loss of humanity is luring them into this dreadful abyss of materialism.

Before illustrating to what point education with sciences and its related values have caused numerous side effects on people and society, let us mention the multiple advantages of this education through science and technology.

A. The benefits and setbacks of scientific education

Education through science as well as scientific formation has brought a great deal of advantages and benefits to people and society. Sciences are extremely useful in the industry to modernize and develop them in a competitive way. Never before had companies been able to modernize at such an accelerated rhythm, and all this is due to scientists, technicians and the scientific formation.

How could we even suspect, just a century ago, that science and technology would one day modernize the environment. Before the Second World War, no industry could produce all the goods available on the market today. The education system met the needs of yesterday. Before science and new social ideologies began to transform mentalities and education, things were relatively stable and simple. **Students were formed with liberal and capitalist ideologies that were still attached to traditional, religious and material human values**. Education available was more or less free of charge and given by religious communities for the greater relief of governments. Trades were passed on from father to son. The power of religion over the people was as important as it is today in several Moslem societies. The young people who went to university exercised their professions and performed their professional tasks with the humanist principles learned throughout their education. Universities were still limited to the education of traditional faculties such as medicine, law, chemistry and physics and the individuals were formed to respect the humanity of man. Universities served their society by teaching the rudiments of science, humanistic principles and human values necessary to the practice of professions. At that time, materialistic ideologies were hardly perceptible, and so the world was inevitably less scientific and there were also much less technological applications than today. All those graduating from college and university practised trades, rudiments of science and the traditional values they shared. Love, kindness, compassion, respect and other humanistic virtues were still found in most individuals and societies. They were less scientific but more human. People may have been poorer, more ignorant and their societies less evolved, but people were content with what they had and what they knew.

Governments invaded education in order not to miss the technological explosion and the development of their consumer societies. From this moment on, however, scientists and intellectuals began putting aside the teaching of the humanities and replaced it by a technological and scientific education. **As this new form of education spread, its material values also spread everywhere and at the same time the teaching of the humanities and the human values were put aside.** Having seized the education, governmental scientists dedicated the major part of teaching time to sciences and new technologies, but the humanities soon became the poor parent of teaching. The accent was put on science, technical training, professions and occupations requiring this new technical knowledge. **Human formation has almost disappeared from education.**

The young people are now formed at the "scientific school of reasoning" very early on in their training. Our societies are now flooded with technicians and scientists trained with this type of education in all levels and all possible categories of human activity. Trades, professions as well as all the domains requiring human activity are overflowing with science. More and more university students complete a bachelor's degree and several new professions that did not exist before this craze for science now abound in all areas of society. Never before had schools and universities been in such high demand and students been avid to study in all possible scientific domains. Spheres of activity have diversified almost infinitely. Several professions have become more and more specialized and so more students complete their master's degree (second university cycle, M.Sc.), their doctorate (third university cycle, Ph.D.) or post-doctoral studies in a growing number of branches of scientific activity. The last decades of the twentieth century were marked all over the occidental world with such specializations and the prolongation of studies in different domains of human activity.

At the beginning of this time of change, people in power had to help schools and universities adapt to education with science, technology and research by injecting small amounts of money. Certain universities were able to accomplish wonders on these very tight budgets in large numbers of scientific domains and in a very short time. Very early on after the beginning of education with science, scientists and experts began to cover all fields of human activity.

By emphasizing science and training through science, education costs increased. Education through science requires laboratories and sophisticated equipment. It also requires more specialized professors to give the best possible training. Rare are the university professors today who can obtain a teaching post without having at least a third cycle doctorate (Ph.D.). Several universities do not even accept professors without a specialization in a particular branch of research, aside from their doctorate. The candidates must also be able to obtain research grants in their respective field of specialization, they must have published articles in scientific journals and be excellent professors...

Governments tried to invest the necessary money to allow university faculties to bloom, hire the best possible professors and have adequate laboratories for their respective fields of research. In several countries, it is with great difficulty that governments managed to subsidize these places of high knowledge with appropriate education. Education with science and technology is much more expensive than education with the humanities. It is much more costly to form a scientist than to form a historian or a theologian. Scientists must learn to use more sophisticated and expensive instruments than the simple purchase of books. For example, the cost of the instruments needed for scientists and specialists to practise their profession in hospitals is often astronomical. Think of the costs of a laboratory in nuclear physics, in astrophysics, in atomic chemistry, etc. Society was quickly overwhelmed by the excessive cost for qualified personnel, necessary research equipment or simply to automate the analyses of hospital laboratories. Everything changed with the education through science and technology. Techniques have been refined, instruments have been refined and research has been refined. Scientists themselves have become more refined. Universities refined their education and professors their knowledge, their laboratories and their equipment, etc. Scientific training soon became a priority for certain countries because it favoured the expansion of new domains of business, new industries and especially the development of new specialities and consequently the increase in jobs. We had already figured, at the beginning of the 1950's, that one Ph. D. alone created one hundred new jobs.

The world of education has changed in many regards at the contact of all this formation through science and technology. Students are now receiving a more technical training. Professors were needed to teach the new and more scientific professions needed by the industry. Students, from secondary school to university, can now study in almost any branch of science they wish. They learn enormously more about the techniques to serve the industries and the consumer societies than their parents formerly did. Professors of the secondary school as well as those of universities must permanently retrain in sciences to be able to keep up.

New scientists flood society. The workers who wish so and who are familiar with all the basics of science can complete their studies in extremely specific fields of expertise. Never before had schools formed so many technicians. Never had so many students embraced scientific careers. All professions, even those with less scientific needs form professionals in science. University law faculties, which had formed very few Ph.D.'s during the first half of the twentieth century are now forming specialists in several specialized branches of the law. It is the same for all university faculties. There are specialist scientists formed in all the domains where science can intervene, which is everywhere.

Never before had our world abounded with so much knowledge and so many individuals endowed with such competence in the specific domains of current science. Never, be it in medicine, in engineering, in biology or in sociology, to mention only a few, were there so many specialists in sophisticated technologies. It is the case in medicine as well as in communications, aeronautics, transportation, etc. Never before have there been so many professions and so many individuals to exercise highly technical and scientific professions. Even the professions which formerly had little to do with science are now forming many scientific experts in their domains and their respective radiuses of action. All professions bring sciences and scientific reasoning in their domains. Even car mechanics, farmers and journalists need scientific elements of training to exercise their professions. People study much longer in order to exercise more specialized professions. Never before had occidental universities formed so many masters and doctorates or as many students in post-doctoral studies than during the last two decades. We live in scientifically evolved societies where science reigns everywhere.

Everything has become scientific around us, from the techniques of "marketing", to economics, work management, training schools, companies, powers and even the teaching of theology is now the teaching of the "science of theology"... There are researchers in history, in archaeology, in politics, in social sciences, in psychology: just name the branch, you will find them. The schools formed scientists and they all teach or practise science in their respective fields such as the science of history, of archaeology, of politics as well as that of social sciences or psychology.

But and there is one but, and it is the objective of this essay: all these sciences, all these professions, all these occupations and all these scientists have been formed at the same school of science and of technology which is regrettably essentially material. So, because of this purely scientific training and because of the permanent contact of the people with the material values of science, each and every person coming out of these schools has inevitably received the same materialistic training. **I have**

**never known a scientific university faculty to give, apart from the
teaching of sciences, lectures or classes on the humanities or courses
on anything but the material values of this scientific method of
education.**

In numerous occidental countries, schools have no confessional
status and education is completely separate from spiritual and humanistic
knowledge. Moreover, most scientists only practise the religion of
"science", and so many became atheists. This is a problem because the
people in our societies are only trained with the material values of science,
and so they can hardly study entities other than those recognised by their
most primitive senses.

What is worse, the number of hours of education in elementary school
is derisory when pupils want to receive moral education principles apart
from the materialistic one. It does not exceed 72 hours per two year cycle in
Quebec. It is too little, and too often taught by individuals who do not often
practise any religion. As far as the choice of studies in secondary schools,
it belongs to the pupils, at least on paper. In Quebec, any school board
can replace this rudiment of humanistic training by a local ecumenical
program of orientation, a program "of ethics" or of religious culture, but
only with the authorization of the ministries and with particular conditions
that civil servants alone determine and authorize – if you please – (*cf. "La
Rentrée Scolaire", le Cahier du Journal de Montréal, le 6 août 2002:
"École et religion: Ce qu'il faut savoir pour 2002" "The Beginning of
the New School Year", Montreal, August 6, 2002: "School and Religion:
What we need to know for 2002".*)

From the beginning of their academic training, students are only
taught the rudiments of humanism and of human values which formerly
governed personal freedoms and the training of people. In a not so distant
past, parents had the freedom to choose the kind of education they wished
for themselves and their children. Today's parents, formed with sciences
and technology, do not have this opportunity. They never took the time to
think about the negative effects of the education their governments decided
to give them and their children. Certain individuals are only beginning to
perceive, after years of practice, that they lost one day the right to choose
the education they wanted for their children, because they were all formed
in the same mould and have acquired, in spite of themselves, only the
material values that came with this education. Very few have received
a humanistic formation. A great majority of these people do not even
realize that there exists, apart from the material values learned during their
formation, human traditional values, social and religious values, the right to
individual liberty and mutual respect, the right to guide themselves or even

the right to be free. Some are just beginning to be aware that the intrinsic materialism of their training with science is responsible for their atheism, their absence of humanism, and their lack of general training because they were never taught that there are the other values than those inculcated by science. There are methods of training other than with science.

This phenomenon arose in all occidental societies and is independent from political, religious, social, philosophic affiliation or from the habits of other countries. **Countries are now governed by the laws of "scientific and social materialism" where the collective liberties and rights prime over the individuals ones.** Even in the political parties formerly of the right in today's supposed democracy, which forever pride themselves in respecting individual rights, as liberals and capitalists still pretend to do, scientists and intellectuals in power were formed almost exclusively with this materialistic education during their academic training. Without knowing it, they have that same political tendency for social collectivism and all these powers are administered with collective laws rather with laws supporting individual rights and liberties. It is because of this training that people complain that even though they elect a different political party to govern them, nothing ever really changes. It is because they all had the same education system and share the same social reality.

They now realize that their colleague scientists in power use these scientific materialistic ideologies and technologies as a means to better lay the foundations of their power. Their colleague technicians and scientists remain in power mainly by using these unproven principles and social ideologies of the scientific and social materialism that they learned. They are just beginning to realize that these individuals in power are establishing all sorts of social and collective standards to control everything: the education of their children, their health, their work, their system of justice, their ways of thinking, etc. They have the power to decide which services are useful to them and which are not, and to take away people's rights and personal liberties in order to appropriate their health, their rights, their schools, their justice, etc. They take into their own hands what belongs to the citizens. They can decide, for example, to put into debt their governments, their cities, their schools and institutions for generations to come; they control all the movements of their lives, reap the fruit of their work and finally exhaust them by monopolizing what people asked them to administer. Even the elected members of governments are subjected to their scientific controls, now and then, because there are now thousands of permanent scientists and intellectuals "haunting" the corridors of the establishments of powers and deciding for everybody. Not only are people slaves to the techniques they use, but they are also slaves to their colleague

169

scientists in power who administer them as if they were interchangeable parts of machinery.

Finally, there are two main categories of scientists: a) those exercising professions and precise scientific occupations to earn a living, to accumulate knowledge and to make economy turn; and b) the pseudo scientists that have managed to slide into power, that subject the people to their materialistic laws and their unproven supposedly scientific social ideologies, taking away their individual liberties by keeping a tight rein on them and refusing to take into account that they are human beings.

We have come to this because we have let these scientists steal, without rebelling enough, our rights and our personal freedoms; when they seized our rights on our children's education, they made us believe that they wanted to make school accessible to all - a collective motive which we could not refuse. They persuaded us that they wanted to provide us, <u>with our money of course</u>, with free health care, cures for our diseases, free medications, and that they would eliminate poverty from society and redistribute equally our wealth. They persuaded us it would be advantageous to leave them all our powers and centralize them somewhere in large impersonal buildings, well shielded from our glances. They convinced us that they had the right to legally possess our powers, even though we have never given them such rights. These non elected scientist and intellectuals now manage what we asked them to administer as if they were the owners.

No, "sirs pseudo scientists" in charge of education, health care, etc., our diseases, our work, our possessions, our natural resources and all what you administer belong to each one of us, the citizens, and not to society and specially not tot you. You will never be able to say, as the Marxist - Leninist communist societies did, that you seized all our rights during a revolution! If they have acquired their legitimacy by bloody revolutions, you, you usurped this power with your socialistic laws. While there is still time, try to behave like good administrators, respect what we are and we will respect you. By continuing to act as the owners of our possessions, we will treat you like swindlers, usurpers and dishonest people... Remember that in a democracy, governments are only the managers of the people's properties: they are and should never be the owners! <u>You no longer have to substitute yourselves to the teachers or to the parents to decide on our children's education. You do not have the right to decide what education we want to give our children or which programs and methods of education we want. You must never again control teachers, parents, institutions and even elected members. Teachers must never again be your slaves forced to teach and to put into our children's head your materialist ideologies and</u>

your indoctrinations. We give our governments our taxes to administer, not to deprive us of what rightly belongs to us!

Materialism carried through education becomes the motor

A panoramic glance on the unfolding of events in the education system clearly shows that the reforms in the curriculum of colleges that assured in yesterday's societies different formations and ways of thinking, have been eradicated to make place for the implementation of scientific education. The main stakes and objectives of all these upheavals were the formation of individuals through science, rather than with the humanities, to allow every country to position itself within the new societies filled with science and technology...

One will remember that at that time, there were stirring debates between the teachers of humanities soon to disappear and the pseudo intellectual promoters of scientific and secular schooling which was rapidly growing. But nobody quite knew where this would lead. In spite of the protests, the powers replaced the secular humanist education methods of formation by the scientific method of education that we now know. It is unfortunate that these changes were made to the detriment of a general training for students, to the learning of the humanities and the various ways of thinking and of living of the people who had lived before.

These intellectuals in power took advantage of this change to take charge, hypocritically and without parental consent, of the custody of education as if they already owned it. There is no doubt that education with sciences became, as illustrated above, the tool to create thousands of new jobs and an experimental field for the scientists and intellectuals in power who took advantage of this to spread their influence on the people, the schools and the teachers.

Numerous data show that this materialistic system has failed in many aspects in the training of citizens. Quebec, for instance, after three decades of this system of education, counted in 1998 only, 35,000 non teaching civil servants for about 1, 5 million students (*Journal des Affaires,* February 1998). This excessively large number of individuals administers only the education of elementary and secondary school. There are almost as many administrators taking charge of education as there are teachers. This excess of civil servants is similar in all levels of government management, be it in health care, in work, in justice, in the environment, in municipal departments as well as in every level of intervention touched by scientist administrators. One could say with irony that in the health care system, there are as many administrators as there are patients in hospitals. It is not surprising that certain governments, conscious of these management abuses (as in British Columbia, Canada), understood that for a better functioning of public administration, **it is necessary to cut by about 50 % the civil servants of the various ministries.** Administrative expenses always weighed heavily on the administrations of socialist societies. France and the Soviet Union, two recognized socialist States have for several decades sinned with an excess of bureaucracy!

To illustrate to what point this method of education no longer teaches young people humanistic thoughts, Marcel Gauchet wrote in "*La démocratie contre elle-même", Gallimard, 2002, (Democracy Against Itself)* that "Communism and National Socialism have played the role of true "laic religions", substitutes of previous transcendent faiths". In these societies, the powers dictate the programs, the methods of learning and of thinking and reduce people to infantilism. The scientists of those totalitarian countries had used the same system of training with science and technology without the humanities.

This materialistic scientific model establishes its culture based on the rigid and mechanical "reason" and the scientific method of teaching as "the ultimate completion of the human thought"! Humanities thus disappear from the school system: students no longer learn how to think otherwise that by the Cartesian method of research, and as for scientists, only the material and tangible matter becomes the truth. After a few generations of this dehumanized education, individuals have become standardized and there is no other truth than the one accepted through scientific reasoning. Material values are the only ones now taught and the traditional human, family, personal and spiritual values are now discarded from education.

During the twentieth century all the occidental countries blended into the same system of education with sciences for the training of their citizens, whatever their former ideological belonging. Philosophic differences which formerly identified Marxists, socialists, neo democrats, social democrats, democrats, liberals, capitalists and others ideologies, no longer refer to the philosophies which differentiated and identified them only yesterday. Since the adoption of the social materialist system in education, each occidental society has a common denominator: science and scientific reasoning. Everybody is formed with science and all comply with the scientific reasoning and the socialist ideologies that guide them. The elected members of government inevitably and automatically follow suit, without suspecting into what monumental pool of materialistic mud they have just stepped. This is the main reason why today, all around the Occidental world and in several other countries which also use science without the humanities to form their citizens, the people and the societies materialize to the contact of this education and to the materialist principles which still maintain an unprecedented popularity.

To make more place for scientific education, technocrats put aside all at once the humanities, history, philosophy and general training. This is when materialisation, secularization of education, education and society began to invade all viable spaces possible once and for all. After a few decades, education through scientific reasoning has become the rule everywhere. **The human and social relations materialized, became dehumanized, deteriorated and became chaotic. This human "disease" has contaminated the people and become the only truth to all societies.**

This little game is spreading all over the world as we enter this new millennium. Science and material values are so well anchored in people's minds today that the scientists and intellectuals in power can make all the inhuman laws they want to legally justify their actions and nobody dares criticise. The "science of the reason" substituted itself to God. Scientists and intellectuals are still persuaded that they possess the monopoly of the truth, science, "and that the Hegelian materialistic philosophy of the "eternal matter" can only lead them to the truth"…

It is in these highly scientific and materialistic societies that we live today. After a century of promoting sciences and material values, nobody dares question his right to freedom. Discussions on the damage done by science are taboo. People, who are conscious of having lost their humanity, know that it is almost impossible to return to the way they were. Alexis Carrel may have been right when he wrote in *"L'homme, cet inconnu"* that: "Men only accept with great difficulty the things they do not want to change from the bottom of their hearts, for it would oblige them to question absolutely everything".

Today, however, water is pouring out of this vessel of science without humanity. In all countries administered with this unique "scientific reasoning", we notice enormous flaws in education and enormous abnormalities in individuals. This is found in the materialization of the people and paradoxically, in the dreadful number of illiterates after a few decades of this education with just science without the humanities. Why is this so? Could it be because this sole training does not adequately meet the needs of the individuals? The following data show the failures of this system of education.

The next data show us more and more that the constant contact of humans with this materialist education and the scientific "reasoning" devoid of human sense make individuals gradually lose their humanity and become dehumanized. People's every move, the way they work, the way they think engendered by this materialism reveal this "humanity lost and this dreadful dehumanizing evil" that saturated them and that is creating chaos everywhere.

Examples of failures in education

Over 19 % of the young people of Quebec are considered illiterate at the end of the twentieth century, after over 60 years of compulsory schooling until the age of sixteen. Must we consider a success the fact that 40 % of the boys and 26 % of the girls drop out of school, or that 30 % are still illiterate after high school? These students hardly know how to write their name and never read newspapers or a book. Nearly everywhere in the Occidental world, 50 % of the jobs require technical training. Even with the staggering changes in the educational system, over 40 % of the students finishing high school do not even get a diploma and only 30 % of them obtain a technical training diploma. Is this not aberrant?

The failure of this education system is common in several countries. In an article published in September 2002 in the magazine "*Science et Vie*" *(Science and Life)*, Philippe Testard-Vaillant reported that 15 % of the French pupils starting the third year of elementary school cannot read. Hardly 17 % of their fellow-students understand the information contained in a text and two-thirds of the remaining children can hardly decipher the same information. Difficulties in reading were found in 20, 8 % of the young people quitting school, in 11, 8 % in those that reached the BEP level (secondary 3), and in 3, 5 % to those who obtained a bachelor's degree. The percentage of illiterate individuals adds its weight to the empty intellectual contents of television programs more often than otherwise without culture. "It is urgent that we rethink the way we transmit knowledge when children use the words "restraining and boring" to describe their school." Is this not incredible?

What scientist or intellectual would dare assert that these academic failures are not connected to the disheartening and generalized carelessness of the young, the stress, the desperation, the disinterest of the *res publica* administration, or to the increase in crimes rates in our society, to alcohol consumption, to the forming of criminal gangs and possibly to gambling? It does not seem that administrators and intellectuals have yet realised that the education system with the materiality of science and technique without the humanities and public-spiritedness is a side effect of this materialistic teaching.

The World Trade Summit held in Quebec in November 1999 reuniting thousands of persons from all over the world, confirmed to what point education with sciences has limited itself to the services and to the technical necessities of the industry. The lack of culture of this Summit also demonstrated the depth of the materialization of education and the blatant lack of humanistic values in technological fields. It seemed very clear that more humanistic and general training would help people find interest in their world and allow better judgment in the execution of their work. There would possibly be lesser side effects from the technological applications if the new scientists and technicians coming out of these schools were more aware of the disadvantages that their minimal training can cause to the environment and to their fellow men.

The powers in the universities

Still, there is worse. Completely incapable of administering elementary and secondary school levels and their hands filled with bitter failures, education administrators now want to take their incompetence to university levels. How blind can they be! This is pure madness on the part of the administrators of the Ministry of Education. Their insane behaviour betrays their objectives: to establish some sort of administrative hegemony pushing its tentacles to impose its domination and ideologies by making compulsory for university professors to sign performance contracts in order to better control them... Scientific governmental administrations constantly try to impose their laws and to subject the people and the societies they administer to such a degrading Machiavellian slavery and to an unspeakable submission. This domination has become blatant proof of scientific materialism in administration.

We have shown how the methodology of training with science and its materialistic culture secularizes societies and institutions everywhere it passes, and how this culture imposes everywhere its hegemony and its principles. This mentality of superiority has fatal consequences on people and society.

The scientists and intellectuals of Karl Marx's scientific and social materialistic school of thought will be further discussed in Chapter 7. It is important to be aware that this school of thought uses this scientific reasoning and training and continues to teach, without legitimacy, that the truth is in science and in its method of research. They teach, for instance, that workers are only the performers of the techniques they have learned, that they must not think beyond the application of these techniques and that they are not even responsible for their actions within their frame of work. These dehumanized powers go as far as stating that individuals are not even responsible for their diseases and that it is society's responsibility to support these thoughts. Should they make a mistake, all they have to do is apologize in the name of some archaic principle of law which claims that people in power are not responsible for the errors they commit during the exercise of their functions and the trick is played. In so doing, administrators remove themselves from all responsibility in the face of the citizens. These same administrators oblige doctors, for instance, to reveal their errors to their patients, knowing full well that in doing so, they become exposed to lawsuits which could have been avoided without this obligation to declare their faults. Two weights two measures. More often than not, it is the technocrat's competence that is restricted and limited to the only logical and mechanical techniques he knows. All these thoughts are expressions of socialist ideologies answering political and social aims without any respect for individual rights. Moreover, these thoughts are the expression of socialist ideologies and are not supported by scientific evidence.

Educated in this dehumanized system without human culture, people are eventually transformed into automated machines devoid of human feeling. The individuals in power are the fruit of this materialistic education. They remain convinced that they own the truth and that their arrogant, rigid and inhuman "reasoning" always leads to "the" truth. All the people that went through this system of education end up transporting this arrogance in all their activities and in all their gestures.

Some decades ago, in "The Peter Principle", Pierre Daninos already foresaw that, because of the automatic systems of promotion, employees would tend "to rise to the level of their incompetence", and "that one day the world would be governed by these incompetent individuals". And here we are in this world of techniques and inhuman machines, a world governed "by oddballs in power".

Stressed by so much dehumanisation and obligation to perform, it would not be surprising that studies would one day demonstrate that individuals can become depressed, drop out of school and society, become addicted to alcohol and drugs, gamble, become discouraged, homeless and even commit suicide, all because of this inhuman education. Nobody seems to be able to give the citizens the services they deserve, after having paid high taxes on goods and services plus income taxes. The citizens pay for mechanical, useless and dehumanized services. And we ask ourselves why the young as well as the adults drop out of society.

The violation of human rights across the planet

The Annual Report of International Amnesty, published on June 14th, 2000, proves that human rights are violated in an astonishing number of countries on our planet. It has not yet been demonstrated scientifically that these social dysfunctions are linked to the changes caused by the dehumanized modification of education by science and technology without the humanities, and by the wide promotion of material values and the rigid linear reasoning of science. In 1999 only, human rights were violated in 144 countries on the planet. Political executions were done in 38 countries. The detention of opinionated people is common in 61 countries and torture is practised in 132 countries. Extreme materialization of people, powers and institutions can be seen through these figures. They demonstrate that this materialistic and dehumanized culture is well implemented in the people in power. China still seems far from having left the totalitarian scientific and social materialism that marked the education of its children for about three quarters of century and it is not the only one.

The decline of human values

With such a school training system, human values have been lost in individuals. Discarding the humanities from education can only help the new materialistic morality that prevails in our societies. For the first time in human history, morality no longer takes its logic in a humanistic philosophy, but only in "pure reason and material values". The cult of the "almighty science" dominates every area of education through science and technology. This cult has not yet reached its apogee and it is alive and well in education and in all the ruling classes. All materialistic institutions like using and abusing scientific and politically correct words: it is "chic"! As it did a century ago, technocrat machines believed that their scientific methodology of research and "reason" could not deceive them and would always lead them to the truth! Intellectuals are so conceited in their reasoning! Those educated at this materialistic school even believe, regrettably however, that nothing is tangible outside of their science. And so, they stupidly reject the non material entities they cannot perceive with their primitive senses and this without any other proof than their "reason". Many scientists and intellectuals do not even acknowledge the existence of what they cannot feel, see, touch, or hear... And this is often the reason why so many scientists are atheists...

In our modern materialistic society, the individuals' morality is inevitably modelled on the values that materialism and "reason" dictate to them. But, if material values are acceptable for a materialistic morality, they are not acceptable for human beings. As proof, materialistic morality established that what they considered bad, illegal and immoral yesterday in our humanist societies, such as alcoholism, gambling, casinos, the exploitation of people's weaknesses and greed, suddenly became moral and virtuous acts, simply because they fell under the yoke of the powers' materialistic morality.

Under these conditions, it is easy to see who decides, what is good and what is bad for man: it is the people in power and it is the "religion of science"! What is the difference between yesterday's religions that decided what was good and what was bad because they held the power, and today's socialist and materialistic administrations who have all the power to dictate to people their behaviour and their faiths?

Materialism modifies the expression of morality

The exploitation of human weaknesses is well set in the materialistic laws. Only the people in power can exploit with impunity and in all legality the human weaknesses by using their materialistic morality. Our governments greatly profit from these illegalities which were formerly known as "incomes of shame" taken only from the "despicable" individuals of the underworld, the mafia or other groups of amoral and depraved individuals. **This is where we find the morality of the scientific materialistic powers: in money, in greed and in disrespect towards people...** Who do they want to persuade, these elected as well as non elected materialistic powers, that bad actions can be transformed into virtuous acts when done by themselves? Such materialistic morality is fooling itself and tries to fool the citizens.

How can we conceive that the odious, criminal, vile and degrading actions that other powers condemned yesterday, can suddenly become licit and legal businesses, simply because our scientists in power exploit them? In an article of *La Presse* (Montreal) published on April 6, 2002, Marie-Claude Malboeuf presented the stages of this progressive materialistic morality to the Quebec Liquor Commission. In less than ten years, the number of authorized agencies passed from 550 to 800. Encouraged by the company's marketing, citizens now consume 2, 1 litres more alcohol per inhabitant, the turnover passed from 1, 45 to 2 billion dollars and the government's profits from 753 to 981 million dollars. Is it in the governments' task description to sell alcohol and drugs, promote gambling, exploit human innocence, sell lottery tickets, own casinos or establish the new moralistic and dogmatic social scientific materialism of Karl Marx?

Sharing the wealth – let's talk about it

Materialistic governments with socialist tendencies, as well as those of our beautiful dehumanized democracies, think that they can better share wealth by nationalizing patrimony: that is by interfering in the leadership of private companies. In 1981, a socialist government took over power in France and nationalizations began. In just a few years, some of the well implemented private companies began going downhill and some even closed their doors. It is a characteristic inherent to these pedantic socialistic powers to consider that they can suddenly become the competent owners of private companies. Not so, dear civil servants! Do you really think that what belongs to everybody is better administered than what belongs to private individuals? You must know that when a company belongs to everybody, it really does not belong to anyone. When there is no real owner to take care of the business, there is danger ahead. When one places at the head of these State owned companies friends of the powers, subjected to the will of elected or non elected members who practise the 9 to 5 working culture, we are far from the culture of the true entrepreneur owner. He or they work hard all day, often in the evening and on weekends to insure the survival of their company and to protect it. The people appointed to take charge of a State business do not have this entrepreneur spirit, innate in owners of private businesses that think about their company day and night, even though they share materialistic ideologies. These new entrepreneurs will rarely be reprimanded for the gestures and actions they take. When these generously paid leaders lose billions of dollars of the pension fund money paid by taxpayers because of ridiculous expenses or badly planned investments, who will reimburse? Will the people responsible for this lose their jobs?

The influence of materialism on communication between people

Technologies in general and communication techniques in particular allow people to be informed as never before. As soon as they occur, events are quickly given media coverage on the radio, on television, in newspapers, in magazines, on the Internet, etc. This coverage is done according to the moral values of the distributing network and those presenting the news. Sensational news is profusely repeated according to the institution's appraisal of its value, and often just serves the purpose of boosting their listening quota or the sales of their publications.

Regrettably, the objectives of the news are not, as they once were, to give out information or to enlighten us. Knowing that some people rush to get their hands on "sensational" news, such as vampires wanting to drink

179

their victims' blood, the materialized institutions profusely use the news to their advantage. The more the journalists and their organizations are supplied with morbid details and macabre photos, the more they quench the thirst of their readers, the more they strike their imagination and in so doing increase the sales of their publications. The manufacturers of this kind of literature exploit to the utmost the purely material credulousness of those that nourish themselves with this kind of horror. Techniques of communication can serve to inform as well as to materialize and dehumanize the people. They also carry the same values because the individuals who practise them have been formed at the same schools of though.

Without knowing why, some people are ill-at-ease in front of all these roles that touch the deep materialistic feelings of their fellow men. Do we care about anything else but the golden gifts of scientific materialism? We feel this uneasiness more or less deeply in the face of the growing materialism of our world. It leaves a bitter taste, even in those that adore the material goods that enslave them. The people who materialized their faith have no consideration left for the humanity of man. The simple fact of asking information to a linear and logical technocrat in power, as on the health of a loved one in a hospital or about a file in a ministry, creates uneasiness in the one that must give the information; the lack of compassion of the technocrat who answers is perceptible. The given information is short and the comments are mechanical, dehumanized, without compassion, but more often than not politically correct. These persons are not aware of their own lack of humanity. They work like machines. Scientists and technicians in power learned that there is no other truth than the one registered in their machines. Have you ever tried to have an error corrected, an error made by a technician while entering your name in the data processing computer of a ministry or on a health insurance card? You will see that the truth lies in the machine. Months after having proven the error, <u>with</u> a birth certificate, <u>with</u> a passport and several other papers, the error will remain in the machine. You will struggle with the lack of humanity of these persons because they refuse to believe what you say!

The materialized scientists and technicians in power have transformed themselves into machines incapable of thinking, of feeling or of showing any kind of respect. Constant contact with this type of materialism in all their every day actions persuades the people that the truth is only in the machine. From the moment civil servants learn to master the series of techniques needed to exercise their profession, the machine dictates the laws. The machine cannot make a mistake. From that moment on, it is the machine that controls the civil servant who does not need to think any more. All he has to do is follow blindly the machine and the standards dictated to him. He has become a slave to his machine. This will remain true in any type of government, weather it is democratic, capitalist, socialist, etc. since all the individuals of these societies are educated with the sciences. After reading *"La démocratie contre elle-même"* (Democracy against Itself) by Marcel Gauchet, Editions Gallimard in 2002, we can easily understand that after the vote, it is the civil servant machine that leads the world.

Armed with this "reason", the people formed with science banish "the supernatural, the mysterious, the poetry, the philosophy, the history, the religion, the humane and the realities of life in a mechanical way without noticing it" wrote Stahl in his criticism of *"Les Contes de Perrault"* (Perrault's Tales), The Complete and Unabridged Edition, Jean-Claude Lattès, Paris 1995). This criticism was probably written in Perrault's time. Education with science stimulates the people who have a positive outlook, but takes away all their humanity. Individuals formed this way are incapable of understanding any kind of truth, if it does not come from the machine. By believing that only their technology is capable of establishing the truth, these mutant human beings ended up like scientists thinking that the machine is their infallible god. "By withdrawing poetry", will say again P.J. Stahl, "Vergil, Homer, Dante, Milton, Goethe and hundreds of others might as well have not existed".... Blind faith in the machine keeps any mutual trust and compassion from becoming established among the people. When only material considerations become the managers of the human psyche, it is that men have already been transformed into non thinking robots by this education. They have become slaves of the machine and of its powers over them. "History", wrote Stahl, "is filled with unlikelihood, and science is full of miracles. Reality abounds in miracles and its miracles are regrettably not all choices! The truth of these materialistic individual lies in an abyss full of the unknown: ask the true scholars. Science can explain the clock but has not yet succeeded in describing the watchmaker. The failure of the "reason" is at the top of the list of all the knowledge and you, positive people (of reason and science), are a mystery. Please give the fairy tales back to the children".

The human beings perceiving this materialism react in various ways. Some will do the "I couldn't care less" thing, others will become compulsive workers, some will live in desolation and evasion, while others will become aggressive, exasperated, commit suicide, and why not add euthanasia which was legalized in April, 2001 in the Netherlands and in Switzerland. Evasion from materialism allows people to escape from the rigidity of this reasoning which stifles them, to escape social tensions created by the absurdity and the dehumanisation it creates, as the individuals of yesterday who were demoralized by the abuses of religion... At the beginning of this twenty-first century, this is where we are at. Why look elsewhere for the loss of our humanity or for the source of the numerous social problems and chaos that are upsetting our world? Let us look at the right place and we will find the source of our troubles.....

The American program *Survivor*, which gathered in front of the small screen over 40,000,000 American viewers during the summer of 2000, gives us a truly transparent demonstration of this confrontation between lost humanism and current materialism. This is a mimic of prehistoric fights for survival, highly improbable in our days. Similar experiments have ended in failure in Great Britain and in France. Is it because the love of tribalism determines a need of wholesome evasion only to the over materialized capitalist populations?

Human relations deteriorate when made through impersonal machines! Is it not paradoxical that in a world abounding with the biggest communications that ever existed, people speak less and less or speak to each other by interposed machines? Speaking to automated answering machines rather than to individuals becomes frustrating to human beings. But the materialized man does not seem to feel this lack of humanity and accepts all that is technology without any kind of judgment.

Another major impact of the materialization of individuals and societies is that this rational and materialized individual stupidly agrees to be controlled by these inhuman machines without a soul, in other words that only answer to the tyrannical imperatives of technology and reasoning.

B. The materialism of technological applications deeply modifies human behaviour

Ancient biblical stories tell us how people behaved towards each other in the days of "an eye for an eye, a tooth for a tooth". If they often behaved as animals without a soul, a certain glimmer of humanity seemed to animate their actions!

Human beings materialized by sciences and technologies have acted towards their fellow men in a very degrading way indeed, without this glimmer of humanity. Several facts confirm this. During this century of material abundance, certain leaders and their people have caused hundreds of millions of deaths and made man regress to the state of wild animal infested by some virus of inhuman intolerance. No glimmer of humanism, however registered in man's superior brain since thousands of years of evolution, transpires in their actions and gestures, and these people formed with sciences and the tangible facts of their "reason" have been using only their primitive brain to guide them.

Completely flooded in innumerable material goods, the people formed with science and technology and without the humanities have not evolved as humans. The every move of the people of the twentieth century is proof of this: a large number of individuals participated in nameless massacres during that century. Let us take a look.

From the beginning of the twentieth century to this day, there have been dozens of wars, ethnic exterminations, social problems, an incredible increase of domestic conflicts and divorce, confrontations among individuals as never before in the history of humanity, etc. These brutal and delinquent acts by the people of the twentieth century show without a doubt that they used the multitude of technological benefits offered by science to satisfy their instincts of destruction on other humans, rather than to use these technologies to evolve humanely. The people who wrote the Bible offered at least a motive to excuse their sometimes massive slaughters. They put the blame on an evil and avenging god whose message was very different from the one God sent to men: "Love each other as I have loved you".

Formed at the school of science, heads filled with knowledge and materialistic ideologies which are still sweeping the world after over a century, the people took the matter as their new god. They elevated it to the rank of "almighty". They made their misdeeds in the name of precious temporary ideological principles that had never been a part of the human vocabulary. **The twentieth century became a century of wars and continuous attempts at extermination of the people by the people. Several million persons were killed on battles fields. People of the twentieth century can't even hide under the dome of a non material god of which they have always denied the existence! So they destroyed themselves even more cruelly than the people who lived there a few thousands of years before.** Where is the iota of evolution in the human nature which basked in so much knowledge and benefits acquired since the beginning of the industrial era? People are entering the third millennium by continuing to kill and hate their fellow men... Love and compassion are non existing entities for the individuals trained only in science, technology and material values.

Conflicts between the sacred and the matter which tore humanity for centuries are still very present today, as if the negation of the sacred by the science had managed only to create such an immense void that the socialist ideologies infiltrated them as water in a river of rubble. In front of the inflexible and mind boggling mechanical and technical powers without a soul subjecting them to slavery, people can only get sick to death of it. The reactions to these facts are seen in each one of us, according to his or her personality. Some acquire more possessions and money or produce for the whole world. Others react by escaping in compulsive work while others suffer breakdowns or abuse alcohol, drugs and start gambling. Others sink in a state of "I couldn't care less", become school or social dropouts or simply bored, work excessively, lose their homes, commit suicide, become members of criminal gangs, become violent, and so on… It becomes more and more clear that these reactions express the human beings' feelings towards the lack of spirituality and humanity in their world lost in disillusionment. Who could deny it?

Useful in our physical life, the material values taught by science, materialism and its method of research were never able to answer the existential questions of the individuals who still consider themselves as human beings. In today's material world, people in search of human values have no other choice than to turn to sects, religions or gurus, etc., to get away from the growing materialism of their dehumanized civilization.

The practice of material values in everyday life inevitably ends up in this world filled with individuals, institutions and seriously unbalanced and totally dehumanized powers. These individuals act as machines, without thinking and without being able to express feelings. **They are lead by their "reason" without humanity. Feelings, poetry, philosophy, magic, the supernatural, the unexplained, the inexplicable, the impossible are at the same moment in the truth and in the imaginary, to parody Stahl's words, do not belong to the scientific vocabulary.** In the absence of humanism and of magic, the individuals' ways are particularly inhuman, not to say animal-like. The materialized people in power cannot establish that a management based only on merciless and rigid, mechanical and linear laws dictated by their "reason" without humanity is doomed to failure. It is with a lot of sadness that we are powerlessly witnessing this growing deterioration of humanity in our societies. These individuals' tangible gestures without human feelings have reached this level of materialization and of dehumanisation in our intellectual societies.

The materialism which transforms student mentality, even in universities

In spite of the wealth of our technical societies and the studies that like to praise an unprecedented evolution of science and technology, the young are disrupted more than ever in the face of this materialism that strikes our occidental world, educated only at the school of scientific and socialist materialism of the nineteenth century.

During the week of the Youth Summit in February 2000, we learned the results of a study done by Doctor Jacques Hamel, a sociology professor at the *Université de Montreal*. This study demonstrated to what point the young people of university level educated in this contemporary materialism have materialized, as well as the degree of materialization they have reached.

The young from 18 to 20 years of age, these half-dependent individuals who have just left home to learn to live on their own, are quite disenchanted. They are lost in all the administrative mazes of their leaders. Indeed, a cat could not find its kittens in the middle of all these administrative technologies as illegitimate as they are useless and to which they are subdued. A large number of the young people coming from less fortunate and less united families, have serious problems and become dropouts, commit suicide, remain unemployed and/or become homeless. Those that come from "baby-boomer" homes (parents born between 1945 and 1965) live in greater abundance and are more spoiled. They know that they will one day have access to more secure and well-paid jobs. "These young people are determined and want to succeed", will say Professor Hamel. Some attend university and often work, besides their studies, about 35 hours a week.

They all dream to one day obtain job security. Once on the work market, access to this security is transformed into a major handicap for these young people and, it is only when they will have reached this material stability that they will be able to think about starting a family and having children. Their main interests are money and comfort. They have to work hard to reach these very materialistic dreams that they consider so important.

Children having lived through their parent's divorce no longer believe in true love and have no intention of repeating this in their own union. They believe in the values of the couple, in faithfulness and in success (*cf. Journal de Montreal, February 23, 2000, p. 7*).

Students, assert the professors, do not want to be deprived of anything. They confirm that money is their main reason to live in today's society. They even have consumer relations with their teachers, which means that the professors have to give them their money's worth. They say "they want to date a girl who has lots of "cash" and that one day they would like to earn 75,000$ a year". These very materialist attitudes are already well anchored in the heads of these university students!

Other possibly less fortunate young people are not happy with their situation. Runaways are found in certain families. Let us take a look at the repercussions of such behaviour.

This antisocial behaviour of young people in general is disturbing. An article in the *Journal de Montreal* of March 29, 2000, revealed the number of disappearances in Canada. There were 60,360 children who disappeared in Canada in 1999. At the end of the year, 337 were still missing. Quebec sadly reported the disappearance of 8,270 children that year. Out of this number, 44 were not found; 47 % of these children were girls and 53 % were boys.

In 1999, 14 out of the 3,651 missing adults were not found.

About 4,000 children ran away from home or disappeared in the Montreal Urban Community. The runaway cases are more frequent in May, June and October. They account for most of the disappearances and according to the police, 96% of the runaways are between 12 and 17 years of age.

Material values inculcated by the materialistic education with sciences and the absence of humanities are also found in university administrative powers. To make ends met, the *Université de Montréal* signed a 10 million dollar contract, over a 10 year period, with the Pepsi (soft drinks) Company. This contract allows the company to advertise on the walls of the University. Professor Jacques Hamel reminds us that we cannot blame the students for over-consuming when they are formed in a university that looks more and more like a shopping center. The "Club Price" University, as he calls it, is a school that gives money an extremely important material value. Many scientists in power who manage the university inculcate these notions to the students. It is not so surprising that on each floor, one can find soft drink vending machines and / or automatic banking services.

The role of materialism in the induction of crime

The young people brought up with this kind of education and its material values know nothing of human values and even less what they mean and their importance. Numerous examples confirm it.

The new "fashion" in the young French people is to burn cars. Young delinquents from large cities and aged 15 to 18 years burn cars mostly belonging to the police, just for the fun of it. Thefts in automatic cash dispensers are done by the hundreds of thousands every year. Magistrates have to resign because other judges, closer to the powers, protect themselves and keep them from doing their jobs every time a high-ranking person is pointed at. In this admittedly socialist country, it seems that juvenile delinquency has reached alarming proportions. In June of 2003, French prisons exceeded their capacities by 25 % with over 60,000 prisoners.

Shooting on our roads is a phenomenon known in capitalist as well as in socialist countries. Murders are committed on the roads for no reason at all. Individuals don't dare look at each other in the face for fear of being yelled at or even attacked and killed. A man stabbed another one for a mere 10, 00$. Another one shot a bus driver that has just cut off his car. A truck driver enjoys pushing cars off the road. Another one follows to scare them both off! We often find out that individuals commit suicide at the wheel of their cars by running into a tree, or worse still, into another car, killing completely innocent victims.

Suicides increase in young depressive individuals as well as in the not so young. In his book entitled *"J'abandonne"* (I Give up), Phillipe Claudel demonstrates the impossibility to change this crazy world. Like Napoleon in the face of Love, some people react by running away. These reactions are not at all connected to the fact that a country is socialist or capitalist, but to its state of materialism and to the values passed on by an education where humanism shines by its absence. South American countries, where violence and social problems live alongside strong Christianity, are there to confirm it. Brazil is the country where we find most killings per inhabitant. It is followed by the United States and then South Africa and Russia.

The fact that psychologists attribute these deviations to stress, to narcissism or to the degradation of values, the cause of these phenomena does not change a thing to the materialism they undergo. Michel Campbell, a psychologist, claims that he is often consulted for "aggressiveness at the wheel". He claims that women have become more and more numerous to present such behaviour and explains these acts of violence by a low tolerance to stress, a high opinion of oneself, an evident narcissism, competitiveness, paranoia or the low empathy of people towards one another.

Violent incidents by parents attending their young children's sports activities and then prosecuted are eloquent examples of these attitudes. Let us recall the case of this American father who beat to death the referee who dared punish his 10 years old son during a hockey game. Another father gave drugs to his child's opponents during a tennis match so that his child could win! A struggle between two groups of parents during a hockey game in a small village ended in several people being wounded amongst the players and their parents. These are some of many daily expressions of frustration in individuals with no conscience who lost all restraint and human value. This phenomenon of violence on the roads was called the syndrome of "get out of my way, uncle, I'm in a hurry", a commercial slogan seen in a car dealer television commercial. Is this not a perfect example of the prevailing materialistic culture conveyed by the aggressive people in that industry?

Criminologist André Normandeau of the *Université de Montréal* believes that these individuals are not criminals and qualifies their behaviour as the reactions of deprived human beings. Why would these reactions not be qualified as "normal" in individuals already in an advanced stage of materialization because they have been submerged since childhood in material values only, and that they have learned nothing of the other human values: could the degradation of the values of respect between individuals explain such frustration? Is putting aside the values of mutual respect a dominant feature in materialized individuals? I am happy to notice that certain researchers are finally beginning to think that there could very well be a relation between the degradation of traditional human values and the social phenomena of spontaneous violence that flood our societies, as one of the elements to be considered to explain the growing crime rate in individuals formed in the materialistic values without the humanities (*cf. Journal de Montréal, March 16, 2000*).

A news story, published in *China Daily* in September 2000, **showed that 34, 3 % of the 86,786 children of pre-school age from two to four years are already affected by psychological problems and 31 % by behavioural problems.** One third of the Chinese children would suffer from psychological problems! The study does not give more details, but the Chinese Press often denounces the problem of the "child king", these children showered with material goods and hyper protective parents who have the right to have only one female child per family. In November of 2000, a simple act like pushing and shoving in the stairs of the Number One school of the city of Shenjian, in China's Henam province, ended with five persons dead and dozens wounded. Let us remind ourselves that this country has been completely governed and educated for decades now, in the ideologies of Karl Marx's scientific and social materialism and that it is still not completely freed of its Maoist deviation.

A growing number of children living in Quebec are labelled as aggressive, hyperactive and even violent. According to the Canadian Child Health Institute (CCHI), about three children in four are considered as being "very annoying" by their parents. More than 170,000 prescriptions of Ritalin were prescribed in 1997 only, to treat this 14% proportion of mostly male hyperactive children. Can anyone demonstrate that this behaviour is not linked to the culture that prevails in our world? Must we blame the environmental pollutants to explain these facts? What must we think of the thousands of children kidnapped every year in the world to serve adult prostitution?

C. Materialism and deterioration of the exercise of power

Citizens are constantly assaulted by the dehumanized interventions of their governmental powers. Have you ever filled your tax report without thinking of all these technocrats who always can find in the mazes of their laws ways to offend you, to condemn you, to make you feel guilty, to destroy you or to reduce you to begging, if they want to? Is the absence of empathy and respect not characteristic features of the materialized individuals who work only by "reason"? And these things happen even if the technician in front of you is smiling, in a good mood and friendly... If these interventions seem normal in administration, it is that they reflect their materialistic culture and does not arouse in them any sense of compassion, respect or feeling. This culture in powers creates dangerous individuals for those that resist the materialization of their society and for those intending to change this mentality.

Modern democracies, if they can still be called democracies, are administered by people who have never received any legitimacy other than that inherent to their job. The large majority of them have never been elected. Their scientific training allows them to experiment on people serving as guinea pigs, the wild imaginings of their disturbed "reasoning" brains. Learning about the human spirit is regrettably not a part of science or of education because the scientific method of research has not yet learned how to analyze the non material subjective knowledge of the human being. Why do we not teach people that there is also a human thought different from the one coming from the reasoning on material subjects and which the people have the right to know?

The rules of only using "reasoning" in human relations have reached their limits because the normal individual is not only reason but also spirit. By wanting to always regulate everything with inhuman laws, the materialistic state inevitably ends up in anarchy and chaos, because administrators easily forget they are dealing with individuals and not machine parts or objects.

People have had more than enough of the multiple legalised materialistic interventions of their governments in their private lives because they stifle their rights and their humanity. Problems seem to reside in the fact that the actual education system teaches too much ideologies coming from the "reason" as if it were the only way to function. Why it is only in the long run that these omissions are uncovered and that the masks fall? Why are governments so slow in understanding that there is a strong lack of humanity in the education system of the individuals and their societies? The extensive use of only reason and material technologies end up by removing all sense of responsibility in man and his powers and this creates violence.

The technologies that make mistakes acceptable in powers and stimulate violence

The powers numb their citizens' minds with words that make no human sense. They hide their misdeeds under mountains of "politically correct slogans", use incomprehensible specialized vocabularies while talking to ordinary people by designing immoral and inhuman materialistic laws only to better enslave and control them. Who are these scientists and intellectuals in power who try to give morality lessons by trying to persuade people that they must always respect their absurd and inhuman laws? What makes justice by "reason" amoral and incomprehensible for human beings, if not this collective feature which elevates social rights above individual rights?

Powers tried so hard to persuade people during decades that all the information they compiled on each citizen in their unique file (cf. the Ministry of Human Resources of Canada, Chapter 4), was necessary to manage the country! How crazy can these civil servants be? Where have they acquired these convictions? References - please! Created in 1996 only, this same Canadian ministry of 23,000 civil servants accumulated more monumental blunders than any other previously. It is indeed this same ministry that, besides this "unique file of information" on each citizen, administered a monumental mess of hundreds of millions of dollars on what it is been advisable to call in Canada the "scandal of the Human Resources". It is this same ministry that was incapable of adequately administering the distribution of 5,000,000 social insurance card numbers. The Ministry of Immigration of the same government lost track of thousands of immigrants in the past year and the judges at the immigration department have already received bribes.

Civil servants of the Ministry of Revenue, though well formed in accounting techniques, attempted one day to simplify the income tax report on one single page (cf. *Simplified Tax Form, Quebec, 1998*). The simplified form represented a substantial increase (a theft) of disguised taxes compared to incomes reported on the usual but more complicated forms (cf. *Nouvelle (News Flash) Radio-Canada, April 12, 1999*). Can one believe that a tax at 2 or 3 different levels would have at least the advantage of removing loads of useless unnecessary complications and legions of cumbersome civil servants, while considerably simplifying the administrative machine?

In another order of powers, citizens were anxious to know what had happened to the individuals who had allowed, between 1981 and 1985, the AIDS and Hepatitis C contaminated blood to be transfused to the citizens. Who is responsible for decisions taken in the big governmental circus today? Everybody knows that love, kindness, empathy, respect, social sense and feelings are no longer in the vocabulary of these people in power.

Whether or not made consciously, these bad decisions coming directly from the "reason" undermine the confidence of the people. Should not these errors make the elected individuals in power understand that technology and machines are also subject to error?

It will soon be necessary that the grey eminences of the political parties as well as those responsible for the decisions learn that the citizens have had enough of these grotesque political dominations of parties without a soul, without a philosophy other than the materialistic one that seems to take pleasure in managing with only what is mostly material: money, weapons, the underworld, with disrespect and ignominy.

I have often wondered if the victories accumulated in the United States by Mr. McCain at the beginning of his election campaign of 2000 for the leadership of the Republican Party, were not due to his humanism or to a return of this nation to more traditional human values. Without the financial support of his party, this man had managed to please the people by expressing human values totally absent in our modern political world. How we would have loved to see the onset of such a change in political American materialistic cultures! I often wonder if, to the material values of money, power, stupidity, scientific and social materialism, the man liberated from the chains of slavery could one day oppose human values. Can we hope for a return of occidental governments with more human managements and laws? We would have been happy to learn that Mr. Bush, the elected representative of materialistic and American capitalist values, would choose Mr. McCain as his fellow candidate once elected as the President of the United States, because the people had recognized in him the qualities of a human man!

Materialism and arrogance of the powers

Why should we accept the errors of these scientific technocrats or their elected people who wrongly put, generations to come, their people into inequitable debts? What is the advantage in merging if citizens do not want it, especially when other cities having experienced this have already shown that such changes do not reduce taxes, administration or management? Quebec also demonstrated this in the years following the application of these municipal mergers. Why then? Is it fashionable to do so? Is it because some scientists and intellectuals in power think that it is time for municipal merging? Is it because it pleases the non elected powers that manage the other branches of government, the syndicates or the unions? If no valid demonstration shows that it simplifies administration, reduces costs or has clear advantages, must we anyhow lose the proximity of the municipal elected members and the territories of the town to the profit of large impersonal mergers too big to be suitably administered? Let us think simply of the dangers in case of union disputes: they will be multiplied by unthinkable factors. To think that certain civil servants stupidly assert with consumed absurdity, that if we regulate today the mergers of cities, it is "because the project has been on the ice for years". If it is the only reason, it is more and more clear that it is high time for the people in power to be changed.

It was thought that by merging the school boards, school taxes would decrease. On the contrary, they have increased each time we let go of superintendents. How do we explain these increases in costs? Would it be possible that the scientific machine of the "reason only" that searches for the truth is unhinged and only serves at favouring certain persons? Don't you think that the salary increases of the administrators immediately after the 2001 municipal elections are an indication that something has degraded somewhere in the administrative machine?

I read one day on the Internet the opinion of a "normal" citizen who was fed up, as we all are, to be considered as an idiot by his own scientists and intellectuals in power. This is what he wrote: "It is time for the government to STOP making decisions for me, telling me what is best for me, deciding whether or not I am responsible, telling me how to raise my family. I am my own person and I know what is best for my family and me. Give me back that power to do for my family what I know is best for them. Give me back the rights that our fathers guaranteed us in the Constitution and Bill of Rights". **Is this not a cry of despair from an individual in front of the excesses of socialist measures that favour only the socialistic powers?**

In spite of all the protests and the evident lack of scientific proof to ascertain the validity of their experiments on people, these individuals nevertheless continue to decide and to intensify their materialistic and collective laws to keep on distorting the truth.

One has to remember that the world one day learned with horror that the communist regimes that had applied the same social and materialistic scientific reasoning in Russia under Stalin, in China under Mao, in North Korea, in Cambodia, in North Vietnam and in a number of the other societies, had remained in power only by killing millions of human beings. Let us remember also that these countries were the first to teach the dogma of this social and materialistic scientific reasoning of god the matter god and also the first to fall under the weight of the nonsense of this materialistic social system. **Whatever the ideological variants of communism, the doctrines and the social ideologies all come from Karl Marx's 1850 "Social Capital". They all used similar totalitarian powers after having stolen from their citizens their rights and their personal freedoms at the point of the sword; is this not so?**

In these countries, the powers operated with reason only by rationalizing, planning and managing the people and the societies as if the matter was god. **It is in the name of this same "god the matter" that the leaders gave themselves the right to destroy, punish and kill all those that did not share their Hegelian faiths in "god the matter".** In Stalin's days, tens of millions of people (putatively over 60 millions) died in Russia or were either deported to Siberia in the name of this Marxist - Leninist socialist system. How many disappeared in China, in Cambodia, in Vietnam following similar reasoning and "directives"? Our world has now nearly come to this point because of our acceptance to educate people with the same sciences, techniques and material values without humanity. And in spite of this, we continue to educate our people without including in education the notions of public-spiritedness, humanism or more human values! We do not know how long it will take to see such pitiful repercussions in our materialised societies because of the use of the same education system. One thing is certain, it will occur if we do not quickly add more humanities to the education system. In his book *Virage Global* (Global Turnaround), Éditions de l'Homme, 2002, Ervin Laszlo thinks that such changes have to be made within a generation and he explains why! This is very little time, isn't it?

A large number of modern societies do not refer to themselves as communists or Marxist - Leninists, **but they teach and practise the same ideologies**. How can scientists and intellectuals think that by continuing to feed on this materialistic education and by sharing the same values, they will not repeat the same mistakes? It is to know History very little to think so! In all countries choosing to form their citizens at the only school of science and technology without the humanities, people have materialized at the contact with this education and its non material values. All these countries suffer today from the same disease of lack of humanity. **As in numerous modern industrial nations, powers pretend not to be aware that they practise the same materialistic educational system as the one that drove so many countries to ruins less than two decades ago!** If it is not due to the lack of knowledge, it is because they share the same social ideologies and refuse to acknowledge, period!

People in power hide behind masks. They know very well that they are green on the outside and red on the inside, and they will not admit it because they are liars. When Castro came into power in September of 1959 in Cuba, few persons who had nevertheless supported him in his revolution knew that he was green outside and red inside. Castro masked his face under false pretences as do number of those in power in our societies. If masks fall in countries that elect leaders educated in the variants of Marxist socialist ideologies, why do so many "social democracies" still refuse to reveal their socialist allegiance? Are the people in power in our supposedly modern democracies forgetting that it is their fathers that imposed this type of education and their values? If they know this and continue to hide the facts from their citizens by using pernicious vocabulary that no one understands, we must identify them as soon as possible and chase them from power, because they are dangerous liars.

Many of our citizens, who were yesterday Marxist-Leninists, now call themselves today moderate socialists or social democrats. Moderate in what way? **Do they want to show that the socialism they practice is different and more human than the one called Marxist-Leninist yesterday, because their people are periodically subjected to elections?** That is far from being different in ideologies. One thing is certain: these individuals govern their countries with socialistic laws whose collective rights take precedence over individual rights. Their every move shows what kind of materialism they teach and practice. The thoughts and ideologies they transpose into the management of their societies show how similar they are to those taught by the social and scientific materialism of the nineteenth century. The change of vocabulary does not modify their thinking. **They are simply individuals who do not want to be identified as they really are.** And so they all wear masks and like hypocrites hide the colour of their ideologies and betray themselves in their actions and in the inhuman materialistic laws they promulgate.

In all the countries where leaders wear such masks, everything is a lie, and is as flashy as in the days of the communist socialist dictators. In our occidental countries also, everything is a lie. If the whole world was unaware of what was happening behind Stalin's iron curtain, in Mao's country or the way Cuban citizens live, it is because lies is the law in countries where humanism is nonexistent. In these countries, as well as in our own, the sacred has gone and the powers hide their faces their behind their masks. People with masks are seen everywhere in power today in our societies: in justice, health, arts, businesses, transport, in many ministries, in unions and this in all Occidental countries.

Materialism and families

Most heads of families today are individuals who, like their elected leaders, were formed with sciences and material values. **They are technicians and scientists who work and think as scientists.** Never before have our societies had so many individuals applying all at once the rational and rigid methods of education or have dehumanized their own children to the point of making them atheists. Let us take a closer look.

The arrival of a child in a materialized family is planned well ahead, logically, not humanly. Mentalities, values and fashions bring parents to plan their offspring according to a Cartesian, linear and mathematical logic. They educate their children by following these same rigid principles and without humanity, learned along their passage to the school of science and technology. For them also, family, economy, careers, prosperity follow the same mathematical and rational rigour of the Cartesian logic they learned.

Children are few in these materialized families. These parents ask a lot from their offspring. As their governments, they do all they can to control their young children. **With both parents working, births are planned to suit their personal and selfish needs. They schedule and plan the birth of their offspring with so much care that they come to believe, as their governments do, that they have all the power over them.** They control their names, their potential, their talents, their physical appearance and even their sex. Their character is often planned according to the horoscope. Having taken so many precautions to select their children, it is with great difficulty that they can let them live their own childhood. Too often exhausted by their work and busy lives, materialistic parents live in constant stress and put their children under stress. They think they can make their children forget their inflexibility by giving them more than they ask, but in return, they demand a lot. **These children are constantly** confronted with the feeling that they must satisfy their demanding parents and their own desire to live their young life. They are put under excessive stress.

Doctor Jean Wilkins, a Montreal paediatrician specialized in teenage medicine, said after 30 years of practice that the teenager has not changed, but the socio-cultural environment in which he evolves has changed. **"Parents want perfect teenagers"** he said in an interview published in the *Journal de Montréal*, on April 9, 2000. **Social pressure, the money needed for schooling and for every day living and the young people's need of autonomy touch the parents and amplify the need to learn in their children. Conflicts arise, evolve, become a reality and often leave indelible marks "of this inflexible materialism" in the young people.** Domestic problems, divorce, school difficulties add up and increase the number of school dropouts and later on their own refusal of this materialistic society which requires, as do their parents, excellence. "I have never felt such distress in parents at grips with the problems of their teenagers", adds the paediatrician. The children often suffer from our "materialistic way of life" and are its "result" without knowing it.

The personal and social image of the family is strengthened through films, commercials and education in both the parents and their children. And all this only adds to the natural confusion of their behaviour that is expressed by anorexia, more so in girls than in boys, plus fear, rebellion, depression, violence, suicide, criminality in both sexes, etc.

"The search for excellence" is one of the many materialistic "leitmotivs" of this dehumanized civilization that the fashionable materialistic values make shine in the eyes of the young people. These "leitmotivs" are transported by the parents, the system, the education, the professors, the journalists, the people in power... The search for excellence is everywhere a materialistic symbol of performance. Even the industry is marked by this excellence which we agreed to call "ISO 9000", etc. Excellence is compulsory everywhere in modern materialized societies, in families, in the young and the not so young, as well as in studies, sports, fashion and social status, etc. **The search for this solely material excellence is harmful to teenagers at grips with their quest for identity, as it is for all these positive people raised in the scientific spirit and governed by the only dictates of this deified "reason".** People must perform as much as their computers. When we think that these young people have only very few notions of humanism, public-spiritedness, human and traditional values to direct their lives, we have to worry about their future... We shall see further on to where this extreme materialism leads in the every move of these young people that feel trapped between school work, parents, domestic problems, money and the vague possibility of being able to work one day in the professions they will choose. "When the school sends home a child with a problem, it kills this young person already in trouble", said Doctor Wilkins.

The materialism of Daniel Kemp's Teflon child

Children model themselves easily on the materialistic values taught in schools and to the daily contact with the material values of their society. Young people are saturated with these material values. They will use them completely, as to try to free themselves from these yokes. Daniel Kemp, in his book "*L'Enfant Nouveau*", (The New Child) describes the Teflon child as the one that does not respect the adult who has been engulfed by the system. "The Teflon child reasons, is already mentally capable of criticizing like a scientist. **He refuses to evolve in a human world exclusively based on tradition, collective customs, emotional bases, acceptance of contradictions and traditional values", in fact all that is part of human nature.** Even before having set foot in school, his parents had already educated him in the new materialistic values that they themselves had learned at the school of science.

Born and educated in this materialistic world, the Teflon child became an individual of "reason", positive, materialized, but incapable of discrimination. He refuses to comply with all that comes from the human part of man which he knows little or nothing of. He considers it normal to live in this world where "reason", intelligence, logic and evidence outdo all things. The Teflon child quickly learns to integrate the data he is taught, which is that intelligence and "reason alone" that allow men to become an individual. And so, he does not have to worry about things related to human nature or the ideologies that come from instinct or emotions and can dedicate all of himself to the only learning of his senses.

The Teflon child is no more intelligent than other children, but he is able to reason as he was taught, like a scientist. Contrary to the children of yesterday who learned to think before they talked, Teflon children learn to argue before they learn how to think and this since birth. Nobody taught the Teflon child that man is matter and spirit, essence and existence: he totally lacks human values. It is an atheist from birth. He is only a material being who reasons clearly and well, some sort of young humanoid robot. The methods of education with science and technical applications have only taught him to perceive through his senses, which are things and material values. Let us remember that the method of research with sciences that his government forced him to learn, can only perceive things that are material and the material values that come with it, things that the Cartesian logic of this method uses to search the truth in the matter. Like the scientist he already is, he can only perceive the data and the truths that are acceptable to the linear materialistic and rigid logic of this method of research. Unfortunately he was taught to use only the most primitive part of his brain to obtain that truth, his most primitive reptilian senses. **He is cold and inflexible, the Teflon child, like a real scientist.** He is the living fruit of this education based on a method conceived to do research, not to form the mind and to think. Science having always denied human values, they obviously cannot teach them. This method cannot study other things than the matter, and human values are not palpable with the senses. **And so, he is reduced to thinking like a scientist as he has been taught, that intelligence and "reason" are the height of man. Science is incapable of evaluating and understanding human feelings such as love, kindness, empathy, respect and only recognizes the beauty, the wonder and the magic of the matter...** Everything is "reason" for the scientist. As a consequence, the Teflon child is, like many adults, educated in this materialism since decades, and so reduced to being a dehumanized intelligent robot that argues but does not know the true values of humanity. The principles of the Cartesian methodology of research and logic are in all his pores and he is perfectly well formed to become another scientist, another materialistic and atheist individual, some sort of humanoid robot.

If the Teflon child "takes himself in charge more easily and cannot stand being told by an adult, which he feels is inferior to him, what he must do or how he must think", says Kemp, it is perhaps because he already obeys only to the linear and mechanical "reason" that he has learned. The Teflon child is already a "green" scientist, a sorcerer's apprentice who doubts everything and refuses the opinions of others if he can not verify them himself. He will always be incapable of spontaneity and for him there will never be any magic. Formed at the school of scientific reasoning, he will have a hard time learning how to use his emotional brain, become a disabled person humanly speaking and an atheist even before even knowing what an atheist is.

Materialism and religions

Religions took a hard blow with the materialization of the people. And so, numerous are the individuals who do not practise any religion other than that of the science. More autonomous, people prefer making their own personal experiences with spirituality. They will prefer practising their religion without intermediaries that impose their way of thinking. Several of the older occidental working adults are former Teflon children, issued from the first groups of individuals formed by science and technology. They are more mature than the Teflon children because they have learned both materialism and remember some traditional values. The age of the oldest materialists varies from one country to another and depends on the time when education with sciences really began to replace education with the humanities in a country.

Those that feel the need to escape from this distressing materialism sweeping their society become members of sects or escape to the country and its tranquillity to get away from the hectic city. Those people are the most aware that something is not quite right in their world without really knowing why. They prefer to get away from the negativism that strikes human beings taken in by the materialistic realities of their unbearable societies. It would not be the first time in history that such phenomena occur in reaction to something that no longer is acceptable in society.

From the Middle-Ages until the rather recent time of the Inquisition, the Roman Catholic Church and its leaders who had power over the people, conferred to their officiating priests the power to decide what was good or what was bad for them.

Religion and priests that claimed the right to take over the power in these terrible times of darkness are in no way different from today's scientists in power in our materialist societies. There is a striking parallel between the powers of the materialistic imperialism and the powers of yesterday's religions. Religion was to yesterday's society what scientific and social materialism is to today's society. Do the expert scientists and intellectuals in power not behave like the modern high priests of the materialistic imperialism? The difference is that they take their power from the good functioning of their consumer societies and from catalogues like those of Canadian Tire, Vivendi, Sears, etc. They are the high priests of economy, of money, of weapons and of other material values. They are the ones who rob and manipulate the personal freedoms of their citizens according to their own extravagances. They are wrongly convinced that their powers are justified because they are conferred to them by truths emanating from their "reason".

The fundamentalist mullahs of Islam take their power in their interpretations of the rules of the Koran. In the end, things do not change. The mullahs are wrong when they abuse the credulity of their subjects and administer their conscience. The materialistic powers are also wrong when they abuse their people, take away their liberties and their individual rights, subject them to their will, make slaves of them and impose their values, their dehumanized laws and their behaviour. **How are they any different from the former religious powers or from the current mullahs?** The power lies in the people's beliefs and is science not a religion that imposes its beliefs and its powers? Who would dare deny this truth today?

Materialism and international powers

The situation of human rights is alarming all over the world, according to International Amnesty. In 1999, two thirds of the countries of the planet violated the human rights. "No continent is spared" (*Annual Report on International Amnesty, June 14, 2000*).

We have already mentioned that human rights were violated in 144 countries in 1999, there that there were summary executions in 38 countries, detention of prisoners of opinion in 61 countries and torture in 132 countries. "For the majority of the inhabitants of the planet, 1999 is synonymous of repression, poverty and war". "The governments of several countries resort to detention, torture and political murder to muzzle the opposition and remain in power".

Children are enlisted in rebel forces or in regular troops in six African countries. China undergoes the worse and the largest repression that the Chinese dissidents have undergone in the last ten years. In Pakistan, authorities still have a poor opinion of women and don't even look into the commerce of women or the murders "of honour" on girls. Most members of the Israeli Forces benefit from impunity after violating human rights. The United States, Japan, Switzerland are criticized for allegations of police brutality or cruel and humiliating treatment against their prisoners.

Materialism and democracy

In what we still call democratic societies, only the name remains: **democracy today is a lure.** It lasts the time of an election, a time for the citizens to vote on a number of propositions made by the political parties which don't identify themselves to any philosophy on global propositions, without any power to judge them and without having thought them over. **After the election there is no more democracy.** The elected members claim that they have just acquired all the powers and the rights to do what they want. Several capitalist, socialist, socio democrat or other governments all deceitfully hide their ideologies under the mask of democracy. **The philosophic distinctions of yesterday's political parties were easier to identify between one party and the other, and between one country and the other.** People knew where they stood and to what human and social principle they referred to. But people are no longer duped and are starting to realise that they all use the same thinking process and **that there is less and less fundamental ideological difference today between a democratic State, a socialist State, because all these individuals have been educated with science and technology.** They were all trained with the materialistic system as a common ground. The socialist government became more democratic and the democracies became more socialist. People knew for what human and social principles they voted. But today's governments teach, act and take by following the ideologies of Marx's scientific and social materialism. **Their legitimacy is acquired by a democratic vote and all practice the materialistic principles of their training with sciences and "reason".**

People know today that citizens of capitalists or imperialist societies have lost their rights and their individual liberties, as in the Marxist - Leninist socialist countries! Education with sciences has standardized the training and more and more countries are sinking into materialism. **The stealing of the citizens' rights and individual liberties and the current materialistic laws standardize the lack of humanity in governments so that there are very few political variants between those in power.** The laws that powers vote themselves to seize these privileges confirm that they feed on the scientific and social materialism and on its values. With all this appropriation of power by the governments, less and less countries on the planet recognize that their citizens still have rights. Judge by yourselves the principles that guide these "false current democracies"! Even though I dare imagine that in countries governed by true democratic principles, powers should refuse to steal the rights and individual liberties of their citizens, we know today that this is no longer the case.

The intervention of powers in their citizens' private lives is justified only by the principles of this scientific training and the principles of the social materialism described in Karl Marx's Social Capital, with some slight differences. In these societies people are always lowered to the rank of material objects subjected to "god the matter" that has all the power over scientists and intellectuals: they are the "new priests". **This is why in these materialized states, people are only considered as "interchangeable machine parts", as Traian expressed in *La vingt-cinquième heure*, (cf. Chapter 4).**

The decrease in personal freedoms and human rights to the benefit of materialistic powers is evident. <u>**Scientists and intellectuals have only replaced yesterday's priests in the administration of the affairs of the State.**</u> Countries governed by the ideologies of scientific materialism all live in states of collective rights, where individuals are just pawns on a chessboard. This way of doing is another characteristic of materialized powers. This is why our telephone conversations, our vacations and journeys across the world, our transactions on the Internet, our ways of life, our intimacy, our home and all our rights and liberties are violated every day by the powers of our "supposed democracies" in this time of peace. Imagine what remains of individual rights in all Occidental countries since September 11, 2001!

The previous chapters illustrated that these violations of rights have increased at the same rate as the development of the new technological applications which also dictate their material requirements to the powers and contribute in changing the customs and cultures of the powers. Let us never forget that all these interventions in people's private lives are done, no less, by "scientists and intellectuals" who use the "reason of their senses" and the mathematical logic of the Cartesian method which bases its truth only on the senses controlled by the lowest levels of the human brain. **And the powers use their social unproven ideologies to subject people to their will, to legally and consciously kill all traditional human values and lower them to the rank of slaves. We must become aware of this!**

The suppression of the citizens' personal freedoms by these powers is a reality, not fiction! One can observe this phenomenon every day in our societies and in all levels of human activity. It is simply necessary to be aware of this to "classify" the actions of our powers where they belong. These infringements on our personal freedoms are so constant that no one can possibly believe that it is a simple coincidence. The astronomical amount of information on every citizen that powers demand by law and then store in their big computers are not justified. And so we must consider that these people in power are blind or irresponsible when they lecture other countries of the world to show hypocritically that they respect "human rights". They should find out how the individual rights and liberties of their citizens are handled in their own country before giving lessons to others. It is high time to stop dazzling other countries. (*cf. Speech of the Canadian Prime Minister in China on Individual Rights and Liberties, in January, 2000).*

The new laws, voted since the end of 2001 in different democratic countries of the world to counter terrorism, push even farther the unfair intervention of the powers in the personal freedoms of citizens, journalists, the press, the length of detention time, etc. It is illogical to continue speaking about people's liberties while at the same time suspending his fundamental rights and liberties. The lies of the powers should be put into light because they cannot stop contradicting themselves and deteriorating societies.

It is because they thought they were the owners of "the rights of their subjects", that the totalitarian governments ruined their countries, abused their subjects and drove them towards the abyss where they were thrown. We have already tolerated for too long the meaning of all these interventions of powers in people's lives, because these powers hid their ideological and philosophic allegiances. These people knew very well that they appropriated the powers of their subjects with unhealthy collective laws that denied the existence of humanity in man. Here, the materialistic laws are planned by lawyers of the same allegiance of power. It is they who make the law, is it not? Not satisfied in having enslaved their citizens in their passion for power, they made them their slaves, stole their rights and liberties, materialized them with education and dehumanized sciences: these are the same powers that now want to interfere even in the homes of their citizens.

When governments plan to spend millions of dollars to install the Internet in the homes of their subjects, when a governmental institution such as Canada Post offers lifetime connection to the Internet, I become very suspicious and avoid these "unwholesome gifts" like the plague. Is it being paranoid to be aware that such measures are sneaky and can easily allow individuals, rooted in their materialistic culture, to get into the bedrooms of their citizens to better watch them, like their colleagues already watching public places, buildings, cars on the road or the traces left around the world by people using their credit cards?... Knowing the dehumanized culture and past actions of these powers, we can not refrain from thinking of the multiple reasons urging them to want to get into the homes of their subjects. It is quite clear that a door wide open into the citizens' homes will enable them to worm their way in, allowing these dehumanized individuals to intrude and further implement their ideologies... The Nazi youth relished on such interventions in German homes by training the children and History shows us what such ignominies have produced.

Materialism and the confidence of citizens

All these interventions of governments in their citizens' private lives, the violations of their rights and liberties, these materialistic and inhuman teachings translate by reactions of disgust of the citizens on the *res publica*. Citizens are disheartened that their governments consider them all as criminals, slaves or even machine parts. They are sick of these collective materialistic laws that undermine their rights and liberties as yesterday's masters undermined the liberties of the slaves that they brought back with them when returning from war. Our fathers embedded them in the "Charters of Rights and Liberties" precisely because they were inviolable human rights. Even those drenched in this scientific and social materialism since birth can no longer accept these abuses, as the Teflon child does not accept that his parents were trapped by political materialistic powers without saying a word.

I read one day in *La Presse de Montréal* of March 2, 2002, a letter to the editor from a certain Serge Riendeau saying **"I can't stand it any more"**. **He shouted that he was no longer able "to look at flowers growing in the manure of administrative governments and private societies"**. "Not a single day goes by," he continued, "without the newspapers denouncing a new robbery, another scheme, another fraud or embezzlement by individuals in power without moral sense. Unless we are perfect imbeciles or "happy" visionaries, this is only the tip of the iceberg". His cry of alarm said, **"Stop this and stop wasting paper, I totally can't stand it any more. Get another job! Become gardeners, I can no longer watch flowers growing in the manure of these administrations. We went much too far!"**

Let us quote as another example the dreadful public thefts of the Enron / Anderson affair, Global Crossing, String, WorldCom, the Cowpland affair and all these manipulations of the Stock Exchange Market by the swindlers of Wallstreet in the United States, the Elf and Vivendi affair in France, the blunders of the Ministries of Labour of Canada, the friends of Ministers, the Oxygen 9 file, the Cinar affair, Vidéotron, Hydro-Québec's 20 to 28 % extra billing and a number of other supposedly "non-profit" organizations or of the special accounts placed in untouchable companies by finance ministers, etc.

As many of my fellow citizens in this world, I have lost faith in these people in power, in these socialist and dehumanized technocrats of our modern countries, in these scientists and intellectuals who want to invade our homes and give themselves the right to think in our place... So, I have personally become a "free thinker" and I will die for this freedom. But I have no intention of dying today for the powers that control my country.

It would not be surprising if researchers demonstrated one day that these feelings of disgust of the citizens, added to the fact of being treated as slaves by their own dehumanized powers, is responsible for the facts that the Canadians army has more and more difficulty recruiting soldiers, that the Israelis do not want to resume military service when they quit the army, that several European countries and Canadian provinces are now turning to the right. People want to live as free individuals and not in countries that seem democratic only at election time.

It is imperative that powers stop annoying their citizens, that they stop thinking in their place, planning their lives, giving them programs they never asked for, stealing the fruit of their labour or getting into debt for generations to come. The citizens ask them to manage and administer like good administrators. **Nobody gives them the right to take control of their lives, their health or their children's education.** Power abuse is responsible for the lack of interest of people in political matters, as are the unforgivable insults to their liberties which result in a total lack of faith of the citizens towards their leaders.

This lack of confidence is more obvious in the young. And so, in the face of the immensity of the dehumanized social device, the complexity of the States and of various and multiple controls, the disgusted individuals drop out of society by any means they can. Nobody believes any more in the impartiality of this robotised reason of the scientists in powers. Nobody can accept their interventions in their private lives as if they were the owners. Nobody believes in the necessity of stealing and violating their rights and liberties. Nobody believes that all the information the governments store in their merciless computers on each one of their subjects is necessary for "their protection". Nobody believes that this information accumulated in their files will be one day used for their protection. They could however be used to condemn them, if one of them dared oppose their inhuman laws.

Let us think in this respect of the disagreeable mentality of the judges of the socialist power in France that went as far as making false accusations against their own colleagues, simply to have access to their documents or to make them undergo lawsuits as odious and useless as those that the Marxists – Leninist communists of Stalin, Mao and others did to their fellowmen.

Governments: omnipresence and incompetence

We have given numerous examples of proofs showing that governments are omnipresent everywhere in our lives. **There is not a step we can take, a product we can make, an idea we can propose, a meal we can eat, a credit card we can use without feeling the omnipresence of these powers. Enough is enough!** The vase is overflowing. If only they were aware, these scientists and intellectuals who decide! They spread chaos in all that they administer without realizing it. All they touch is transformed in over-administered structures. Take the time to look and you will see.

Poverty has never been so present in our cities, even though certain organizations try to make us believe the opposite. We only need to look at the hundreds of voluntary institutions that collect from door to door and in public places for the deprived of the society in all Occidental societies. We only need to become aware of the growing needs of soup kitchens, endless demands of help on the radio, on television, in the newspapers, in phone calls from companies which assail us every day to collect some money for one thing or another. All these demands for money, food, clothing and furniture are only one demonstration of the social failures of our socialist and materialistic powers towards the growing number of deprived people. Must we talk even louder about the failures of our social societies in front of this omnipresent poverty to demonstrate the incapacity of the powers "of logic and of reason" to appropriately administer our inhuman societies?

On December 13, 2001, I was listening to a radio program where the commentator asked a university scientist who had prepared a report on Medicare for the Québec Health Ministry, to answer the people's questions. **Well, even though health is first and foremost the doctors' business, <u>this professor of economy</u> never thought of mentioning the importance of the doctors' participation in the management of Medicare.** For him, only pharmacists have the competence to control the prescription of certain medications to cure certain diseases. Mr. Monmarquette, it is not because pharmacists manage the distribution of medication and are connected to the governmental computer systems **that they should be responsible for the disease of the patient,** or that they should decide if such or such a medication is adequate for a particular disease. Only the medical doctor who prescribes the medication knows the case history, the exam results, the laboratory analyses reports and number of other information given by the patient that he is the only one to know and that motivated his prescription! My God, let us come back to earth and give to Caesar what belongs to Caesar and to God what belongs to God. **Too many pseudo-experts believe that they have a role to play in the management of <u>my health</u>!**

It all happens in the heads of these materialistic people as if my health had to be managed by an avalanche of paramedics and absolutely incompetent civil servant scientists and technicians involved in the administration of MY own disease. What is the part of the medical doctor in all of this? **Why form doctors if societies do not need them anymore? When did I give all these non medical and non paramedical professionals the right to choose the medication needed to cure my disease?**

Damned, let's be logical. To each his job and we can all be happy. **If health systems no longer work, it is mainly because too many such**

scientists and paramedics, knowing only a small part of MY disease, exceed their competence. All these individuals want is to persuade themselves that are playing a role in <u>my</u> disease and <u>my</u> health, and that they are competent to do it. **Too many paramedics get involved in giving, down from the height of their powers, medical opinions which in no way concern them.** Would doctors have lost their role in the treatment of diseases to the profit of these intermediaries in power in Health Ministries simply because are they administer the money? **People no longer want to be treated by intervening politicians or useless para-politicians which delay all the decision-making on My health.**

When an administrative power has reached the point where he must hide behind just anybody's absurd ideas to back up his decisions, it is that this power has reached its level of incompetence. When such powers speak to other incompetent powers to justify their administrative incongruities, it is that these technicians in power are not at the right place. All businessmen who managed to get their own company going have occasionally asked for expert opinion, but never ever have they left to the experts the task of taking decisions in their place!

Is a revolution the only thing that could free us from these socialist and dehumanized interventions? How can we protect ourselves from these useless transgressors? Will it be by opposing their villainous materialistic system through civil disobedience or by filling the prisons? Will it be by voting for the return to a more sensible and human power? Are the people who commit crimes against humanity by psychologically killing the human side of man less guilty than those who physically kill their neighbours? **Are we going to keep quiet because saying things as they are won't please the sires in power?** *"Silence, je tue"* (**"Be quiet, I am killing!"**) is the title of a Riopelle' painting.

The principle of sovereign immunity, behind which hide the governmental decision-makers to remove all sense of responsibility coming from their decisions, comes from a medieval concept of justice which claimed that the sovereign upholders of the law are not responsible for the errors they committed during the exercise of their functions. It would indeed be necessary to review these evasions, so convenient to protect all the bad decision-makers which abuse power and try to convey that they are good administrators, good leaders and good heads of family. The leader that does not know what his right hand is doing does not deserve to be a leader. Period!

D. The influence of materialism on the degrading of man's work

Any kind of under the table work is not a fault against the state, as "virtuous" powers would like to persuade their citizens. **It is a fight for survival which answers a hard but fair natural law.** History teaches us that when monarchies or other powers required from their citizens more than a certain percentage of their work to support the excesses of their leaders, it is the instinct of survival which dictates the means of survival. Some simply stop working. Others secretly work for themselves. Others go back to barter or stir up movements of rebellion to release themselves from these yokes. Others take advantage of social welfare. People have always survived the excesses of their greedy governments by using these means of survival, well anchored in their genes and the superior levels of their brains where their conscience seems to lie…

The desire of freedom in employees and employers

How can we possibly understand the burdens of rules as pointless as they are disastrous that powers legally impose on storekeepers and companies? The creators of work submit themselves to certain standards, inspections, multiple charges or to the very numerous interventions of their powers in their own business. These interventions discourage the entrepreneurs and these constraints go against the creation of work by private individuals and companies.

Knowing full well about the difficulties and controls to which their fathers' companies had to submit in these omnipresent States, the children are little inclined to carry on the family business. Some become discouraged under the weight of such laws and binding standards. The young people are less ready than their ancestors to suffer at the expense of all these legalized interventions, all these committees of this and that, all these numerous controls required by the powers of cities, states, syndicates and governmental civil servants who never stop putting dead wood in the wheels of the companies. **These are nevertheless the same companies which turn the economy, which assure work to over 80 % of the employees of a country and which are responsible, besides, for taking off taxes and income taxes at the source for the governments. I will only give an example of this burden.** As I write these lines, an entrepreneur has been waiting for nine months for the inspector and the municipal council of the City to accept as a "gift" thousands of feet of road made and paid for by himself, in order to obtain the right to work on the land adjacent to this road and for which the municipality will receive taxes for hundreds of years…from the road built entirely by the entrepreneur.

All people naturally strive for freedom in their work and everywhere else. Nobody has a natural inclination to slavery. The employees know that the governmental management of the State affairs is often more expensive than that of private companies. Furthermore, the social benefits of the States' employees are often superior to those that the small private companies can afford because they are in competition with one another.

To boost its declining economy of the last decades because of binding laws and standards, a new Irish government has to cut into the mountains of rules and laws which were keeping away investors interested in investing in this country. The inspectors of the Ministry of Agriculture of Quebec have just forbidden, for instance, an eight year old child to sell the eggs of his 10 hens to the corner restaurant! The simple preparation of a label for the ratification of a product takes to a few dozen civil servants to apply a law! Those of the Ministry of Agriculture of the Canadian Federal Government also **impose ridiculous annual fees,** a form of disguised tax, in order for the producer to have the right to maintain <u>a label on a bottle of inoffensive "insect repellent product"</u>.

Every person who owns a business, as simple as a garage, must necessarily work for hours and hours every week for the powers, as if these businesses belonged to them! To open a simple breakfast restaurant is laborious work. It requires an incredible amount of paperwork for the only purpose of keeping armies of civil servants busy checking you. Should one of them go bankrupt, who do you think will be the first to receive the due sums: the suppliers or the governments that make them work at collecting their own taxes without paying them? The powers that often cause the bankruptcy of a business are also the first to help themselves in cases of bankruptcy. The one that risked all and worked for years, without a salary, loses everything. Why do the powers have the right to lead these companies to bankruptcy because collecting their taxes does not adequately meet their imposed requirements, standards and recommendations? They should collect their own taxes without requiring that private companies assume this responsibility besides not being paid! In the name of what cursed authority can these civil servants transfer their own responsibilities to the business owners, as if they had all the rights of life and death over these storekeepers? Who will create the wealth of the country when these people stop complying with these insane requirements?

Stupidity

Is it possible to have reached such a level of human stupidity? Ah yes! The examples of such governmental interventions in the lives of storekeepers and people are legion. **Years are needed to simply justify the desk work for a simple label of ratification affixed on a product of consumption at the Ministry of Health!** Why? Because a whole "bunch" of scientists and useless technicians in power like to persuade themselves, with their university degrees in hand, that they are useful to something. In order to approve of a natural product, the Ministry of Agriculture of Canada, recently merged with Health Canada, can take four years and more. These people ask that analyses be made sometimes at great expense to the companies. Furthermore, once the editorial staff have approved of the label, these sad sires are going to charge, every year, an important royalty on the total amount of the sales, simply to obtain the right to place this label on the bottle. Such weighty and useless abuse! We must accept the disgusting robbing of our energy and our money by these greedy powers! They are "hidden and abnormal taxes", thefts through disguised taxes, abuses of power and extortion of companies whose costs always echo on the products...

How much is the hidden tax on 4, 50$ hypertension pill? Why do you think that medications cost several millions of dollars and must to wait ten to fifteen years before receiving ratification and permission to market after having already been accepted in numerous other countries? Because these governmental agencies demand impressive files, multiple expensive analyses besides the hidden taxes on every tiny pill sold and this is always to protect the citizen... **In the name of this great protection, tens of thousands of civil servants in power put their noses in private enterprises to watch them and make them responsible for the administration of their own taxes.**

To have the right to produce a simple product, small and medium-sized companies must comply with all this heavy pointless bureaucracy that costs hundreds of million dollars to society. The impact of these millions of dollars is often fatal to the industry. The Ministers and the elected members rarely know about these thefts in procedures, because even their own inspectors don't know that their leaders reach into the company pockets. How much power abuse and disguised taxes does this represent?

Here is another example of a fiasco. During a transfer of power in management in the Labour Ministry of the Québec Government, this ministry managed to make, within six months, a bankruptcy bordering

90 million dollars. Nevertheless, this government had been wanting for years to control the money and the power related to labour. Now, once in power, these technocrats that claimed to know everything about labour management techniques have succeeded in making a dreadful fiasco of it.

The government's role in democracy is not to interfere with the management of private companies and to administer them as if they were the owners. Their role is to set up simplified structures and the simplest of rules to favour the development of small and medium-sized firms which supply, let us no forget, between 75 and 80 % of the jobs in a country. Briefly, <u>too many</u> programs of surveillance, <u>too many</u> taxes and <u>too much</u> administration are damaging to the employees, the companies and the progress of the States.

E. Materialism is invading the professions

All schools of education from law to medicine, as well as administration and engineering schools, faculties studying society, religion, psychology, etc., have been reviewed either to confirm or to destroy the "truth" of what they were teaching. **The method, as we mentioned earlier, consisted in reviewing education in every domain by subjecting all their teaching to the principle of the methodology of scientific research.** This scientific wave was so strong and powerful that it managed to influence all the professions that wanted to find their place under the sun "of science". For reasons we will discuss in Chapter 7 on the saga of the development of scientific ideologies, the teaching of science and technology has spread like wildfire, invading all university faculties and professions and materializing them. **<u>In order to reach this "truth" all the domains had to comply with the logical, mathematical and linear exercise of the scientific experiment and with the "reason" of their intellectuals.</u>** Who is not in favour of the search for the truth? Even the sciences that do not lend themselves to this mathematical logic such as psychology, theology and history started to think that in order to be recognized as true sciences, they inevitably had to examine closely their contents in the light of the methodology of scientific research.

Let us gave an example to better understand this. The fathers of behaviourism, those psychologists who pertinently knew that psychology is the study of the consciousness and human feelings, have unsuccessfully tried to scientifically **<u>document the existence of the human consciousness in laboratory rats</u>. Not having found it after over 20 years of laboratory research on rats, they proposed a new "materialistic psychology" of human behaviour, based on a tangible** and measurable logic of the conditioned reflexes observed in rats: **"Behaviourism was born".** To deny what we cannot understand with the knowledge of the moment is very common in science. Descartes had never considered denying the existence of the things that could not be studied with his method of research, as scientists and intellectuals did later on. The positive thought of the "reason" and the logic of the scientific research method not having allowed researchers to demonstrate the existence of the human consciousness, **they simply denied it as an existing and provable entity and proposed to replace psychology by their theory on behaviourism.**

In order to be labelled with the prestigious word "science", all the university domains had to play the game of this materialist methodology of research and of its "reasoning". All the entities which could not be explained and verified by the mathematical logic of this method were dismissed as if not being part of the existence…

Accepting only the measurable, visible, audible and tangible facts perceived by the senses, scientists could only study things that were material. This method of research serves admirably well material sciences such as chemistry and physics, accepting all the material evolution and accumulation of knowledge, but on the other hand denies all that the primitive senses cannot perceive, like feelings, human questioning on existence, the spirit, etc. It is not surprising that all the knowledge acquired and the certainties that science teaches are of a material order: science only considers concrete, material and visible things that can be studied with their method.

Being able to study only concrete things, this method is thus inevitably incomplete and deficient to study non material entities. So far researchers and intellectuals have rejected all the entities that their method cannot confirm with the primitive senses of comprehension. These entities were then abandoned, and in time were considered as non scientific and non-existing. This is why all the faiths, which could not be based on concrete data, palpable and material, have all been rejected by science as non-existing. It is very simple for scientists and intellectuals: what is not matter does not exist! **But this "certainty" is not a certainty at all, because it relies only on the perceptions coming from the most primitive part of man's brain: that of his reptilian senses.** A lot of "certainties" taught by pseudo scientists are often totally false because they are absolutely not proven by scientific experiments. They nevertheless continue to study only the material truth. **This proves their insanity, because they reject what does not fit in with their own principles of truth, without proving it!**

This is what drives many people to atheism: the acceptance as the truth of only entities recognized by the primitive senses and this, without any scientific proof or experiment. They keep on believing and **teaching that only matter exists and all what is not matter does not exist.** In the end, what they ask is to have faith in their reasoning without any proof! How can we believe these pseudo-scientists and intellectuals? They do not even consider studying the entities which do not cross the realm of their ways of thinking...

A truth can not be qualified as scientific if it cannot lend itself to the materialistic game of experimental methodology, and that is how psychology got its letters patent of nobility through behaviourism, by "materializing" human conscience, feelings, love, kindness, empathy, respect, joy, the supernatural and the soul as simple material conditioned reflexes.

Even though today scientists are starting to recognize that there are more evolved levels in the human brain than the only level controlling the senses and on which they still base all their scientific decisions, they keep on relying on their primitive senses to deny entities that their defective method of research does not enable them to study. Why do scientists not accept modifying their method of research to enable them to study other things than only material entities and why do they still only base their hypothesis only the perceptions of their senses?

What should we think when theology is only taught today with the scientific technique of theology, history with the scientific technique of history, psychology with the scientific technique of psychology, administration with the scientific technique of administration, as if these pseudo sciences were only material like physics and chemistry. **Are these branches really sciences or simply domains which cannot be studied with this material methodology of research?** Should we believe that spirit, consciousness, history and theology have become materialized and dehumanized to the point of denying their own existence, because science obliged!

When science is imposed in an education system and only uses what is tangible and material to train the individuals, there is a big hole in such an education. It is not surprising that there is a great lack of humanity and human values in such an education system.

This scientific methodology of research which gave humanity such important material knowledge must evolve one day and become more universal. This method should henceforth take into account the "data" that today's knowledge does not yet allow us to understand, simply by stupidly denying their existence. For example, we had to wait for Newton in the seventeenth century to understand the unknown and imponderable forces, nevertheless very material, of the moon on the tides, already proposed by Copernic on the attraction of the matter by the matter. It has been necessary to convince many tortuous minded powers of yesterday that the earth was round, turned on itself and around the sun, and was not, as Ptoleme claimed, the centre of the universe. It is always pretentious for scientists and intellectuals to deny as nonexistent less tangible entities, simply because nobody yet understands the mechanisms with the knowledge available at that given time. Who did not hear these supposed intellectuals, atheistic and conceited, deny all that is new or that "the reason" still cannot understand? Regrettably, and even though this behaviour is childish, it was and still remains fashionable in science, because it well serves the conceit of certain individuals who praise themselves as being intellectuals or scientists when they don't understand something. There is a difference between the scientific doubt that requires the Cartesian methodology of research and the doubt due to the lack of knowledge at a given moment.

On the other hand, one has to wonder if all inexact sciences can really be called sciences, especially when their objects are non material such as in theology, psychology and history. Can they be taught as if they were real sciences when they have never been proven with scientific experiment? As if they were scientifically proven sciences such as chemistry, physics, mathematics? Are they sciences just because pseudo intellectuals and supposed pseudo scientists make them pass through the mould of the scientific methodology of research and "reason"?

Example of the materialization in the medical profession

To better concretize the impacts on people and on societies by the only material knowledge and values without the humanities acquired in the education with science, let us examine further the influences of certain dehumanized decisions in medicine as well as in schools of administration, engineering, and even religion and psychology. Let us see, for instance, how the influence of scientific and social materialism has settled down in the exercise of what one formerly called "the Art" of practising medicine.

Medicine is an Art because it cannot be only a science even though modern medicine bases its vision of the world on only the scientific method of research! **Medicine must remain an Art because it will always deal with individuals that are at the same time matter and being.**

We already talked about the administrative techniques used in the exercise of the medicine and demonstrated that medicine and doctors had also modified the exercise of their profession, after their forming with scientific training. The materialism linked to the scientific knowledge only and its values is felt as much in medicine as in doctors, educated as everybody else at the school of science without humanities and material values. As for all the individuals educated at this school of science without the humanities, the doctors' gestures transpire with materialism in their everyday work. Doctor – patient relations, the exaggerated confidence of doctors in material computable laboratory analyses, X-rays, new and precise techniques of scanning, nuclear magnetic resonance as well as the multiplication of the numerous other techniques which they use each day in all the medical specialities, from surgery of the otic bones of the ear, to the treatment of cataracts, vascular and cardiac surgical technologies, brain surgeries, etc., deeply modified the physical exercise of medicine. **The modern scientific vision of practising has taken away the human aspect of practising medicine.**

The blind trust of new doctors for these tangible technologies has made them less prone to listening to their patients' complaints... **The new doctors trust the more tangible results of the laboratory analyses and of the scientific techniques than the subjectivity of the case history, so useful in establishing a diagnosis.**

The decrease of working hours to conform to other professions governed by the same teachings, group practice and the almost complete decrease in house calls, the loss of the family doctor, etc. all added their weight to the new ways of practicing modern medicine. By so doing, however, doctors have dehumanized "their art" of practising medicine. It is less and less necessary for doctors to see a patient in order to make a diagnosis. Reading the abundance of lab test results is enough to make the majority of the diagnoses. Modern doctors confirm their diagnoses by asking their patients if they really have pain in such a place or if they present such or such a symptom, etc. It is reverse medicine for the older doctors. **The latter practised a deductive approach of diagnosis, while the doctors formed at the school of current sciences practise an inductive method.**

As technological applications became more and more sophisticated, a deep ditch was created between the traditional humanized art of practising medicine, characterized by a doctor who knows his patient and his family, and the dehumanized scientific medicine which is characterized by a lesser knowledge of the patient and a greater knowledge of all that is technique and laboratory analysis. Gradually, patients have lost the spontaneous contact with their doctor and are complaining about it.

The former art of practising medicine was, in a sense, saved in the early stages of these changes because medicine was slow in accepting novelty. Medicine has always been, like religion, too prompt to react to novelties, often inequitably and without the proper knowledge. Medicine denied what it could not understand, like scientists who acquired professional habits doubt of everything. This has possibly undermining their secular bases. Conservative by nature and strongly encouraged in this manner by their medical associations which are prompt to punish those who break the tradition, medicine has also mistrusted for too long the technological novelties and the scientific progress.

An unhealthy medical conservatism

Regrettably, because of this conservatism, medicine did not know how to raise itself in time to the level of the scientific research. During decades, it wasted precious time fighting its opponents rather than trying to compensate for its own inadequacies. As wrote repeatedly Doctor Charles Godfrey of Toronto in *The Medical Post* among others, "Medicine has wasted too much time denying the medical novelties, instead of trying to understand the advantages and the disadvantages of the other types of medicines".

"Medical powers preferred raising sterile debates on complementary medicines rather than trying to study and understand them, because medicine risked seeing its authority undermined. As the religious powers did in their time, medicine denied for a long time all that was new. It did not care to form doctors, did not try to acquire more knowledge or to do as well as its adversaries with new therapies". **It is the case of the epic fights of the medical profession against chiropractors, naturopaths, homoeopaths, etc.**

For having placed incompetent people in charge of management, "The Colleges of Doctors and Surgeons were set on disgracing their own medical doctors who were interested, more intelligently than they, in new medicines and in new ideas. Those that saw farther than their peers were reprimanded and often wrongly condemned".

The gratuitous assertions of medicine

It is unfortunate that medical authorities were content to assert, without any scientific proof, that alternative approaches to medicine had no value. It is also unfortunate that doctors can still slander and ridicule, says naturopath Jean-Marc Brunet (Montreal), things they know nothing about. "It is this narrow-mindedness and this childish behaviour on the part of the conventional medicine", wrote Doctor Godfrey, "that opened wide the door to alternative medicines".

Medicine also underwent considerable cuts on the financial plan. Over a period of just a few years, Quebec has been traumatized by a cut of 2,443 beds of short duration, the closing of 9 surgical units, 7 emergency rooms, a decrease in hospitalization stay from 8, 2 to 6, 8 days, with expense cuts of 16 % in hospitals and 350 million dollars elsewhere in the system, besides the departure of 15,000 experimented employees and several hundred doctors... It will take a long time to work adequately after the implementation of this administrative incompetence because great inconveniences have marked these changes. Waiting lists have lengthened in emergency rooms. The long waiting lists for exams, blood tests, X-rays, tomography scans, mammograms, etc., made people charge to their private insurance companies expenses what actually belonged to the health system. Treatments and operations are delayed several months. It is a major shambles in all health systems in several countries. "Medicine and doctors lose credibility with their patients who opt more and more for alternative medicines" wrote again Doctor Godfrey.

"By acting so, medicine missed the boat of scientific revolution and medical training has been deficient for several years in these domains". It is because of "this deficient medical training that patients suffering from back pain very often consult chiropractors rather than doctors", or still use "healing plants to replace numerous synthetic medicines". Even today, studies demonstrate that the exercise of the medical practice is far behind scientific novelties, over one generation in certain cases...

"If conventional medicine had kept the most effective approaches in treatment, alternative medicines would have had a lot of difficulty being born and surviving" wrote a naturopath in his health chronicle (*Journal de Montréal, April 8, 2000*). It is absolutely certain that medicine should not have turned its back to all these novelties during so many decades. History shows that medicine often monopolized treatments belonging to others and eventually made them its own when it suited them, and continues doing so today. The recent acceptance of aspirin as a very useful medication in the prevention and treatment of cardiac diseases, atherosclerosis, vascular cerebral accidents and certain cancers prove it. The usefulness of "glucosamine" in the treatment of arthritis, the use of antioxidants and more powerful doses of vitamins than those recommended for the prevention of certain chronic diseases are other striking examples of this phenomenon. The role of oligo-elements in the metabolic balance has not even reached traditional medicine. But, on day it will come...Very often these therapies are simple, natural and less toxic than those used in medicine. They sometimes offer real advantages in the prevention of certain diseases. Curative medicine, which is ours, is only beginning to acknowledge the virtues of disease prevention through exercise, nutrition, meditation…

The example of Jenner's vaccination

Official acceptance by the British medicine of the inoculation against smallpox is a good example of the acceptance of novelties in medicine. It is only in 1895, over one hundred years after its discovery that the famous Royal School of Medicine of England finally accepted to officially recognize the great discovery of English veterinarian Jenner, made in 1788: inoculation against human smallpox with the virus of the bovine smallpox.

Jenner had observed that farmers working with cows developed bovine smallpox in the endemic state, but never presented human smallpox during epidemics. He then imagined inoculating people with bovine smallpox to protect populations against the much more dangerous human smallpox. Smallpox is an epidemic disease which devastated human populations for generations. It killed thousands of individuals and ravaged complete cities when it appeared in Europe in the thirteenth century. We now owe this English veterinarian the prodigious discovery of the inoculation of human beings with the bovine smallpox. This inoculation effectively eliminated this terrible infectious disease from the surface of the earth in less than two centuries.

Why wait for over 100 years for medicine to recognize that thousands of lives would have been saved, had arrogant doctors deigned to officially recognize the extraordinary merits of this great scientific discovery?

Let us recall that, although this observation was one of the greatest medical discoveries of modern history, prestigious medical journals of the time, such as *The Lancet* that still exists today, as well as others just as well quoted, repeatedly refused to publish Jenner's discovery. Discouraged by these obtuse and repeated refusals by medical journals, the veterinarian published in 1798 a monograph of 61 pages which became the rage everywhere in Europe and America, except in his own country, Great Britain. Is this not a striking example of medicine refusing scientific data?

Once his monograph published in 1798, Jenner's inoculation was at once accepted and practised all over Europe and America by foreign doctors, without waiting, as it is case today, for the permission of medical authorities. History shows that this inoculation instantly saved the lives of thousands of American soldiers during the 1801 war, for instance.

If certain simple treatments, known for millenniums, must wait even today for decades before being accredited by medical authorities as for the case of smallpox in Great Britain, it is not surprising that medicine lost some credibility with the patients who are more and more informed and capable of consulting other types of therapists.

Consequence: medicine misses the boat of scientific revolution

In medicine as in other domains, what was to happen did. Because of the excessive slowness of the medical profession to accept scientific discoveries that did not come from his milieu, "medicine did not adapt to the quick scientific revolution which spread over the world in the twentieth century", said against Doctor Godfrey. And so, it fell so far behind that, in science, it has not yet been able to catch up to this day. Scientific novelties are taught late to young doctors and, because of the costs, the practice of the medicine regrettably remains decades behind other scientific novelties in spite of numerous scientific researches.

The absence of medical researchers

Another problem inherent to this behaviour of the medical body was that very few doctors worked in research until the end of the twentieth century. It was only in last half of the twentieth century that we found a few doctors starting to show some interest in scientific research. Today, after decades of research by the non medical scientific community and related sciences, medicine can count on the fingers of one hand the medical researchers earning their living in research laboratories.

Sometimes, during their studies in a speciality, future specialists do "stages" in research laboratories. Few however will turn to this less profitable practice than the practice of medicine. So, specialists quickly abandon research to concentrate on their medical specialty. Doctors are often nevertheless born experimenters. Because of the numerous constraints imposed by scientific powers today, it is almost necessary to be "delinquent" in order for a doctor to dare change one iota of a technique in his specialty or to not follow the sterile lines of conduct established by his peers.

Lines of conduct

Lines of conduct are sort of "changing protocols" that doctors must know and follow to palliate to different medical situations. So, when facing a heart attack patient, for instance, doctors are almost obliged by their own authorities, to follow a series of predetermined techniques to treat their patient, independently of the data and observations collected at the bedside. Now, these lines of conduct are not scientifically proven medical data. They are established for the greater part by statisticians, epidemiologists, laboratory data or other individuals who may have never seen a single patient. They nevertheless represent the "directives" on the procedures to follow in the face of a given medical situation. Doctors have too often been judged by their peers on these technical procedures, non standardized and based on what one agrees to call convincing data, disregarding the objective data obtained at the patient's bedside. **Lines of conduct can be multiple in front of the same problem, especially when the pathology overlaps several specialties. So, in front of a tuberculosis problem, lung specialists, bacteriologists and other health specialists each have their own lines of conduct which vary from one specialty to the next.** (cf. *Evidence's Weakness, by Charles Godfrey, The Medical Post, October 8, 2002).*

Adopted and accepted as "standard of excellence" by the evaluation committees of medical procedures, these data are inequitably required without on any scientific proof by these committees to judge the medical acts of their colleagues! A study done on over 900 such "lines of conduct" demonstrated with supporting evidence **that hardly twenty out of these 900 (less than 2%) deserved to be retained on a scientific basis and that over 95 % are simply not confirmed on a scientific basis.** Nevertheless, a doctor who would neglect to apply these "non scientific lines of conduct" in a given situation exposes himself to the wrath of these medical evaluation committees. A doctor can even be inequitably made responsible in front of a court of justice for not having followed these "lines of conduct". Is the intense teaching of these lines of conduct creating well-trained robot doctors?

Experimentation victimized

It is unfortunate that modern medicine does not remember that anatomy, for instance, would have never been able to acquire its letters of nobility without the "violation of corpses" in cemeteries. In today's current conditions of clinical research, experimental vaccines for rabies, smallpox, diphtheria, whooping cough or poliomyelitis would still be plagues for the humanity, if audacious doctors had not first experimented immunizations against these diseases. Organ transplants might still be waiting to be experimented if researchers had been obliged to wait, like today, for the "Malthusian" approvals of ethic committees and their too often incompetent peers...

If doctors had not dared practise the first transplantations of kidneys, heart, liver, lungs, bone marrow or other organs as a measure to save their patients, they would have died, period; and all the knowledge acquired on organ transplant would not exist. Nevertheless some of these treatments are done today and allow medicine to save thousands of lives and/or considerably reduce medical expenses. Let us think, for instance, of a simple kidney transplant to understand this! Such research is almost not possible theses days. The researcher who wants to experiment a technique has to obtain beforehand: a) the necessary funds from committees or support agents; b) the approval of his peers and c) of the hospital committees. He must almost demonstrate in advance, which is absurd in research, the results of the experiences of his research, before obtaining permission to make the research! Research is thus nipped in the bud. It must come from other countries and we must be content to repeat that of others. With the new laws of the Québec Medical Code of Ethics which stipulates that doctors must henceforth inform patients of any errors committed, we will need decades and a number of deaths which could have been prevented before research becomes a natural reflex in doctors.

All the controls made on the medical practice by the medical profession and governmental powers only favour materialized medical care and accentuate the dehumanisation of the care. Teaching groups of rigid techniques, the practice of medicine within the "lines of conduct" not scientifically proven do not allow doctors to exercise the Art of medicine and take into consideration the fact that the patient is also a human being. A certain freedom is essential to the exercise of good medicine. This freedom and humanity are not found within the numerous bureaucratic medical constraints of the management of medical acts which are compulsory for doctors and current researchers. Because of all these constraints, medicine has lost its Art and its humanity.

These excessively severe measures of control inevitably lead to a more mediocre and less effective practice of medicine. And so doctors are more careful in their practice: they double the tests, the X-rays and the lab analyses to protect themselves from medical errors. By killing the freedom, administrative authorities have destroyed at the same time initiative and creativity, which are the motor of efficiency. To stupidly follow like robots those "lines of conduct", besides the fact that they are not scientifically proven, and the amazing quantity of new technologies that deprive doctors of their freedom, prevent them from analyzing the data collected from their patients and even to act in accordance with their training. Such exaggerated controls on medicine have become a major obstacle in the exercise of good human medicine and contribute even more to the dehumanisation of the medical acts and of health care.

If the medical profession and the non medical socializing professions that politically stole health care from citizens, consider health care and medical acts so simplistic that it should be enough to simply apply mechanically pre-established "lines of conduct" to treat a given medical situation, why not simply program them in automated computers and doctors will no longer be needed! **To practice this "politically correct" medicine directed by non medical higher instances is similar to removing front line doctors and eventually replacing them by dehumanized robots.** As long as all these political "layers" of powers and all these non medical professions persist in wanting to administer medical care as they presently do without any medical knowledge, medicine will remain bureaucratized, dehumanized and on the decline. The administrators of the medical business must be aware of this.

Patients are quick to understand

In their natural wisdom, patients understand that if there are few alternatives to surgical treatments, it is quite different for chronic diseases or for other medical conditions where traditional medicine has nothing much to offer. Giving more freedom to doctors who wish to investigate these radiuses of action would be more intelligent than condemning those who try to help their patients by offering them at least another alternative. Powers as well as non medical administrators who manage health care, diseases and treatments do not know that it is not with their socialist ideologies that knowledge is acquired in medicine, but through scientific experiment.

The socialist speeches of the powers infiltrate medicine

Those that work in the fields of social medicine betray the most their socialist ideologies, by trying to impose their views and their thoughts through the political route. **Certain professors of social medicine and paramedical academicians, working at the various levels of medical education, now try to teach their false socialist and materialistic ideologies in medicine.**

A book entitled "*How Political Correctness Is Corrupting Medicine*", recently written by Doctor Sally Satel, M.D. BASIC Books (Perseus Books Group), Copyright 2000, ISBN 0-465-07182-1, states that these materialistic and socialist ideologies are invading medical education today. Analyzed for "Medscape", on April, 24, 2001 © Medscape Inc., Doctors Ceccoli and Kinderman wrote that "Doctor Satel presents evidence that medicine is being infiltrated by social activists, academicians and health professionals". Contrary to the principles of traditional medicine which is based on the scientific data of the disease and not on fictitious or political ideologies, **these people are trying to persuade their students that it is politically correct to impose their ideologies and their socialistic way of seeing disease.** "If it is left without defence in the hands of this minority of extreme activists", writes Doctor Satel, "**medicine will soon pass from a medicine of evidence and proof to a discipline based on purely political socialist ideologies. Consequently", she writes, "standard medical care will diminish and patients will become the ultimate victims of a politically correct medicine**".

To support her words, Doctor Satel demonstrated, by means of a series of data, the problems of this "politically correct socialistic medicine" which is settling down in public health services, in mental health, in clinical research, in nursing and in medical education. Here are some of the abnormalities which confirm this tendency of the socialist politicians and doctors to socialize and to politicize medicine.

1- "Academicians in public health services are trying to extend their mission so that the prevention of wounds and diseases becomes part of a reform or change in medicine. This medicine is based on the fanciful unfounded assumption that social conditions are the first causes of disease. These socialistic political concepts oppose the scientific concepts which demonstrate above all that it is the people's individual conditions that are at the origin of their diseases."

By attempting to invest in such a grand mission, these activists are trying to divert the traditional mission of public health services. **These socialists' claims that society and not the individuals responsible for the people's health and diseases have no other foundation than a political one.**

2- In the United States, former psychiatric patients and their lawyers started campaigns to give their clients more freedom in the treatment of their mental health problems. These lawyers claim that patients suffering from severe mental diseases should not be forced into taking antipsychotic medication or being hospitalized. This unfounded opposition, claims Doctor Satel, results from a deep denial and a refusal to recognize that there are treatments and effective pharmacologic medication to treat these patients.

3- Dissatisfied with the current medical system dominated by men, nurses rebel against "the medical establishment" and attempt to dissociate themselves from the conventional practice of medicine. This deviant way of teaching medicine is found in nursing literature. Nurses support alternative therapies, without any scientific evidence, particularly those by the touch. "This rebellion of the nurses", says Doctor Satel "is expressed by inappropriate application of medical treatment. They perform actions that do not correspond to their professional training. This abnormality should be corrected at once by a return to the basic principles of nursing", thinks Doctor Satel.

4- Women working in health domains cry out that clinical research is always in favour of men. To support their demands, these women present valid and scientific arguments. They show, for instance, that this biased thinking towards men engenders bad decisions in the priorities of national research. Doctor Satel uses arguments to explain these abnormalities.

5- Civil law activists cry out that there is racial discrimination in medicine and that it would try, according to them, to infer racial preferences in the admitting departments of medicine. Doctor Satel demonstrated that this approach is false and non scientific for several reasons:

- Because the current data show that this assertion of a racial discrimination is not proven;
- Because there is no scientific evidence at all that the relations doctor / patient / race produce better results;
- Because certain actions in this sense have already allowed a minority of students to begin a career for which they were not intellectually well prepared.

    These examples demonstrate that current medicine tends to be gradually "infiltrated" by social activists, academicians and health professionals with socialistic political tendencies. These tendencies even place socialist political ideologies above the medical practice which is based on scientific evidence and facts. Doctor Satel's book also mentions several other examples of this socialist "infiltration" of medicine. She draws our attention and goes against these activists, because these ideologies urge doctors to practise a "politically correct" medicine, by opposition to a medicine based on scientific evidences. This phenomenon creates a deviation in health professionals, which consists in considering patients as victims of social circumstances, without the proof of scientific data.

Doctor Satel also draws the attention of doctors on the legitimacy to continue considering scientific proofs of evidence because these proofs clearly demonstrate that it is the individuals who are responsible for their diseases.

Forcing doctors and patients to believe that it is either the circumstances, the race, traumatic events or society that are responsible for diseases, is an insult to the intelligence.

Doctors Ceccoli and Kinderman, who analyzed the book, write that occasionally Doctor Satel simply seems to react to the activists with whom she disagrees without bringing any solution. They also write that sometimes the conservative views of Doctor Satel seem to be as dangerous as those of the social activists'. The analysts however strongly recommend the reading of this book as food for an interesting reflection.

We would like to stress that **this publication has been written with the precise purpose to draw the attention of doctors and the medical world on the political and socialistic opinions of the powers on disease and medicine and not to propose solutions to the problem.** It is clear that certain politicians in power would like to exploit their socialist policies to prove and to better anchor their influence on health, patients and health care.

Hector D. Ceccoli, M.D. is a board certified cardiologist practicing with Cardiovascular Associates of East Texas, and Karla Kinderman Ceccoli, J.D., L.L.M. is a founding partner of MEDLAW, LLC, and a former American Medical Association Health Law Counsel.

The principles backing the administrative techniques in governments are of material order and were developed to concretely administer factories and garage yards. These principles are not universal nor are they scientifically demonstrated and as such should never support the principles of managing the administration of the people. A non socialist government that respects itself should never oblige by law, for example, doctors to transform themselves into civil servants in one day. A law which dictates to every doctor, as the Québec Government did in August 2002, **when, how, at what time and in which clinic or hospital a doctor must practice medicine in a given emergency room is a demented and degrading law that takes away the freedom of doctors and is in no way democratic.** It is only another example of the authoritarian management of those powers governing with the "reason" dictated by socialistic principles, not at all scientific and totally deprived of humanity. A democratic government cannot transform with impunity the members of a liberal profession such as medicine into civil servants, by subjecting doctors to the authority of the non medical civil servants of the State!

In spite of the prodigious quantity of knowledge brought by science and necessary to the practice of a profession, numerous individuals in positions of power regrettably reach their level of incompetence in the ministries they manage, because of this too rapid evolution of science. Those that are chosen to manage a ministry should retrain into the domain or not accept its leadership. Professions presenting a more rapid evolution become more quickly obsolete, and so those occupying these positions must not only be well informed, but also know what they administer. The acquiring of knowledge is made more quickly in certain professions than in others. It is unfortunate that this notion is not yet well understood in "administrative sciences". Continuous education is necessary today in all the fields of activity. There is nothing stable in science. Individuals in positions of power are incapable of following the speed of these changes in their respective domains. This is true as much in governmental administrations as well as in other domains. An airplane company administrator cannot become a Health Minister in one day and impose, by law, his socialistic conception of work by taking a profession such as medicine as hostage. What can we make of all this?

F. The influence of materialism on other professions and institutions

Several other examples show that training with science, technology and material values have not only affected the medical profession: many other professions were affected as well. Let us see how it affected them.

Justice

In several occidental countries, judges, politicians, bureaucrats, manufacturers and many others are involved in shameless scandals in absurd and unthinkable ways and by sordid political authorities, in order to cover up certain persons. Demonstrations of this are numerous.

Lawyers, for instance, drag causes endlessly to increase their wages to the detriment of their customers. Their fees are boosted to the maximum. Their colleagues working for governments draft absurd laws or defend causes so stupid and useless that it would have been preferable for them to have never completed law school.

In order to empty prisons, political powers force judges to give less severe sentences or to serve time in society with almost no surveillance. French judges complained, in April 2002, that 33% of the judgments they gave were not executed for all sorts of incredible reasons. These attitudes are disastrous and are not without creating chaos and social injustice. Sentences sanctioned by judges then become farces and contribute to make light of the justice system. The vocabulary of the justice system has also been transformed into an incomprehensible political language. Imprisonment for life can also mean 15 or 25 years; a ten year prison terms can mean three years, sometimes less; what must we make of all this?

During this time, in the United States, justice seems to have lost all sense of decency and measure. A fellow named John Marquez was condemned to life imprisonment in Oklahoma because he spit on a policeman. In California, sentences up to 25 years are given for shoplifting or drug possession: this is out of proportion with the crime. How can the American Supreme Court rule that 50 years of prison for the theft of some video cassettes is not too cruel or unusual a punishment?

Justice has been transformed into shameful bureaucratisation and too often into an institution of lies similar to that formerly used in totalitarian countries. "Welcome to the circus", was entitled an article in *La Presse de Montréal* on July 29, 2000. The journalist was commenting on a judgment made in British Columbia (Canada) that had just authorized the presence of TV cameras in the lawsuit of seven Koreans accused for having helped a group of illegal Chinese immigrants enter the country.

A judgment of the Québec Superior Court one day condemned Social Security and its pitiful civil servant lawyers for power abuse, in what was called the P. Brunette case. A 64 year old handicapped woman, confined to a wheelchair, lived in an apartment with an intellectually handicapped person. These two persons helped each other, as only human beings know how. In 1996, three years prior to the judgment, the materialistic technocrats and lawyers of the Ministry of Social Services and Welfare cut off, without any consideration or humanity, their social security cheques. They demanded that the 64 year old lady refund the 56 626, 00$ received during the last 10 years for this ridiculous motive. The affair was judged by the Québec Superior Court in January 2000, which took advantage of the situation to bring this injustice to light in regards to the unacceptable and inhuman procedures of these dehumanized lawyers in power. The judge wrote: Social Security (and its technocrats in power), proved to cruel and stubborn towards these two handicapped persons who lived together to mutually help each other and to be as little as possible to the charge of others.

If citizens allow their governments to distribute some of their tax money to the ill or needy individuals, it is not so that civil servants can stupidly justify their jobs and spend the governments' money on

numerous and useless judiciary costs, and even less to teach these people how to live in society. The fact that 2 or 10 people share the same house is not the business of governmental powers that like to interfere a bit too much in people's private lives. At least give the citizens the freedom of power to help each other mutually, even though the technocrats in power have already lost this human sense and seem to have forgotten that mutual help is a non tangible, non material, and non visible human notion.

Can the civil servants of Social Security or any governmental lawyers allow, with impunity, to justify their jobs by continuing to commit such stupidities? If so, it is that their administrative culture is not only materialized, but completely dehumanized. It is necessary to quickly change these laws and to have all these high and mighty persons retire as soon as possible. What a golden opportunity for a Finance minister to recuperate millions of dollars from jobs as useless as they are stupid, directly paid with the taxpayers' money.

They should also be reminded that it is these same materialized, socialized and dehumanized powers that have confused everything! They regrouped under the name "Social Security" the deprived people of our society as well as perfectly healthy people who, for one reason or another have decided to live this way, and also, and this is much worse, individuals suffering from multiple chronic diseases which our society must support financially. I want to speak about these people who are medically ill, physically or mentally handicapped and who will never be able to work normally. What are the powers waiting for to definitively separate these two classes of individuals, as doctors have asked them to? In no way is it the business of these technocrats and scientists in power to contradict medical diagnoses, to put patients under undue stress by leaving them in uncertainty or to keep requiring medical report after medical report from doctors, a procedure which increases more and more medical expenses!

Let us remember that France, a country officially very socialist since 1980, is going through a "delinquency crisis" amongst politicians, judges, bosses of big companies as well as ordinary citizens. Magistrates of this socialistic power have made up, for example, false actions in order to protect certain politicians. In a book entitled *"Vengeance d'état"* (Revenge of the State), edited by Alban Michel, in November, 2001, Alain Guilloux tells how his lawyers' archives were searched by upholders of the State law, with the only purpose of seizing his documents.

It seems that in the French country, as well a in other countries, there are new sagas implicating judges. **After having condemned sex, delinquency, crime and family crises ending in crime, the French judges now give themselves the power to impose fines to "publishing**

firms". One of them has just taken the rap of a substantial fine for having published the revelations of a retired French general, in regards to the atrocities committed by the army that tortured or killed Algerian prisoners during the war in Algeria over 40 years ago. In a French television broadcast, (*February 17, 2002*), one of the commentators mentioned that the journalists from the *Journal du Monde* and from other newspapers, had already published these same facts well before the publication of this book. Why then was the publishing firm the only one condemned by the French justice?

Judges brought racism charges against Houellebecq (*Platform, Flammarion, 2001*), while intellectuals shyly talked about these new events, if at all. It is the "thought" that is now brought to the bench of the accused by judges and politics, claimed the commentators to this newspaper.

We have already reported that France, a socialist country, accounted for 168 crimes per thousand inhabitants in 2001. A short newscast on Channel 15, on March 4, 2002, supported by two studies, showed that this increase in crime was neither linked to unemployment which decreased while crime increased or with higher family income, as crime reached summits. Mr. Jospin's socialist government admitted that there was a 16% crime increase during its mandate. How so? A big part of the 2002 presidential campaign was in fact about finding ways to reduce crime across the country.

In April 2002, the Magistrate Union announced in a short story that 30 % of the sentences given by judges were not executed. It is completely inadmissible that in a system of justice, one judgment out of three is not executed, whatever the reason!

Some are convinced that the courts go too far while others speak of a "government of judges". The elected members must respect the will of the people and must answer to them, not the judges, mentions an article in *La Presse de Montréal* on April 13, 2002. Judges, claimed the article, demonstrate so much zeal and enthusiasm to judge the constitution that it is obviously these high-ranking civil servants that manage today's justice system. "We simply exercise the powers that the legislator himself gave us" said the Québec Court of Appeal Chief Justice. "Only the courts have true independence which enables them to face controversy, unpopularity and to sanction the individual rights and liberties of the most vulnerable. This responsibility", added the magistrate "justifies the fact that we can ignore the legislator's agenda, enhance the responsibilities of public service and modify budgetary priorities".

Julius Grey, a veteran lawyer of the Charter of Rights causes, said that "the courts do not put democracy in danger, they save it because "the legislative power is becoming more and more illusive nowadays". It is globalization that gives true decision-making power to the investors and the multinationals who threaten to move away". Where is this leading us? I think that it would be preferable to lose such investments as to let foreigners lead our countries....

At the beginning of 2002, Québec judges unilaterally imposed to the Government to pay Legal Aid lawyers and members of the jury rates clearly exceeding the norms. How could this happen?

A Québec individual was arrested for speeding in the state of Louisiana in the United States in the beginning of 2000. He was deported to Quebec because of an unpaid traffic ticket 20 years earlier. How has justice come to this?

The objective of justice is have laws respected and to protect citizens against people's maliciousness. **Now, nobody understands how the administrators of justice can degrade themselves into condemning individuals to only serve one sixth of their jail sentence?** To decree that crimes committed by the young must be considered more or less offensive, even if they kill their own parents while declared sane, is an unthinkable social vice. **To accept that policemen, condemned in criminal courts, be reinstated in their work without having the time to think about their acts is also unthinkable, especially when studies have demonstrated that 16 % of the American policemen are criminals.** This means one criminal out of five policemen! The responsibility of this manipulation of the justice system is attributable to groups of community activists, social committees of good people and inequitable laws as much as to the lack of prison space. How can justice representatives degrade justice to the point of freeing criminals after only a few months of punishment for their crimes? The responsibility for this lies in the materialized system and the training of the justice representatives, also educated with sciences and material values without any humanity. Is it a particular trait of a justice administered by materialistic laws if criminals have more rights than their victims? The occidental world is administered today with the same materialistic justice that guided the totalitarian socialist countries with inhuman laws that fed on the same material values. **"By giving priority to a full and complete defence for criminals, the people in power completely forget their victims and their future victims"**, said the chief inspector of inquiries of the Montreal Police Force in an article in *La Presse* on April 13, 2002. "Crime is more attractive to big gang leaders", it read, "because these individuals know that by paying a battery of lawyers they will manage to find the small breach in the system". "It is no longer a Superior Court, it is a place to discuss the philosophical cause of individual rights versus the State", said an ex-judge of the Supreme Court. In Canada, three quarters of the cases refer to the Charter of Rights, while only 12 in 15 causes a year treat the real problems of the people. Marcel Gaucher, in *"La démocratie contre elle-même"* (Democracy Against Itself), **tries to explain why, after the failure of the Hegelian philosophy and of Karl Marx's capital sociologic principles of the Charter of the Rights of man, the socialist States are destroying the democratic conception of democracy by interpreting these rights to serve themselves.**

It is abnormal that the judges of the Supreme Court can reverse established practices by giving the right to vote to all the prisoners detained in the federal prisons of Canada (*October, 2002*), that is to criminals who lost their right to vote after having been condemned for murder, violence, rape, pedophilia, theft and/or other crimes. It is abnormal that the judges of the Supreme Court can decree, from the height of their praetorium, that the public has no right to know how the elected members spend the public funds (*Quebec, end of October, 2002*). It is extremely urgent that government limits the powers of these civil servant judges who behave like elected governments.

When laws and the Charter of Rights and Liberties give more protection to the criminals than to the victims, it is urgent to look at what is terribly wrong in the law. Maybe is it time to humanize the laws, to make them more adapted to the administration of the human beings they address? Maybe we went too far in the socialistic left-wing ideologies?

Having secularized their States to the point where we are at today in Occidental countries, what government would be ready to consider that "the humane", the sacred, the feelings and the respect of the others are still real and human necessities? When activists and community organizations choke on their words of "goodness" to protect criminals to the detriment of victims, when, in front of the opposition of certain influential immigrants, the Mayor of Montreal activated a general outcry of protest from his citizens because he agreed, rather stupidly, (*December, 2002*), not to put up the symbolic Christmas tree in front of his City hall, with the excuse that it represents a religious symbol in a laic State, **it's time to worry seriously, because people have reached the point where they confuse legality and morality with the traditions of the people of the country**.

We would need much more positive actions such as those of George Ryan, Governor of Illinois. **Before leaving his post as Governor of the State in the beginning of 2003, he released four persons condemned to death who were and finally found innocent. Moreover, convinced**

that corrupted policemen without scruples do not hesitate to make up proof, preferring vengeance to justice and condemning innocent victims without valid motives, the Governor stated that this judicial system is arbitrary and rotten and that lawyers are often incompetent, and he released the 167 condemned prisoners in his State.

Correction institutions

There is more concerning justice. In foster centres in Quebec, 52 % of the young delinquents have already tried to commit suicide (R*adio News, April 13, 1999*). The great majority of the prisoners released from jail, and even after having received specific training by the technicians and experts of the penalty services, return to prison for similar offences. These bureaucrats were nevertheless formed with the most advanced and most rational new specialised technologies... Are these new scientific techniques effective or are they not simply experimental measures and enactments of education? Could we possibly have not identified the real problems? The fact that Quebec possesses 10 civil servants per 1000 persons, while nearby Ontario account for 4 per 1000 persons, could this mean something?

The Police system

It has been demonstrated that since 1995, crimes have increased even in the police corps. Over this period, prison sentences of American policemen have increased by 500 %!

In a radio news report presented in April 2002, we learned **that an American study had just demonstrated that 16 % of the policemen were corrupt criminals.** That means that one policeman out of 6 is a criminal. Some steal, others participate in crimes or create proofs to condemn innocent victims by deliberately adding drugs in clothing or in cars, etc. It is rather disturbing to be confronted to "this police that won't forgive", to use a euphemism from a T.V. commercial by a car insurance company!

Policemen have the responsibility to respect and to make individuals respect the law. But their training with sciences, technologies and material values without the humanities seem to have helped police training techniques become dehumanized. They believe to be above the laws that they have the responsibility to respect. They have also lost, during their training with sciences, the sense of traditional and instinctive human values that have always guided people during their evolution. This materialistic model of forming and their materialistic values make them incapable of recognizing that there are values other than material and that people are more than pawns on a chessboard: they are also human beings who think and who have feelings.

This materialistic phenomenon is regrettably international, because many countries opted to reform education systems by modelling themselves on science. Anthony Garotinho, Governor of the State of Rio de Janeiro, launched in March 2000 Operation Clean Hands to study the assertions of a security coordinator who had claimed that the high spheres of the State Police were corrupted and engaged in all sorts of crimes and misdemeanours (*La Presse, Montreal, March 18, 2000*). Charges against policemen would also have increased by 400 % in that country in five years. Documents seized in 1993 at the home of a Rio mafia member, had enabled to bring into evidence the fact that several policemen worked for this underground organization. Lawsuits and "arranged" inquiries set up to prove that the police was really involved in criminal activities, proved to be false and investigators had to drop the charges. Do these inquiries not resemble those of organized crime in Quebec or those of 1997, 1998 and 1999 in the Montreal Police and in so many other countries?

Societies have endowed themselves with polices services with the objective to prevent individuals from taking justice into their own hands. But for all sorts of reasons, these objectives were diverted by the social powers that followed. We ask ourselves today if the policemen, formed with the most modern police techniques, feel at a loss because of the peculiarity of their profession, among the most dehumanized. One thing is certain: the police corps were not formed to give out fines or to fill the cash registers of the administrations that pay them!

Expertises and interventions of the police corps in criminal cases are often rotten, deceitful and degrading as much for the human beings as for our societies. The Mattix affair, the Barnabé affair, the Poitras Commission in Quebec, the Dutroux affair in Belgium, the massive dismissals of French judges, the disappearances of children in series in certain French departments, etc. etc., are these not dazzling proofs?

"*El Universalun*", a popular Mexican newspaper, wrote in the beginning of 2003, that Machiavellian **students can join crime schools for 200 Pesos (20$)**. The objective of this crime academy is to give an intensive course in teaching students how to steal cars, such as BMW's or Mercedes, equipped with all the possible alarms. The professors teach how to manipulate the "class instruments": jimmies, hooks, blades, knives, brass threads, 22 calibre pistols, etc. The pupils also learn to sequestrate a motorist at a red light at gunpoint, hold him to the ground, steal his belongings, empty his bank accounts with his blue card and ask his family for a reasonable ransom so as to be paid quickly. They practice on stolen cars in the schoolyard. The whole thing is filmed on video to correct dexterity faults. The exercises then continue in the streets on parked cars. How can we not consider these schools of vice dehumanized?

Policemen were recently filmed beating citizens. Others are accused of using excessive force to control people already arrested. **Even though police techniques teach it, is it necessary to humiliate a detainee for not having paid a small fine, for a simple discussion or because of his skin colour, by putting handcuffs on both hands and feet?** What nonsense to handle all individuals brought in for questioning as if they were already condemned criminals! Deaths have been reported in several countries following the use of brutal force towards the accused. How must we consider the gigantic police deployments, set up for the G 8 summits on globalization in Québec City in April 2001 or in Genes, Italy just a few weeks later? And these are our materialized cultures...

When the materialistic culture influences sports and State actions

The Olympic Games and sports in general have become monsters of materialism where money and deceit are the main motors. The virtues of the extreme competition of scientific and social materialism are very visible at these levels. There, as everywhere else, athletes are forced to reach "excellence", a purely materialistic value, or to "win at all costs". In order to demand the monstrous salaries given to great athletes, the young people, their parents and the coaches do not hesitate to urge children beyond their capacities and even to commit indecent acts in their names. **The current culture of sport is a strongly materialistic feature of our civilization in search of excellence.** It is legalized and in great progress. Have our good materialistic governments and their accomplices not installed gambling areas, visible everywhere on their territory because it is paying? A very materialistic measure, isn't it? Let us recall the sad story of the miserable American father of a young ten year old who killed the referee at his son's hockey game. Do you think that these are human feelings on the part of individuals?

Influence on people in power

Thieves, liars, money launderers, overbearing, these are a few words to qualify some modern politicians. Political scandals spread in broad daylight in a number of countries confirm these facts. I agree that there have always been dishonest politicians. But the representatives of democratic political parties formerly had the decency to claim a certain humanistic righteousness. They tried to keep away from these unsightly manners of doing politics and mainly to make of lies their only warhorse. It is not so anymore. It does not seem that today's politicians and scientific technocrats, elected or non-elected, have learned at school or university afterwards, the humanistic education and values others than material. These people do not even know that there exists, apart from the scientific method of education, forms of education and thought other than their own, that take into account traditional or human values. They simply have not learned this and few are the ones who have done enough reading to be aware of it. It is not surprising to note that the people in power are also materialized and practice a dehumanized culture that transpires in all their actions. **They give free reign to their inhuman and unhealthy passions in the making of laws and norms that do not stop hindering people's freedom.**

The Canadian federal elections of 2000 were marked by a systematic misinformation of the kind, worthy of that used in Stalin's totalitarian regimes in Russia. Negative advertisement is not new in politics. But it is more often than not the weapon of desperate parties in loss of speed. So the Party in power started calling one of his opponents a "demon". It directed all its election campaign on the demonic aspect of his opponent, a Protestant minister. The attacks against him never ceased to be unbecoming, negative, unconstructive and deceitful throughout the whole campaign. Moreover, they tried to make "politically correct" the negative and personal advertisements that have always been considered in politics, by the strong parties at least, as unacceptable. Politicians will also go as low as using misinformation to try and gain terrain on their opponents. The individuals planning these dishonest acts have been doubly misinformed. First, these materialised people were generously paid to plan these insanities. These dishonest attitudes are a result of a complete misunderstanding of traditional human values. They can be classified in the same category as these unreliable scientists who work in armament or at the sick destruction of their fellow men. Regrettably, they are more and more such scientists deeply dehumanized to the marrow of their bones. These are other striking examples of this materialised political culture.

Hypocritical scientists in power try to persuade their subjects that "a game must remain a game", that "moderation has much better taste" or that "under the table work is a theft". These powers treat "under the table workers" as swindlers, while their co-workers transfer billions of dollars of surplus money towards fake companies or non-profit organizations, to escape the controls of their own supervisors. They hijack millions from the tax money of their citizens in all sorts of manipulations.

Is it not their colleagues of the former materialist government that got into debt, with impunity, their country, their provinces and their municipalities with such enormous sums of money that will take generations to be paid off?

Why two weights, two measures? **Before giving lesson to their subjects, maybe these governments and their scientists and intellectuals should practise what they preach.** "Swiping is stealing" civil servants keep telling us. They should give their citizens the example by correctly administering the public funds without diverting it. Should they not put some order in their own useless expenses, reduce the unnecessary programs people never asked for, pay off the dreadful debts they make on their citizens backs? Elected members and mainly you the non-elected members, you have to be aware that the people have had enough of your cavalier ways of administrating, of your useless travelling, your useless studies, your useless meetings, your commissions made to reward yourselves mutually for your lies. Stop adding programs that people do not need and pay off the debts you contracted on the back of future generations. Imitate Boileau in his *Art Poétique* (Poetic Art, 1674) who suggested: "cross off, erase and often start over again". Would it not be time to retire these scientists, intellectuals and elected as well as non elected civil servants?

The Québec Liquor Board, a governmental company that has legal monopoly on alcohol distribution, was careful to eliminate all competition and especially to remind us hypocritically of the virtues of moderation. Alcohol is an overtaxed product, as was *La Gabelle* (40 different taxes on salt) in Europe, abolished in 1790. Theft is this 100 % of taxes added to the other taxes on certain alcohols, before the taxes on the value itself successively added by various levels of government as it is done with so many other products! It is not because these thefts or disguised taxes have been a part of the materialistic practices since decades that they are acceptable. A theft is a theft for everybody! Why are you surprised? Why be offended and make such a fuss when faced with citizens who make and/or sell alcohol or traffic their cigarettes? And yes, it is that same dehumanized materialistic culture that you practise: this culture that you received at the school of sciences without the humanities that you teach and that your peers chose to teach as "the ideal model of education" for training!

Working under the table, adulterating alcohol, smuggling cigarettes, boycotting the high price of fuel, criminalized street gangs, the culture of cannabis, etc. are only normal demoralized reactions from individuals against the very numerous abuses of their materialist administrations. This will continue as long as powers will continue to steal their citizens or to treat them as slaves who owe them obedience. **The best way of stop all these "crimes" would be to stop oppressing citizens and to decrease these exaggerated taxes. One day you will have to open your eyes...**

Is it in fact the role of governments to tax all the products used by people, to own gambling halls, to sell and control alcohol everywhere, to encourage people to hyper-consume in order to turn economy instead of trying to understand "why" chaos strikes everywhere in the world? These leaders have only immortalized their materialistic culture. Such hypocrisy on their part! The socialist powers of the Marxist communist countries had the same vocabulary empty of sense to explain their schemes. Trying to spread the idea that "your citizens are richer, that they travel more, than they live better and than they are best in the world" does not change a thing to your dishonesty. It is blackmail and the promotion of only material values. All this is eloquent proof that the materialistic culture was hammered into your heads by the education your predecessors decided to impose on you one day to the detriment of a humanistic education.

Your inhuman materialistic habits, Sir civil servants of the State, multiply the occasions to incite people to gamble, and contribute to dehumanize people and to accentuate their weaknesses. The hundreds of gambling houses legally established on our streets because they belong to the powers, have no other purpose than to encourage citizens to maintain this culture and to play their hard earned money. Families receiving social security spend, in these game arcades that their governments conveniently placed at every street corner, their allowance for survival which these same governments give them hypocritically to live on. The expression of this culture is seen also in the implementation of other places to spend their money: in bars, inns, restaurants, shopping centres, arcades, everywhere a gambling machine can be installed. These games bring in a lot of money to materialistic governments. *"Salesmen of Dreams or of Nightmares?"* was entitled an article in a newspaper one day...

The situation of compulsive gamblers and that of their families is alarming. According to statistics, pathological gambling makes for 2, 1 to 5 % of the adult population. Worse still, 4, 7 % of the young people in Québec are already gambling adepts. Compulsive players increase by 5,000 each year. Statistics indicate that 75 % of these individuals also suffer other types of dependences (alcohol, drugs, tobacco, etc.). In 1999, the office of the coroner related 31 suicides to gambling problems. An average of 436$ per person per year is spent on these games of hazard, out of a population of approximately seven million people.

The time when people in power gain the control on drugs may not be so far away. In Québec, we already think the name of *The Quebec Liquor Board* should be changed for "*The Quebec Liquor and Drug Board*". **These people will soon try to control the dozens of kinds of pills that young people use in their "Rave" parties or have the control on sex which still escapes them.** Nothing stops these scientists and technicians specifically placed in these posts for their natural capacities to manipulate the money. Their absence of humanity, their love of material values and their dehumanized greed well characterize these individuals. There are so many distressing situations demonstrating that powers are rotten to the marrow of their bones. They help create individual dependence which brought them in 1999, 1, 2 billion dollars. Now, believe it or not, these irresponsible individuals of gambling enterprises in Quebec push the indecency into planning a gambling house in Chinatown knowing that the Chinese are, for some reason, very susceptible to **becoming** addicted to gambling.

In *La Presse de Montréal* January 26, 2002, Foglia reported the words of a lady who wrote: "I worked for nearly a year as an "executive hostess" for the Casino. One of my tasks was to call players at home. **When a "high stake" player leaves the casino, his "executive host's task" is to make a follow-up call. I was afraid the wife might answer the phone on even children, I was afraid of a tragedy".** How can we describe such Machiavellian behaviour on the part of our social leaders? I hardly dare believe that such administrative decisions can be accepted by normal people!

There is even more: the leaders of the Québec Society of Gambling Clubs and Casinos have invented casino games, intended to entail the children from their young age to play "the casino" when they get older. The objectives address the performance and skills of these young people at "video" games that they master since childhood. These new games are modelled on habits formed while playing "Nintendo" or "Super Mario", for example. Who do you think are the most experimented in these games of "Nintendo" type games: the children naturally, and much more than the adults. **It is extremely odious on the part of these scientists and technicians, heads of family besides, to be as blatantly dehumanized and unconscious as to design games of chance that children will love to play once they reach adulthood, I mean those games of chance for which they have already acquired dexterity in their childhood with the "Nintendo" their parents gave them as presents.**

The dreadful Machiavellian tricks of these retarded leaders must urgently be modified. Somebody in these governments must have the guts to deactivate such infernal machinations, made to bring children to play casino games later on in life. I cannot conceive that there exist such materialistic spirits!

Greedy like no other, these criminals in power only give back a mere 3% of the gambling money, not out of the total expenses but of "the net earnings", for the treatment of individuals that became compulsive players. One again, the population will pay for the balance with still other taxes. Who can still convince us that such behaviour in our leaders is human? Here are other irrefutable proofs confirming the disembodied materialistic culture of these individuals educated in the rules of scientific materialism and socialism.

Numerous other domains eloquently demonstrate what degree of materialism people have reached. The decadence of worldwide sexual tourism, as Michel Houllebec described in *Plateforme* (Platform), *Flammarion 2002*, is one of the examples of the depravation of atheists' "human" nature in our modern societies. There are millions of *Sades* today in our dehumanized world. Houllebec's book and its depraved ending, much more than the racist comments for which he was quoted in a lawsuit, are examples of the outcome of these individuals fallen in the bottomless despair of materialistic nihilism.

The numerous financial dirty tricks done by dozens of American, Canadian or European company leaders (WorldCom, Anderson, etc.) embezzling their companies, forging reports on the stock-exchange markets for years, says a lot on the state of corruption, deterioration and materialism of these leaders. The too numerous embezzlements legalized by people in power in governmental ministries or by administrators of private companies are other examples of this.

G. The influence of scientific and social materialism on poverty, criminality, violence...

Can we establish a link between materialism and dehumanisation on one hand, and poverty and violence as well as the increase of shortcomings in the people of our occidental societies on the other hand?

Materialism and poverty

The increase in poverty is one of the immediate consequences of a deceived, disrupted, dehumanized society. Overexploited by powers and the dehumanized people managing companies, cartels controlling the prices of food around the world, manipulations on the price of petroleum, the price for conveniences, the necessary possessions for everyday life, the commodities such as coffee, wheat, meat, cereal, fruit, to name but a few, add their sad side effects on the financial imbalance between the poor and the rich. A drop in prices for coffee in a country where coffee is the main exportation product soon becomes a disaster that echoes at once on the economy of the country. Drought, heavy rains or any other imbalance of nature only add to the weight to their problems. Let us think of the impacts of drought on food prices in a country and consequently on the people of that country!

Within the framework of a broadcast on the reality of poverty of September 21, 2000, Jean Lapierre of CKAC (Radio/Montreal) mentioned that in a society as well off as Quebec, 350,000 students could not afford a good breakfast before leaving for school in the morning. This is more than one student out of four. On January 9, 2000 another broadcast reports that there is 1, 4 million Canadian children in this same situation. If statistics are right and we live in one of the favoured countries, can you imagine how the dozens of other less favoured countries of the planet manage to live?

A study done in the New York area showed that between 1979 and 1999, psychosocial problems in children had tripled, poverty had increased and one third of the school absenteeism was found in this less advantaged group of society. How can we understand that all this poverty is living alongside the promotion of scientific knowledge and the profusion of tangible assets?

In several other countries of the world families have more and more difficulty supporting the family burden. The weight of social constraints has become too heavy for a growing number of people. Citizens working to earn their daily bread are far from being able to discuss the rights of their children to education, to health care as well as to elementary social measures. Not only are the parents deprived, but the countries that tried to give themselves elementary social measures can not support them any more. How can they equitably share the possessions and the repercussions of globalization?

Without money and at the end of her rope, a Montreal mother lived in her car with her two school aged children. She was then arrested along with her children for breaking and entering. They needed the basics (dishes and pots and pans). Even though poverty has always existed, can we think in front of such poverty and misery, that powers and institutions that abound with numerous possessions have reached a point where they cannot even recognize that they are administering human beings?

Materialism and suicide

The waves of suicides among the young and the not so young are sweeping across the richer populations more than the poorer countries. Research has not yet enabled us to establish the real causes for this. Many experts underline that these disturbing social facts strangely arise at the same time that countries develop and that cries of appeal from the young progress. Several have left letters crying their refusal to live in their possibly rich but dehumanized societies where they don't belong. Rare are the studies, as far as I know, that mention the materialized customs spread by the educational system with science and technology without the humanities as one of the possible causes of these suicides. Speaking about the wave of suicides that swept over a small region of Quebec where seven young people between the ages 16 and 23 took their own lives between 1999 and 2000 a Montreal journalist observed that this phenomenon is generally worse in the poorer classes of the society.

On September 5, 2000, we could read in the newspaper part of a testimony of a "fed up" young teenage girl who had just committed suicide. In the note she left her parents, we could read: **"Life or rather today's society makes me sick. We do not live, we more or less survive and we are always fighting for I don't know what"**. The rest of the letter, it seems, was a charge against the powers and was not published. This type of testimony goes a long way in telling of the despair of the young people caught in the nets of their materialised societies. What state of despair has the atheistic education inculcated in the young!

One day, a 15 year old boy asked a former journalist of an "advice column on love and psychological problems", (Solange Harvey), why 15 to 20 % of today's young people consider suicide or suffer from severe psychological problems, while there are fewer teenagers than in the past. The close-to-70 year old journalist replied that "her generation had formed mentally strong but spiritually weak individuals. **A few decades ago, she wrote, parents were strict, discipline was exaggerated, religion was anything but flexible, and families of 10 or more children could more easily accept sacrifice. But people were good workers, resourceful, full of projects and dreams, and did not want their children to live the life they had lived.** Once they became parents, these "former children" relaxed the discipline, stopped making efforts, and bit by bit lost all sense of family. Then, there were divorces, heartbreak, children torn between their separated parents, etc. Brought up with many opportunities and without too much effort, children often reach adulthood lacking the necessary resources and incapable of renunciation. They become easily discouraged, drop out of school and later of society, incapable of finding solutions to their problems. (*Le Courrier de Solange, September 27, 2000, Journal de Montréal*).

We hope to have demonstrated that our Occidental societies are sick of this materialism and of the loss of their traditional and human values. They are also sick of their mentally strong but spiritually weak children they themselves educated. We could all tell hundreds of anecdotes demonstrating that the main problem is intimately linked to the lack of hope and to the lack of spirituality in people because they were educated and formed in the last decades with sciences, technologies and a lot of material values devoid of human feeling. Such disturbing facts abound in our societies. They are the prerogative of our dehumanized institutions and of the education dominated by the distribution of materialistic values taught and the rejection of human values. **Maybe we will never know the true role of the additional weight of the dictatorship of these mind-numbing technologies, the loss of individual rights and liberties and the multiple interventions of powers in the lives of these individuals. But who can convince us that all these things did not play a role?**

Powers should begin to think about the causes of this lack of humanity in people's actions. Can they? Scientists and intellectuals in power have been brought to the point where they are unable to recognize the existence of a certain form of humanity in people as one of their fundamental needs. We will have to realize that one day!

Abuses leading to violence

The animal instinct of "an eye for an eye, a tooth for a tooth" has reappeared in people because former acquired traditional human values have been replaced by material values. Lost in this materialism, people got bogged in the chains of dehumanisation. When human beings stray from the traditional values acquired with great difficulty during the course of history, materialism resurfaces, controls their behaviour and so they regress. Could aggression, violence, evasion and several other deviant human behaviours be symptoms of this absence of humanity in man?

"Violence is not a hereditary trait", wrote one day Robert H. DuRant, a researcher from Wake Forest University in the United States, "but a feature acquired through contact with violence" (*Journal of Paediatrics, November 2000*). The results of this study illustrate that violence is not an innate reflex but an acquired behaviour, suggesting the existence of a close link between the exposure of young teenagers to acts of violence and the temptation to use violence thereafter as a means of defence. The study also demonstrated that the chances for a given individual to become violent presents a certain degree of correlation with drugs, street gangs, smoking, being a male or depression symptoms. The strongest correlation seems to be linked to exposing the physical act of violence itself. **These same scientific data inversely prove that certain more human behaviours, opposed to violence such as religious practice, reduces the risk of developing violent behaviour.**

Defence mechanisms of individuals assaulted by their leaders

As materialistic systems go on spreading their tentacles on people, they also modify their ways of living and acting and the traditional human values begin to fade. People have dehumanized at the same time their actions materialised and their human values faded. Must we be surprised of in the place of the traditional order promised by science, it is crime that has grown, and governments legalize it, exploit it and monopolize what was bad in yesterday's societies? Plato wrote over 2000 years ago that in decadent States, laws multiply. This is why we observe more degradation in the States where human values do not exist, where chaos has settled, where people kill each other at a younger and younger age. Criminals, born and educated in the materialistic morality of our education system, feel justified in killing the very persons that chose to teach and propagate this materialism, such as people in power, elected or non elected. This is the case in Italy, in France, in Austria, in the United States, in Colombia as well as here in Canada, amongst other places.

How can a simple citizen warn his powers that they went too far? That such or such a law is inequitable for them? The acts of despair individuals express in those moments must be read by powers as cries of despair. The strongly armed anti-governmental American terrorist groups of which Thimoty McVeigh was at the same time hero and victim, the suicides of the Waco people in reaction to governmental violence, the actions of the Palestinians or those that sacrifice themselves by fire on the public square are examples of these gestures of despair on the part of the citizens. Is not the American Islamic terrorist Lindh, found and arrested in Afghanistan during the American war in 2002 and judged thereafter in the United States, another gesture of despair?

As I wrote these lines, a Radio-Canada journalist was trying to make governments understand the violence that justice arouses in men, when it does not treat men as it treats women in divorce cases regarding child custody. He went on a hunger strike for several days to try and draw the attention of powers, so that they would set up a crisis centre for men and for governments to establish a board of inquiry to study the fate of the men deprived of their children after a divorce. He shouted on the rooftops that society killed him by making public his personal difficulties with his ex-spouse.

The governments made discriminating laws against men to further protect women against men's violence. Some of these laws even created more domestic violence. It is understandable, though paradoxical, that laws can cause a greater intensity of violence in men against women. Why, for instance, following the simple complaint of a wife, can the police arrest the husband, tie him up, put him in handcuffs, beat him up and then imprison him?

The Québec Federation of Women acted with undue violence one day by condemning inequitably on the public place the candidate to a political party during a by-election, while no charges were laid against him. Why do not judges give the same rights of access to the children to both father and mother?

No one responsible for having these inhuman laws respected is shielded anymore from the reprisals of unsatisfied citizens, journalists, political men, policemen, prison guards and so on. All kinds of crimes, as imaginative as those in sensationalist films, continue everywhere on the planet because of this unfair constraint. Disorder is everywhere and no one seems to be able to put the finger on the causes of this chaos affecting people and societies. Is it surprising then that violence and crime explode in all forms in our dehumanized societies?

This is a worldwide phenomenon of generalized violence. It happens in countries where only the education of people with sciences and the ascent of material values count. Even though there are still no valid figures to confirm these phenomena, data demonstrate that violence is increasing all around the world. This phenomenon first drew attention in communist socialist countries where deaths by political murder could be counted by the dozens of millions. Today, the same materialized and dehumanized evil that affected these Marxist socialist countries touches the whole world, even capitalist or democratic countries. How can we not become conscious of the possible repercussions of an education system devoid of humanism, an education that only promotes material values and ignores the controls that assured traditional human values? These problems of course evolve in all countries, in parallel with the social and scientific growth of this materialism without human sense, without spirituality, religion or God. "They would not all die from it" said La Fontaine in his fable, "but all were affected"... Instead of constantly putting the blame on childhood problems, on poverty, on the poor neighbourhoods where people grew up, on bad-tempered parents or on sexual perversion, maybe researchers should think of including also in their hypotheses of research, the role played by the dehumanisation of people and society at the contact of materialistic values transported by the education system? **Poverty, childhood problems, difficult social circles, inadequate and non educated parents have always existed, without causing this "dreadful evil current of materialization" in people and their societies and without all the crime and violence today.**

The French government published in the beginning of 2001 alarming figures on crime in France. Over 3,771,849 cases of fraud were reported to the police in 2000 only. This represents an increase of 5, 72 % compared to 1999. Many people ask if there is a relation with the values of the socialist government in place, even if Lionel Jospin's socialist government seems to have made good social work and has tried to establish measures to counter this phenomenon. Crime shot up all the same during his mandate. Would it be possible that the material dehumanized values, transported by the socialist system, have something to do with it? If in 2000 there were 64 offences per 1,000 inhabitants, there were 168 per 1,000 in 2002! This delinquency is mostly economic and financial and does not seem to be connected to poverty. It represents 130 millions dollars or close to one quarter of the operational costs. Crime is intimately linked to anti-civil acts, formerly better controlled, such as the theft of bank cards, portable computers and cars or even setting fire. In the weeks preceding the presidential elections of 2001, the government felt the need to present a law on daily security to try and restrict the opportunities for the commerce of weapons, to fight against violence and to limit swindles on bank cards. The insecurity of the French also showed at the end of January, 2001, by a violent fight between rival bands in a shopping center near Paris. The generalized dehumanized materialistic habits, the social ideologies interested in groups rather than individuals, the frauds made by the ruling classes could not be foreign to this almost catastrophic increase of crime in this country.

In order to manage the rougher districts where violence exists to the extreme, the city of New York gave itself in 1993, enough policemen and a special budget of several billion dollars, intended to reach "zero tolerance" in violence. Blue civilian patrols pursue mercilessly breaches of public-spiritedness, as common as urinating against a tree, drinking from an unwrapped beer bottle, drawing graffiti, jumping on underground train rails, in order to add more weight to the safety measures of the city. In seven years, these measures managed to decrease by 57 % the number of delinquent acts. Would students and citizens disrespectful of social values not also benefit from lectures on public-spiritedness and of collective human notions in their capitalist education and materialism?

Materialism and violence in parents

Experts of all kinds are beginning to connect actual human deterioration with the disappearance of the "virile" domestic education that materialistic governments condemn with laws preventing parents from raising their children as they wish. Through training with the disembodied sciences of humanity, modern education is done without public-spiritedness, without religious rules, without morality other than the materialistic one and without human principles. **Should we be surprised then that young families and modern societies are flooded with a generalized "I couldn't care less" attitude.** We can observe that in countries practising since much longer this scientific social materialism, education and dehumanized moral customs are deficient and crime has reached unprecedented summits. Now that our societies are trained with these materialistic values, such customs are legion. This phenomenon seems to be more in parallel with the ascent of the scientific materialism in the educational system than with any other cause studied so far. The examples of materialized actions are numerous. Here are some of them.

In June of 2000, two students of a Russian aviation school having failed their exams killed their professor. The Russian mafia imposes its laws all over the country and is expanding here and there around the world. Salt Lake City's games in the United States were, it seems, marked by this Russian mafia. One Sunday of June, 2000, 44 women were assaulted in New York Central Park by young delinquents. A teenage girl was killed by two young racists of 14 and 15 years in Quebec. A man shot his wife with several bullets in a shelter for battered women. Patients in an emergency room in a Montreal hospital die because of administrative blunders when powers decree that the hospital budgets must not, by a law, show a deficit. This same government accepts, in spite of opposed medical advice, to pay 5 000, 00$ for mammary prostheses for a 15 year old girl who claims to need this for her mental health. When will they start paying for facial reconstruction, for thighs or anything else that children want? During this time, emergency rooms of hospitals must close due to lack of financial means and those that remain open know crises after crises. In a Québec town, two young people of 15 and 16 set fire to a secondary school, wreaked two elementary schools and then destroyed several school buses. Citizens united their voices to austerely condemn these two young birdbrains. But, two days after their appearance in court and over a million dollars of damages, they are set free, just like that! They were not, it seems, in their "normal" state and their upset parents promised to keep an eye on them. Hello justice! This is unforgivable. **Besides their very heavy tax burdens, society will once again pay for this as well as for compensating for serious domestic and social deficits and drug abuse by young delinquents that the "I couldn't care less" society contributed to educate!**

Another striking feature of these social materialistic societies is precisely to take the blame for their citizens' errors, to become responsible for their diseases and even for their health! **Such nobility for a society to take away all sense of responsibility from the individuals they have already reduced to infantile levels and to take the blame in their place!** Is it normal for high-ranking servants of the State to give themselves 20 % raises, while they confine other employees of the State to a slim 5 %? Exhausted governments submerged by mountains of deficits still give 40% tax credits on the salaries of each new employee of a private company specializing in the production of videos on violence! There is something quite distorted in these dehumanized governments, don't you agree? The left side of the brain does not always know what the right side is doing! The deceitful, hypocritical, and demonic behaviour of certain politicians during the 2000 Canadian election campaign was unbearable. Have we ever seen such hypocrisy and dehumanisation in people who had nevertheless run the country?

The rarity of traditional human values and the "spiritual and/or moral impoverishment" of the individuals are so blatant that the Pope denounced the generalization of this materialistic life in occidental societies. In his public message of January 1, 2001, the Pope was extremely preoccupied, as are a lot of people, by the materialism propagated everywhere with the techniques of information. **"Information technologies and their progression", he said, "behave as the shameless servants of a "servile conformism" in sciences by accepting, for example, cloning and the use of human embryos in research".**

Instead of mentioning scientific and social materialism and its materialistic values, the papal denunciation used more politically correct words, by talking about "models detached from their Christian origins, inspired by an approach marked by the secularism, atheism and the plans of a radical individualism". He added that this "phenomenon of technology of such great proportion is supported by powerful media campaigns, intended to propagate a way of life and the occidental points of view (materialistic) which damage other cultures and other civilizations".

All this materialism was perceived by a person who printed one day on the back door of an elementary school where the professors entered, this graffiti in five inch tall letters: "I am sick of living" (*Foglia, La Presse, January 26, 2002).*

Materialism and street gangs

The family is the individual's first gang. This is where the child learns the sense of the security, of belonging to the clan and mutual protection. In adolescence, the child meets other clans: the gang at school, the weekend gang, the hockey club gang, the shopping center gang, etc. Young people join gangs to fill their solitude at home or at school, to protect themselves from attacks from other gangs. According to experts, it is at this stage of their lives that young people learn to define their feeling of belonging, their security and their ways of living in order to be able to stick together in the face of danger, whatever the cost.

Today, this gang phenomenon has taken unexpected proportions for all sorts of reasons and in several countries: **the growing number of divorces, separations, the incapacity of children to adapt to new reconstituted families, the uprooting of immigrating families, children lost in these new social masses, new schools, new streets added to already built up areas, the contact with ethnic groups and different cultures.** The progressive loss of the traditional human values by people and societies is not foreign to this gang increase.

Wars between street gangs cause dozens of deaths every year in North America and in several countries of South America. Several individuals are victims of murder attempts. Hundreds of persons of different background belong to these street gangs in all the large cities. Police consider them more dangerous than criminalized bikers because the acts they commit are unpredictable and spontaneous. In spite of the arrests and charges brought against gang members and their accomplices, this phenomenon is still growing all over America and Europe. Ethnic gangs form spontaneously among immigrants in their new country.

In the year 2000, we observed a 50% proliferation of neo Nazi groups in the United States. Because the constitution of this country allows citizens freedom of expression and is part of the North American culture, individuals using Internet abuse the American Constitution to protect themselves from their hateful slogans against Jews and their ideas of superiority, for instance. By so abusing the privileges that the people of the country had given themselves, these violent individuals only manage to rob us of the few liberties we have left, to promote racial wars or proclaim insanities such as "Christianity was invented by the Jews as an instrument of destruction of the white race".

The Charters of Rights should be reviewed to specify that all visible religious signs (kippas, chadors, turbans and others such signs) should be worn at home, in places or worship and religious ceremonies, but not in public. These visible signs pointlessly stimulate the hatred of "sensitive" people when worn in public. Freedom of religion and especially the Charters of the societies of adoption should not permit to each group to flaunt in schools, in the streets or in public places their religious or racial symbols to encourage rivalry between the people. It should also not prevent the inhabitants of the country of adoption to stop expressing their own traditions as usual or infringe on their rights and thus aggravate traditional or ancestral rivalries among the people, so that non elected judges use this to decide who will wear what symbol of his religion in everyday life.

The newcomers should know that it is stressing enough for the societies of adoption to open up their schools to them, to teach them their language and to give them a thousand and one other considerations. Such warnings should be written black on white in the conditions of immigration.

Materialism and violence

Humanity has known, during the century of science and technology, unprecedented violence between individuals and/or races as never before in history. This violence marked the twentieth century with wars and the exterminations of people as never before. These acts are now recorded in the registers of History as the most bloody that humanity had ever seen. This century of science and technology has been marked with new types of violence between races, countries, as well as in families, schools, work places, streets, public transport, cars, governments, institutions, etc. **Is it only a coincidence that during this century of science and technology peoples suddenly lost all respect for one another?** The constant hatred that characterizes today's individuals appears to be the reflection of this materialism and dehumanisation pushed to the utmost. **Whether or not these acts were committed in the name of God or for scientific reasons, hatred always remains an act of pure barbarism.**

Torture, hatred, war and systematic exterminations spread like wildfire and unfortunately often in the name of "scientific logic and reason"... These acts of violence extended throughout the twentieth

century from South Africa (war of the Boers), by way of the Russian Revolution in 1917, the extermination of the Armenians in 1915, the systematic and rational extermination camps of Jews during the Second World War, until the other innumerable death camps in several countries, in Russia under Stalin, in China under Mao, in Cambodia with the Red Khmers, in Bosnia and in Rwanda to name but a few. Millions of individuals died during these numerous human conflicts. Their blood reddened the surface of the earth. Hundreds of millions of human beings were thus killed by other human beings for all sorts of motives, fanciful ideologies and rational logic as merciless as they were distorted and with the most sophisticated technologies in the history of humanity. **Never before in the history of man has a century been more murderous than this century which was the witness of the explosion of scientific knowledge.** Even medieval cruelties and murderous biblical wars remain weak in comparison with what happened during this century of "reason", of "positivism", of "rationalism" or of the systematic and dehumanized "materialism". The monumental human slaughters of the twentieth century spread on all the surface of the world and will forever mark the History books!

Materialism and criminology

For how long must society tolerate that organized powers muzzle judges, restrain the actions of governments, deputies and ministers or execute journalists? Criminalized powers have replaced the bit of democracy that remained in certain people. **How can we accept that a criminal act can be debated in the name of the liberties defended in the Charter of Rights and Liberties? We must stop warped lawyers from twisting the words of legislators.** Those are aberrations worthy of the rational materialism and its values. A crime is a crime – whoever does it – period! The powers must rapidly confirm that man himself is responsible for his actions and his diseases, not society! These socialistic ideologies, stemming from the scientific and social materialism without humanity, were never confirmed by scientific data and are only the wild imaginings of the perverted minds of certain perverse political powers.

Societies, justice, police, the media and governments have tried to control the explosion of violence in criminal groups by all sorts of means that proved to be more or less effective. **We continue to observe in our "so-called civilized world", inhuman criminals attempting to monopolize powers in order to take justice into their own hands.** They kill politicians who oppose them, policemen who refuse to go along with them or judges who want justice to prevail. These dehumanized criminal gangs act as if they were another government in the State. They crawl into the organs of powers through electoral coffers, through the corruption of authorities in place or by direct infiltration. Other criminal accomplices attempt to influence governments and steer them by the power of money. **It is so easy for criminals to jump into the multiple administrative workings of the "state entropies"**, said Mario Roy in an essay in *La Presse de Montréal* on November 2, 2002.

We learned in this same article that "without counting municipal and school administrations, **Quebeckers are now assisted, advised and restrained by no less than 390 ministries, agencies, committees, secretariats, offices, state control, councils, societies or specialized courts answerable to these two levels of government!** The same newspaper has calculated" continued Roy "that this bureaucracy produced 10,000 pages of regulations per year." Is this abuse or useless over-administration? "Impossible to slow down the movement", added the article. **By trying one day to eliminate half of the 204 organizations answering to this government, the Facal plan (a Minister of this socialistic government) found himself four years later with 205...** Must we blame these sound-of-mind individuals for trying to cut down on this dreadful "state entropy" that the civil servants of State and Labour Unions have weaved liked spiders? Moreover, there are in Quebec 10 civil servants by 1000 citizens compared with 4 per 1000 individuals in the nearby province of Ontario! Is it not proof of hyper administration?

In an interview (*April, 2000*) by Stephane Bureau, a reporter – interviewer from Radio-Canada, the Jewish humanist philosopher and Nobel Prize winner Elie Wisel asserted that in front of all the horrors of the world, he keeps a blissful optimism that I have a lot of difficulty sharing. Mr. Wisel thinks that the positive interventions of groups formed to fight all the ignominies and the free violence of the people of the twentieth century will finally succeed in stopping these ignominies. But, they are so few and scattered the positive demonstrations of these groups of individuals against violence that they look like the anaemic interventions of countries against greenhouse gases. **They are drops of water in the ocean compared to the number of problems and interventions that it would now be advisable to make to only begin to contain these pitiful side effects on the environment and the individuals.** Countries have been trying for 15 years to settle the Kyoto agreements and they are still not settled today. To think that over 25 international pacts such as the one of Kyoto on pollution would be needed today only to begin confining these other striking problems! And time is running short because in one generation, as we will further see (cf. Chapter 8), damages on nature and human being could indeed have already reached a point of no return. In spite of all these dangers the destructive ideologies of our socialistic and materialistic education system are always in full effervescence and continue destroy people and societies.

Materialism, violence and overpopulation

Violence is in progress all around the world in families, schools, on roads, cities, in sports, everywhere, etc. We just have to look to see. Could all this violence be connected to the animal phenomenon of self-destruction, well known where there is an overpopulation of animals in a restricted territory?

Let us further look at what is happening around us. Never during the history of humanity were there so many people on the surface of the earth. The first billion was reached at the beginning of eighteenth century. The earth now counts more than six billion individuals. Since two centuries, men have multiplied by six. It is what one calls a geometrical progress. Technology has multiplied at the same time by geometrical factors that I don't know. But during that period, the humanity of man declined concurrently by geometrical factors.

Whatever the cause, we must face the fact that humanoids have generally regressed to animal level. The attitude of the people in the American show *Survivor* during the summer of 2000 is a striking example of this. Can we suppose that the reasons we find so many beached whales, dolphins or sharks are linked to the pollution of sea by man? There are for instance, according to the international movement for nature, over 300,000 whales that die every year because they are caught in huge fishing nets and as mention above, over 60% of the ocean fish have disappeared in the last half century.

Materiality and other activators of violence

Materialisation of people and societies is responsible for the violence we see everywhere, as well as the other types of violence that disrupt man. Let us consider the violence of man against man, the violence of unemployment, the violence of human cannibalism, the violence of prisons that cram together the good and the bad guys of our societies, the violence of this 14 year old teenager who tried to kill and burn his young 7 year old sister, the violence of armed teenagers who kill their friends for no good reason, the violence of the men and women who dominate and exploit the sick and the elderly as well as other disabled persons in centres for persons in loss of autonomy, the violence of motorists who kill the driver of the other car simply because he dared block the road, the growing violence of the modern, cruel, dehumanized man who considers himself more civilized than his ancestors, the violence of the materialistic civilization, the violence of poverty that societies deliberately maintain, the violence of the inhumanity of materialistic ideologies, the violence of the cruelty that appears to be a process of civilization, the violence of all these primitive beings which from the height of their powers activate wars, the violence of ethnic purges or the use of the most sophisticated technological applications to perfect their cruelty, etc., etc. There is also all this more subtle violence of the powers taking legally, from the pockets of taxpayers the "sweat" of their work, the violence aroused by the violation rapes of personal freedoms and the thefts of the individuals' rights and liberties and what else? Disguised taxes by governments, school boards, useless supra-municipal structures, city groupings, unions, etc., the injustice and abuse of power by police against drug-addicts, prostitutes, beggars, citizens they have the duty to protect or justice with dishonest judges are still other acts of materialization which arouse violence everywhere.

Never had animal been as cruel towards his sort as the intelligent, materialized human being who acts only with the intelligence of his "reason", never taking into account the intelligence of the heart. The intelligence of the spirit tries jealously to by-pass the imperturbable laws of nature and only listens to the call of the linear logic adopted by scientists and the intellectuals and that they consider superior to instinct!

The materialism of violence in children and teenagers

The repercussions of these dehumanised actions are found everywhere. Within eighteen months, between 1998 and 1999, the United States registered eight school shootings by young individuals aged 11 to 17 years. The young people wholeheartedly kill fellow students and professors for childish motives. These children from good families use technologies they understand more or less to commit unthinkable crimes. Neither the school, nor the already materialized parents taught them to respect other traditional and human values. The only training the young receive is mostly if not only passed on to them by their materialistic education, schools of science, television, films and material values, etc.! Parents cannot seem to be able to control this. People in today's societies are not capable of renunciation like their predecessors. In laic schools, there is less and less, not to say any religious practice and no one knows anymore the meaning of the word "sacrifice". The people of "reason" claimed one day the only true values that man had to learn and respect were the material values justified by their senses!

These materialized individuals are younger than ever. In February 2000 in the United States, a 6 year old young boy killed a 6 year old girl with whom he had a fight the previous day. What principles had the laicised parents of these young people inculcated in them for them to act this way? As I wrote these lines in June 2000, <u>an 11 year old young boy</u> hung himself in the garage near his parents' house. Two 15 year old teenagers killed a 15 year old friend for racism and a garage attendant was shot down behind a counter. **What edifying social values the school of sciences, disconnected from traditional values, sows in the young of our societies!**

People so formed will kill for a few dollars or simply because they don't like the face of the other person. They kill to show that they are adults, that they are powerful or for trinkets. **Are they normal, abnormal, sick, mad or the fruit of atheism in this inhuman, worthless materialistic education? It is rather terrible to notice that every hour in the United States, 13 minors are killed with a weapon.** That makes a lot a day… and a year…doesn't it? (*Newsflash, Radio-Canada, February 29, 2000).* South Africa had to deplore 21,108 killings in 2001, compared with 16,110 in the United State only, (*Paul Von Eeden, Kitea, December 19th, 2003*). **In Brazil, there are also 13 killings an hour as in the USA... It is a generalised phenomenon!**

This phenomenon of materialization is international. In June 2000, the weekly newspaper *Veja* reported in an article entitled *"Au secours!"* (Help!), **that in Brazil as well as in the United States, there are 40,000 murders a year, which is more than the total of adult murders in nine large Occidental countries.**

Teachers are in danger in their own schools. Pupils practising this materialistic morality take justice more and more into their own hands. The teaching profession now comprises real dangers in this materialized age without human conscience. Professors are often wrongly and unjustly accused of having beaten or sexually assaulted young children, others of having committed actions of sexual nature on infants, because parents thought they had decoded the rudiments of their child's first words... Because badly formed social workers and psychologists "asserted" to the judges that they could scientifically decipher and interpret the language of toddlers from one to three years of age! The lives of these accused persons are thus forever broken, even though nothing really happened, even though there is no culprit and only the invention of babies in diapers, manipulated by imbecile incompetent technicians in need of glory. In causes called "sexual", parents have all the powers until proven otherwise. It is law reversed. "Justice" still plays here an unacceptable role. Young criminals receive mild punishments which only encourage them to continue their unbecoming actions rather than to correct them.

Materialized parents are too soft in front of their children and will be blamed by their children later on. **"Shrink" literature, scientifically unproven and written by pseudo scientists after the war, taught them to never correct their babies and their teenagers, to let them do what they please... Supposedly intellectual laws, based on "reason only" have sanctioned this unhealthy education of the children and the uncertain values of their parents.** Disorientated, these parents are afraid to act and to cry their outrage at not having the right to raise their children their own way, to shout on rooftops what their societies are doing wrong because the "shrinks" close to power can easily accuse them of ill-treatment. Always the opportunist, these powers take advantage of this to take away the parents' rights and to raise the children in their place in order to protect them! Come on! It was in order to seize the parental powers that they were missing in order to make them more quickly into submissive and less "anti-establishment" slaves! It is not surprising that children educated this way are dehumanized to the point of killing so recklessly parents, colleagues, professors, everyone and anyone including themselves: it is the rule of this materialistic morality that legally guide young people, parents and even the powers towards atheism.

History teaches us that the individuals formed with sciences and material values are hardly aware that there are also human, family and social traditional values and other methods of educating people. And so, people act as if everything were permitted. Psychiatrists, authors of these pseudo scientifically unproven methods of teaching, admitted their errors years after having spread their teachings. This is not a new phenomenon!

Stemming from this materialized education, the young of these societies now believe that they have even the right to physically correct their parents. To the question: "What should we do with parents who drink, who drive while under the influence of alcohol, what should we do with disillusioned professors who do not take care of their pupils and\ or manhandle them?", the children answer "Beat them up!" Some even claim this right in front of the courts and Parliament. What a beautiful and tangible side effect of this materialistic education!

The explanation of the psychologist

Leslie Charles, an American psychologist, believes that the phenomenon of aggressiveness in the people of modern societies is a disease which is born, develops, grows, explodes in each one of us and is transferred to others. The objectives of her research were to try and understand the causes of mood swings, overreaction, stress, aggressiveness, frustration and of all this despair translated onto verbal violence, violence at the wheel of a car, violence against the absurd imposition of inhuman laws by governments, violence against schools, city streets, families, between couples, in separations, etc. The psychologist's research demonstrates that this evil comes from the fact that scientists, governments, intervening parties and civil servants ask more of man than what they can give. A study on modern culture, published in an article entitled "Why is everyone so cranky?" concludes that this evil comes from the fact that science and technology promised more that they could deliver.

If the hypothesis of this psychologist is true, we will have to read that it is "the constant expectation of scientific and technological realizations that creates stress in our daily lives, as well as frustration, despair, aggressiveness and disappointed hopes".

H. Attempts at restraining violence

In tying to control violence and aggressiveness in materialistic societies suffocating under the dehumanization and the practice of material values without humanity, researchers can only offer other material scientific experiments. These researchers are already so materialized that they can only advance other material solutions to try and stop the problems of their societies. And so they can only offer other insignificant materialistic techniques and solutions of evasion, solutions of comfort and "restaurant" solutions...

Material techniques of evasion

In response to the stress of daily life, to violence in consumer societies armed with their materialistic "credo", researchers can only offer what they are capable of offering: other material solutions. Conflicts will continue to progress because man strives, consciously or not, for something bigger, to parody *Blondel* in his "*Apologétique du seuil*" (Apologetic of the Threshold). Man always strives for something bigger that science and technology cannot offer and the solutions proposed by these types of research are all of material order and powerless to relieve the dreadful evil of materialism in today's societies.

The "comfort" techniques

Materialistic ideologists believe, in accordance with their training, that by increasing material comfort they will ease these problems. This is to know very little about human nature! How can a spacious home with gigantic bathrooms, majestic dining rooms, large areas for television and home cinema decrease the need of humanity that lies in the deepest part of man? The accumulation of material assets cannot and will never be able to stop the materialism that causes violence, aggressiveness and even less allow individuals to express their humanity. History is rich of such examples. Neither secondary country homes acquired to avoid the noise of hectic cities, nor journeys to insure relaxation, or even the comparison with other people's ways of life will ever replace the quest for the "infinite" etched in the heart of every man, whether he feels it or not. If it had been possible, the young American Lindh who became an Islamist and was found fighting with the Afghans against the Americans would have never been a terrorist. It should be possible to understand that when a young terrorist sacrifices himself by blowing himself up to a certain death, it is that he is ready to die and that he aspires for something much greater than material life.

The "restaurant" techniques

No marble restaurant looking like a royal palace, no gargantuan meal or large modern shopping center will ever be able to satisfy the deep immaterial aspirations of human beings.

The "shopping" techniques

Imagine the degree of materialism of these marketing researchers who go as far as trying to create "the shopping solution" to palliate for stress and violence. In an article entitled "*Je consume, donc je suis*" ("I consume, therefore I am") in *Chatelaine* (magazine) of June 2000, Nathalie Collard talked about shopping being a source of personal development. "Shopping could also be a solution against stress", say the big shots of advertisement. "We go shopping and we consume", according to these people, "to pass the time, to relax and to evacuate our frustrations. People shop to reward themselves and to feel good." Such phenomenal and satisfactory aspiration for the human soul! We live in societies where flea markets have replaced the Sunday mass.

Nevertheless, the marketing industry claims that it has well understood people's needs and goes as far as claiming "that we go shopping to satisfy a need and pass the time". **"Shopping has become a gesture of emancipation in women"** wrote Undrehill in an article entitled "*Why we Buy: The Science of Shopping*". "A "cult book" for the publicists", said the article in *Chatelaine*, "it is that humanity has lost much more feathers than we thought"! Man is more and more conscious "that he is a victim who consumes to reach an objective, that of consuming". If somebody dares to claim that this behaviour is not a materialistic one which leads people to nihilism or to suicide, then he should redo his classes.

Techniques "to change the vocabulary"

The masters teaching scientific and social materialism and who adapt it to all spheres of human activity have only created verbal acrobatics which create inconceivable intellectual lucubration! These powers that change their vocabulary for another one that only they can understand, who are they trying to persuade that they are effective, good and excellent? **This is what happened in antique Babel!** History has shown us the result of this. Now, to solve the problems which they caused themselves, scientists and technicians feel the need to invent new vocabularies or to change the meaning of certain words to give them a different meaning. **These vocabularies are so numerous today in science and in politics that neither scientists nor politicians can understand them.** They create new "Towers of Babel" with what they call "politically correct" expressions. If powers have always loved slogans, it is that they see in them a simplistic method of teaching how to think!

Problems became envenomed when, down from their powers, governmental scientists began to distinguish themselves from the "people below" by modifying their vocabulary. So, since a poor person became a "deprived" person, a patient a "beneficiary", a reform school a "youth center", or yet a "delinquent centre" and so on… Nobody knows what anybody else is talking about anymore, and so problems increase. Did civil servants really think that they would solve the problems they had created by simply changing the words of their vocabulary? How childish! "The ultimate hope of power", wrote Mario Roy in *La Presse de Montréal* on November 16, 2002, **"is placed in the magic power of communication which leads to the paralysis of action and to the morbid reproduction of words and the inefficient build-up of red tape".**

Individuals and materialized societies are thus subjected to avalanches of words that no one understands. No one can speak directly to these Sires, elected by power, without passing by the intermediary of "experts" (houses of consultation) that have learned the particular vocabulary of these powerful persons. In certain circles, it is to talk for nothing, in others to talk so that only the initiated individuals will understand. Even individuals who work in related domains do not even understand each other. Certain powers have managed to build themselves private niches to reign as masters. We have seen the disaster on the financial plan when the American Stock Exchanges crashed in 2002 because the investors could not understand anymore the vocabulary used in the financial reports of the companies. They tried to distinguish themselves form the others by using a particular vocabulary, book-keeping systems that only themselves claimed to understand and "thought" ideologies extracted from their rotten reasoning.

Are there human solutions?

We shall try, in Chapter 9, to enumerate some elements of solutions. There are after all very few advanced human solutions that I know of to defuse the violence and the aggressiveness that took hold of the people. **So far, very little individuals will acknowledge that it is the dehumanizing materialism and the method of education that is at the origin of these numerous adversities.**

In certain Québec schools, conflicts are so numerous that parents have tried to establish programs of resolution of conflicts, by having the young meditate. A group called *"Vers le pacifique"* (Towards Peace) is a community organisation created in elementary and secondary schools of Quebec. Their objectives are to prevent violence or quibbling among pupils. Human processes such as empathy, communication, respect, listening and independence are used as tools. Quibbling often degenerates into battles for the boys and of verbal bickering in the girls. In 1999 in a Montreal school, young twelve year olds managed to defuse 420 quarrels among pupils, some of which could have turned quite bad.

I. Conclusion

Somewhere something changed during the century of scientific reasoning. People have not only uprooted themselves from nature and from the earth that always fed them, but they materialized and dehumanized at the contact of the materialistic knowledge brought by science, technology and the multiple material productions and applications surrounding them. **By teaching only the sciences and the technologies to the detriment of the values taught with the humanities, art, philosophy, or a general training of the individuals, powers strongly favoured the implementation of materialism in the people and their societies.** These changes are greatly responsible for this plague of dehumanisation which marks today the people and creates chaos everywhere across the planet.

Not only do people lose contact with mother-nature and destroy their environment, but during this materialistic revolution of sciences and technologies, they have lost their traditional, family, social, religious and human values, and so deteriorated their humanity. No, the values of our world are not what they once were. The traditional values of yesterday have been replaced by the material values which are disconnected from all that made man a human being.

Intellectuals and scientists must carry the burden of all these adversities. They are the ones who seized the powers, formerly possessed by religion and it is also they, once in command, that formed the people with their education and passed on their material values, leaving aside any lessons of humanism, of art or of non material philosophy. For decades now scientists and intellectuals have been raised to the heads of governments and impose on their citizens their teachings and their ideologies. They have even exported and propagated their scientific, technological and political knowledge to the four corners of the world, while imposing on the people their new materialistic and social conception of the world. They even gave themselves the right to remove the humanities, the arts and the traditional values of education and oblige professors to teach their materialistic and social ideologies giving priority to the collective rights by eliminating the individual rights. And so the people lost the power to think by themselves and to go where they want. Their governments took advantage of this to better subject them to their powers and to infantilize them.

Constantly spellbound by sciences, technologies and the irresistible charm of the ease of material life, materially muted people sank even more deeply into this dehumanization. Knowledge acquired on material domains allowed the leaders to raise themselves as uncontested leaders in government headquarters. It is from there that they replaced the humanities and the traditional human values by their training in material values, thus creating a void of humanity in people. The abolition of general forming, traditional human values, the sacred and finally of all that could resemble anything human, ended up dehumanizing the people and undermining their last fibres of humanity. This void is wider and wider in people educated at this school of materialism. It has been aggravated with the rejection of human values. **<u>Today, the poison of materialism is felt in</u>**

all the activities and the professions of the people and perspires in all their acts and gestures.

How did we manage to reach such a point without realizing it and without reacting? Will we still be able to make the right choices before the night comes and swallows us up?

CHAPTER 6

How do materialized people manage to live in their soulless world?

Introduction

We now know:

That scientific knowledge and technological applications have greatly transformed people's material life while paradoxically inducing harmful side effects which threaten the environment, life on earth and the humanity of man;

That technology has the unpleasant intrinsic power to impose its rhythm and its material rules on those using it;

That by using only reason without human attributes, as well as science and its method of research excessively useful for the study of the material things of the universe, scientists and intellectuals have acquired powerful powers of domination on their fellow men and placed themselves at the head of governments;

That after having replaced the traditional humanist forming methods with education through science, technology and their material values, **these powers deliberately rejected the humanities and the traditional human values found in traditional training, as well as all that was non scientific and had only human values to their eyes;**

That scientists and intellectuals went as far as proclaiming the supremacy of the material values of the "reason" over the traditional, human, family or religious values under the pretext that they could not study them with their method of research because they were not perceptible by their primitive senses;

That the mixture of science, technological applications, material knowledge and especially the ideologies of the <u>Marxist scientific and social materialistic system</u>, added to the absence of all true human notions in education, have shifted people to a materialism where no trace of humanity can be found;

That armed with these powers, **the sorcerer's apprentices in charge of governments stripped bit by bit the citizens of their civil and individual rights and liberties in order to strengthen their powers and to become the uncontested masters of the governments;**

That these people administer institutions and their citizens by using only material values deprived of any sense of humanity and administer them as if they were interchangeable machine parts;

That these scientific and intellectual powers enslave the citizens in deceitful social ideologies never proven with sound scientific data, and <u>in shameless slavery by subjecting them to inhuman social and collective laws which take precedence over civil and individual laws</u>;

That after a few decades of this education with only science and the propagation of the supremacy of the "reason", people and societies are affected by the dreadful evil of materialism which is found in their behaviour, their speech, their ways of life, their professions, their occupations and their every move; "although all are not affected with this evil, all will suffer".

That this evil is found in the culture of a materialistic society of non-respect of the values of others and a rejection of the traditional human values, creating in people this lack of humanity which will mark most of their actions;

That these changes of dehumanized human behaviour are associated to the materialism of science, technology and the material values dictated by their ways of thinking and their materialist reasoning;

That, according to the affected individuals, this generalized materialistic evil is revealed by a total indifference of the people towards their institutions and their politicians, as well as by abuses in work, alcohol, drugs, gambling, dropping out of school and society, suicides, crime, gang forming and the exacerbation of the feeling of implicit violence which characterizes today's occidental societies, etc.

Let us now try to see how the people and the powers of our societies, affected by this evil, live this materialism and this lack of humanity in their everyday life.

A. Let us examine how the people and the powers of our societies live this materialism and this lack of humanity in their everyday life.

While weaving the historical background of the social changes marking the twentieth century in the walk towards materialization, people have used all the knowledge acquired with science to set up the technologies of the industrial revolution. Susceptible of facilitating everyday life, this new found knowledge is immediately used as the solution to the physical necessities of the individuals and their societies. **People use this materialism that they identify to happiness!** Since decades, they have been feeding on these tangible assets with such bulimia that they never took the time to think of integrating this knowledge to improve their humanity. On the contrary, their materialistic education helping, they exchange their traditional human values for values almost uniquely material. The fact that their dehumanized traits determine their behaviour is absolutely normal for them.

As they have done with education, scientists and intellectuals in power try to impose the hegemony of their "scientific reasoning system" to all societies of the planet, as religion formerly did. The absence of human values in their ways of being removes all barriers of ethics, maintained so far by religion, humanist philosophies, the arts, the difference in cultures and human principles. Watching the way people live is enough to see the expression of this growing materialism in their every move and even more in those that manage the powers.

In these societies transformed by scientific and social materialism, we see in their history more crimes than ever before. Slaughters have become common currency everywhere in the streets, at school and/or at home. For instance, one person is killed every 13 minutes by a weapon in the United States and in Brazil, as much per capita in South Africa and a little less in the other occidental countries. Wars between countries transformed by this education continue across the world. Between 100 and 168 delinquent acts per 1,000 individuals are committed each year in France only, depending on the cities where the studies were done. In this country where socialism reigns since two decades, delinquent acts have increased by 16 % during the last five years, (between 1998 and 2002). Prisons have more inmates than they can hold and social charges are increasing...

Day after day, the media, the newspapers, radio, television, the Internet of the Occidental world overflow on the subject of war between individuals, families or clans. Journalistic current events abound with stories of hatred, extermination of people, quarrels within couples or with friends ending in tragedy and fights opposing employees and employers, riots, suicides, power fights between the oppressed and the leaders, between governments and trade-union groups, between governments and the people they manage.

There is so much negativism reported and repeated non-stop on all sides by written or broadcast media that try to widen their hegemony and their materialistic ideologies, that the people have come to think that the modern media techniques have been invented to display and to repeat the horrors of this world of ours. A media that would not repeat the same horrors several times would not be able to survive.

Formerly considered as "the sacred duty to inform", journalism and journalists have materialized to such a point that it is difficult to recognize them. Like their fellow men, they write or diffuse for reasons of "highest rating". The paparazzi scrutinize the private lives of people who make the headlines. No well known personality can make a single move without being invaded by a bunch of tabloid photographers, equipped with the latest technologies enabling to photograph a person from kilometres away and to immediately transfer these pictures across the world via the Internet. Equipped with performing devices, these "wolves" of journalism photograph everything without scruples. **They try to seize the most intimate movements of their victims with unprecedented immodesty. They appeal to the right of the community to information, without taking into account the individual rights of the persons they pursue as in a hunting game.**

People are thirsty for horrors. They are obsessed with them. And so, newspapers and other media, property of large impersonal companies, give them all this literature almost more pornographic than real pornography. Horror films succeed one another and people drink them in as if they take pleasure in all these blood baths. Internet and other computerized media propagate news immediately via satellites right across the world. **People are obsessed and hungry for the misfortune of others as if it relieved their own miseries. In the midst of the horrors and the atrocities of September 11, 2001, American television even showed the delighted sinister looks on the faces of the enemy!**

Horror delights and saddens at the same time. Just look at the frequent shows on "laughter" using scapegoats to arouse it. To display on the public place the most personal gestures to promote laughter often exceeds the borders of decency. Human defects are displayed without humanity on the public square. **Reputations are destroyed, just to make fools laugh.** The technology of laughter aims mercilessly at people who are at their lowest and activates the greed of the population eager for sensations. Mockeries on one person's wig and the other one's crooked mouth, the stupidity, the lobotomy, all this is low, disabused, neurotic, cheap and well paid on top (*La Presse de Montréal, March 18, 2000*). **"These broadcasts are looked at by young people already known for their great "materialistic" culture, their lack of historic knowledge, their lack of respect and of spirituality in these times of human rarefaction".**

It is more and more difficult today to teach beauty and art and to show people the virgin landscapes that man has not yet massacred with his machines to destroy the world. **And so, we show them the horrors of human decadence. People love horror.** Look at the current films, at the television shows, at the newspapers, at society, etc. The Council of radio and telecommunications understands that people prefer "that"; so it gives them "that": acts of violence, slaughters by shovels, horror films, sex, acts made by unhinged people and all sorts of unbecoming acts that come to mind to a producer as disordered and decadent as the film. Art, public-spiritedness, empathy, love, kindness, respect are rarely a part of the values propagated by the education system and even less by the social powers and the dehumanized institutions of the individuals who manage them. Education through art, the humanities or the study of various cultures were put aside to leave the place for the propagation of the sole materialistic culture and its values.

B. Materialism has destroyed the human being

The progression of attacks between people is common in all the consumer societies that one day exchanged their human values for material ones. **Proof of this is that violence is seen everywhere in these societies: on television that shows dozens of violent acts each day to children, in horror films, in the written or spoken media, in families, in neighbourhoods, in schools, in public transports, in city streets, etc.** The wealth of our societies in no way prevents the people drowning in material assets to hate and kill each other. **It even accelerates the human heinous tendencies that bloom in the midst of all this material wealth.**

Violence has been transformed into a "normal" way of reacting and of living in our current societies. **The individuals in these societies are more and more denatured and inclined towards atheism at a very young age.** As proof, in November of 1999 in Germany, a teacher is killed by a 15 year old young boy who bet that he would kill her simply because he did not like her. In the United States, we can no longer count the deaths caused by young people almost "still in diapers". An American teenager living with his mother made about twenty craft bombs during the summer of 2000, leaving them in his mother's living room, **without his unsuspecting mother even knowing that she was living with a murderer. How many mothers and fathers taught the children they had once held in their arms, the hatred of others or the false religious principles they believed to be the absolute truth?** In March of 2000, a young American boy of 6 or 7 years killed a 6 year old girl with whom he had had a fight the day before. A few days later, a seven year old boy killed a five year old girl, just like that. In England, two young people of 10 or 11 years enjoyed torturing to death a 5 year old child. A German student who failed his exams killed 13 teachers and his former secondary schoolmates. Murders and suicides are prepared with "recipes" and with propagated domestic violence. **How can we explain otherwise than by an absence of humanity and of human value these gratuitous killings by children who do not yet know how to read?** Who allows children to witness dozens of acts of violence each day on television and at home, when it is not through the authorities in place, the owners of large communication companies or their own materialized parents who know and practise only this materialistic morality? All these inhuman acts have nothing to do with the poverty behind which societies hide to explain this march towards decadence.

These phenomena of human decadence are generalized in the Occidental countries. We can count a dozens gratuitous murders a year solely in this big city of ours. We need to see these acts of violence as witnesses to this absence of human morality, and the spreading of the materialized values in individuals educated with this social materialism propagated through the forming without the humanities. We kill each other at the wheel of the car if we are unhappy with the other one's driving. Taxi drivers are killed by people who want their few miserable dollars to buy drugs, to smoke a joint, to "sniff" their dose of coke or to inject their drugs of happiness...

Contact sports are more and more violent. Not only do players without human values express their dissatisfaction by wounding their adversaries, but we even saw on television the coaches bludgeoning crowds with hockey sticks because they were dissatisfied. Wounds in the young and the not so young practicing contact sports have increased at a stunning rhythm. We have also witnessed parents of players molesting officials, persons in the audience and a father even killed the referee who had punished his young son during a hockey game.

Other changes can also be linked to this materialistic fashion which foregoes all moral restraints to express its lack of humanity. I would not be in the least bit surprised if researchers demonstrated one day that these changes in behaviour, that some attribute to poverty and that abound in our ways of living since decades, are the expression of this materialistic culture which is also seen in the wearing of miniskirts, bikinis, monokinis, jeans, drug use, hippy movements, rock music, feminism, abortion or the so numerous divorces, etc.

How can it be otherwise? "Shrinks" of all sorts have been teaching parents for decades to let their young children express all their desires since childhood without reprimanding them. It was the lucubration of pseudo university scientists and intellectuals that once again were not at all based on any scientific experiment. **Parents of baby-boomers (born between 1945 and 1965), more well-off than their own parents, were forced to give their children all that they wanted to be forgiven for their faults and to be loved.** The young people educated this way have become the adults who continue to satisfy all their desires without depriving themselves of anything! **Are not the graffitists on the walls of cities and houses of our occidental world alarming examples of this education of generalized "laissez-faire policy" or of this morality without humanity and public-spiritedness guiding the people today?** Teenage games where they bet they are going to kill the first person turning the street corner, or individuals who kill for pure pleasure, without any motive, are eloquent of this type of amoral education given to the young people of today.

This materialized education eventually persuaded the already infantilized parents that they were incapable of educating their own children and that they must inevitably leave these privileges to the powers of the States. Imagine the deception in the face of this propaganda by the political powers as well as by the communication powers! **It is now impossible for parents and even less for teachers to correct these children who were never refused anything, nor to try to teach them the rules of family life and life in society, without groups of socialist activists publicly protesting to give them good conscience.** These good people incite governments to proclaim collective laws and to remind parents and professors who would like to make their children respect the simple rules of propriety, that they are responsible for their forming. These powers have voted collective laws forbidding parents to raise their children according to their own principles. This is another major example of a theft of the parents' individual rights and liberties by their socialistic governmental leaders. Judges were recently obliged to advise these powers that they had gone too far and they gave back to parents the right to correct their teenager protesters. But, adolescence is a bit late to start raising a child!

The "shrink" scientists who had these laws voted have never demonstrated the scientific truth of their theories on education before confirming them with laws. But their colleagues in power nevertheless imposed and sanctioned them as being scientific truths! Since they have been managing education, scientific and intellectual powers have also taught their materialistic values as the truth and deliberately neglected to teach children the traditional and human values. Eloquent proofs show that this system of forming put up front to educate people was wrong, because the results show that this system has only contributed to forming good scientists and valorous technicians, while at the same forming atheistic delinquents without any notion of humanity.

Humanistic traditional values were banished from schools and from programming in the name of a standardized secularization, and eventually lost bit by bit the sense of the human being, religion, the sacred, the respect of others, etc. **While education with science and technology flooded the people with their values and their tangible assets, human values, in loss of speed, increased the lack of humanity in man. And we didn't see it coming...** as the soldiers guarding Caesar had not seen anything coming in the night, because they were sleeping...

The individuals who received this education practise today the philosophy of the life and the rules of this materialistic morality. The human moralities which yesterday took their source in philosophies other than that of the rational scientific materialism were abandoned. Many parents educated in this materialism find themselves powerless to teach their **own** children about family and human traditional values, because they do not know them and do not practise them any more. In the stride of all these changes, powers threw the baby with the bath water, that is: the sacred, the human values, religion, mutual respect, beauty, poetry, the humane, the fabulous... With the lack of human values, decline settles in and spreads dehumanization and chaos everywhere along the way. Not unlike this absence of humanity which settled down in the Marxist totalitarian countries, the morality of the "reason only" settled down everywhere in the Occident and is used today as a guide to the people and to their dehumanized societies.

The large diversity of opinions and cultures that formerly marked the character of the people has become blurred, consumer societies have standardized, materialism has spread as the scientific truth and intellectuals proclaimed the superiority of their values over others. This morality without compassion offends that of others as it settles down everywhere as the absolute master. All the ethnic groups, the philosophies, the occidental religions now give way to the sacrosanct "materialistic morality of the reason" which scientists and intellectuals elevated to the rank of God. We begin this third millennium under the aegis of this merciless domination of the people's "reason", mutated into inhuman persons. The rules of ethics of this morality spread chaos everywhere. Striking examples of these Towers of Babel are on the board of ethic committees in university and hospital research, in committees of this and that, in meetings between governments, in law or medical schools and even in all the secular institutions that adopted them. Those that sit around these "consultation" tables where questions of ethics are discussed may not share the same ideologies, but, even though they come from different ethnic groups, cultures, religions, or atheistic philosophies, all come there to laud agnosticism inculcated by their forming through science and technology. The "matter" has reunited them all and they all speak the language of the matter.

Supported by their scientists and their intellectuals, governments impose this moral code to all professions by saying that it is they who possess the truth and who decide what is good or bad, as religion formerly did. The world is thus administered today by this only force of the "reason" of scientists and intellectuals who only practise the values and the ideologies of the scientific materialism which reigns everywhere as master. When elected members finally notice that these people of science and their infernal bureaucratic machines actually seized the reins powers, these Stalin-like dictatorships of "reason" and those of his accomplices will in turn burst in broad daylight. Must we think that the "political turn towards the right" in several occidental countries is another indication that people have had enough of being driven by those powers which lead them like children incapable of thinking on their own?

If the mutants of the "reason" have learned that "happiness" is found in tangible assets, why have people never laughed so little? Why have they more than ever before "disconnected" so much from society and in such large numbers and why they are they so little enthused by their leaders? Why have people never felt so useless in their societies fed only on tangible assets, why are human values so absent from their lives, why have they never felt so betrayed by their powers, by science and the knowledge used to such degrading purposes?

C. Something, somewhere, has been broken

The wheel of life does not turn round any more. Individuals have lost the joy of living their parents and ancestors had, although much poorer and much more deprived than they. The contact with nature and with traditional human values seems to be what was broken in man and his environment. But some prefer not to look and continue burying their heads under mountains of material technologies and inhuman laws not to see that their loss of humanity has chained and immobilized them.

They prefer to keep on believing that the acquired knowledge has enabled man not to work for hours just to get water from his well, to pick his daily food and not have to work for hours each day to earn his daily bread! But in all these changes, how does he live in his material world, he who now has so much spare time and who hardly works 25 or 30 hours a week? He plays. He watches other people playing. He travels. He reads.

He watches television, discusses, travels on the Internet, etc. He entertains himself. But all these gestures are soaked in dehumanization and societies are more and more marked by material happiness, drop-outs of all kinds, drug and alcohol abuse, gambling, drugs, sex and voyeurism or with the other personal tendencies... Let us see.

Yesterday's man spent his time working to feed himself, to defend himself against the adversities of life and his fellow men. His preoccupations were centered on ways to tend to his primary needs, such as finding a place to live, fighting against the forces of nature, struggling against diseases, etc. Physically tired at the end of the day, he went to bed early and rose with the sun. He lived by day in contact with nature that dictated his ways of life. Nature was there, inviting and loving, said the poet. Even the poorest of men found happiness by living in symbiosis with nature, the earth they knew, respected and loved. They lived in harmony with the rough and demanding life of yesterday.

Regrettably, technology, science and material assets made people drift away from this precious contact with their mother the earth and with nature! Because of his enthusiasm in wanting to use all the scientific knowledge to relieve his daily burden, the physical world of the people shifted into ease. He evolves today in the midst of all these excessive tangible assets. But in less than a century, he has uprooted himself from the earth, his natural environment, and only feeds on science, new possessions, technological applications and material values. The people who live in contact with nature are inevitably closer to the things of life. They are more human, simpler, more religious, less complicated, less Cartesian, less linear and less logical and less uniquely "reason". **"Reason" without the humanities is what separated these mutants from their contact with the earth.** The people who live with nature possess besides the "reason", the intelligence of the heart. In contact with nature, man is less cutting, less complicated, less conceited and is content with the real things. His actions are integrated into the environment because nature makes him more humble. **Those that consider themselves as an integral part of nature are more respectful of their environment and more aware of their humanity and their humility.** Scientific materialism and technologies have robbed the people of their humility, so necessary to man, to catapult him in an incredible material modernism. **The sacred as well as the arts helped him accept the quiet simplicity of the things of life. By straying from the things of real life and by plunging head first into the ease of tangible assets, a break happened between man and nature. This loss of contact with the earth accelerated his dehumanization.**

D. Materialism disillusions man

In the play *"Sous le regard des mouches"* (Under the Glance of Flies), author Michel Marc Bouchard asks **"At what are we playing when there is no more love in the home, when everything has become predictable, when we are lonely, when we swim in an ocean of possessions, when we have everything we need and more?"**

At what are we playing "when the child is a daily witness to hundreds of attacks against the person... is this expected to be without effect on his behaviour"?

At what are we playing "when a man kills his wife, his children and then himself? Directions for use are known and the media go on and on repeating it... The macabre is the recipe".

At what are we playing "when virtual tribalism performs an autopsy live, a crime is shown live, killings live, prisoners executed live, sexuality live...", asks Bouchard?

"And so, the merchants of death grow richer through technologies, authorities such as the CRTC (Control of Radio and Television Commission) prefer them to Art, and we even manage to make sexuality macabre (piercing, banding, "ecstasying", humiliation). **We had already lost our souls (to materialism) we only have our bodies left to lose".**

How morbid are the truths of this world where we live without thinking! Caught in the midst of this decadent turmoil, many of us do not see anything, anything at all coming. **I love the lucidity of the poet lost in the crowd! It is again he who rings the alarm. Others will murmur in other people's ears, as did the prophets of the Bible, to get up and to look in front of them not to put their foot in the giant abyss which is right in front of him.**

"Man yearns for the obscure, for failure, for tyranny, for servitude, for hatred, for fighting, for war, for blood, for death" wrote Francis Fukuyama in his book *"The End of History and the Last Man"*. Materialism only increases his yearning for death. **Wars, the millions of deaths on battlefields, the atrocities in concentration camps, the ethnic purges and the conflicts which marked the twentieth century, are they not eloquent testimonies of this excess of materialism in the people?** So self-sufficient and full of themselves, they use Knowledge to destroy themselves without really realizing it. Is there not a striking resemblance between the greed of individuals for material things and the greed of mother "Eve" for the symbolic apple, presented one day to Adam?

<u>"Liberalism has established its superiority so clearly that it is susceptible to end History and to define itself as the ultimate discovery</u>

of humanity as a social organism" wrote Fukurama. How very close we are to this! In an article in *La Presse de Montréal* of January 29, 2000, Mario Roy wrote that "Fukuyama had also noticed that liberal democracies cultivate the paradox of having been historically the least bad political system on the globe, but at the same time, the most hated".

By working only with the matter they can see and touch, and rejecting all possibility of recognizing the humanity in man, scientists, intellectuals and the dehumanized mutants they formed have come not to see the inhuman monster they made out of yesterday's man. They do not understand that their dehumanized laws are cold and hard like a rock, because they know nothing of the existence of the brain of feelings, the brain of arts and the brain of the instincts **and get their certainties only from the primitive reptilian brain of their senses.** They do not even feel that their laws do not show any glimmer of humanity and that they are deprived of love, compassion and feeling.

The almost compulsory adhesion of the people to the scientific and social materialism and their systematic refusal of the traditional human values in their life, as if they did not exist, brought them to live in their world as automated machines empty of humanity. Should the Frenchmen's vote for Mr. Le Pen's extreme right-wing party in the presidential elections of April, 2002, or the Austrians, the Italians, the Quebeckers and others just before, not be interpreted as cries of despair of individuals continually "trapped" in the unconsciousness of their stifling governments and their intellectuals who will not stop subjecting them to the shameless slavery of their dictatorship? **How can the powers' civil servants not realize that they cannot always with impunity control people like puppets?** In the name of what right do they assume the power, for example, to make out of doctors treating patients, of pharmacists giving out medication, of university professors, of municipal elected members and many others, puppets under contract subjected to their authority? People have a visceral need of freedom. But the social laws that direct them are binding for the individuals and have no other purpose than to use them to serve the powers in place. Can they continue to waste, at the speed of light, what remains in them of freedom and humanity? And so flooded in all this shameless materialism which gives them food and games to amuse them and to keep them from crying out, people and their decadent societies disintegrate in the midst of the lofty pretensions of the intellectuals and the scientists of the governments which lead them.

They managed to transform people into inhuman mutants and standardized robots. **The materialized man is now incapable of thinking and acting with his heart. He is just a logical individual incapable**

of recognizing his own humanity. And to think that these mutants of the "pure reason" without humanity push the boldness and the conceit to think that they are "the ultimate find of humanity". Since they have covered themselves with technology and science, these humanoid mutants regressed on the human plan to the point of worshiping the golden calf of the Bible.

Many of their descendants are now machines incapable of making the difference between what is a law and a natural morality, a law and a philosophical morality, a law and a social morality or a law and a materialistic morality. Have scientists, intellectuals and their muted descendants put too much faith in the "reason" and in their deficient method of research which ignores all human aspects of man? **Regrettably, as glorious as it was, "this reason" and this method of research relies only on the approval of the senses controlled by the most primitive reptilian senses.** Must we be surprised that after having thrown people into this kind of world, this century of science, technology and "reason" ends filled with material technology, possessions, applications and humanoid robots, and at the same time more and more empty of human beings.

Lost in this morality and in accord with ethic principles filled with "reason", and believing to be full of wisdom, **these mutant humanoids and atheists have become more powerful bombs of devastation and horror than the live bombs of the fundamentalist Moslems, because they directly threaten the heart of man!**

I received one day on the Internet a quite concrete description, although not scientific, summarizing the physical situation of this today's dehumanized world. "If we could reduce the population of the world to a village of 100 persons while maintaining the proportions of the existing people on earth, this village would consist of 57 Asians, 21 Europeans, 14 Americans, 8 Africans and 2 others. It would include 52 women, 48 men, 30 whites and 70 non whites, 30 Christians and 70 non Christians, 89 heterosexuals and 11 homosexuals.

Six persons would possess 59 % of the total wealth and all would be American. For each person dying, one would be born. One only would possess a computer, only one would have a university degree, 80 would live in poor houses, 70 would be illiterate and 50 would suffer from malnutrition".

"If you got up healthy this morning", continued the author "you are more fortunate than the millions of persons who will not pass the week without a problem. If you have not already felt the danger of a battle, the solitude of detention, the agony of torture, the pain of hunger, you are better off than 500 million persons always the prey of these dangers. If you can go to church without the fear of being threatened, tortured or killed, you are luckier than the 3 billion persons who do not have that chance. If you have food in your refrigerator, clothes on your back, roof over your head and a place to sleep, you are richer than 75 % of the other inhabitants of this world. If you have money in the bank, in your wallet or change in your pocket, you are a part of the 8 % most privileged persons of the world. If your parents are still alive and still married, you are a really rare person, and if somebody sent you this message you have just received a double blessing because somebody was thinking about you".

The message ended by this proposition of life: "Work as if you did not need money. Love as if no one had ever made for you suffer. Dance as if no one were looking at you. Sing as if no one were listening to you and live as if Heaven were on Earth".

E. Conclusion

Taken in the whirlwind of uncountable tangible assets and unconscious of their gestures, people start this third millennium drowned in the pollution of their environment, the flaws of their technologies, the paradoxes of their "scientific reasoning", the absence of human and traditional values in their lives, materialism and a dehumanization (of fact), and the indescribable ups and downs that undergoes their planet. We must all quickly become conscious of the responsibility that the paradoxical side effects of science, technological applications and the impacts that materialistic powers have imposed on man, asides from good realizations.

Discussions about the adversities of science and technology on human beings should not continue to be taboo subjects not recommended to study. It is not surprising that few intellectuals really seem to be aware that most of the ups and downs in our world today are bound to the materialism of the individuals and to the lack of humanity left by a materialistic scientific education without humanity. **Powers have contributed in hammering into the heads of our children this atheism that is blooming in our dehumanized societies.** It will indeed be necessary to recognize one day that the source of the dreadful evil of materialisation that swept over the people of our planet is intimately linked to the materialism of science and technology and that this plays a non ambiguous role in the propagation of this materialisation.

In the next chapter we will try to understand how this great ideological saga led to this monumental influence of new socialist ideologies in the Occidental world, and how these ideologies were wildly propagated by intellectuals and scientists in the education by science and technology. In other words, why have science and technology so quickly enthused scientists, intellectuals and the people and how they managed to change the course of events and to give birth to the ideologies of the scientific and social materialism that served to usurp the powers of religions by taking them as their own and imposing them on societies in the formation of the people?

CHAPTER 7

The history, the interest and the ideological impact of scientific and social materialism

Introduction

The ideologies of scientific and social materialism and the promotion of the material values of this system originated in the nineteenth century. They arose from a new way of thinking quickly adopted by scientists, intellectuals, men and societies. This system will quickly invade education, then successively the universities, the trades, the professions, the mentalities and almost all other aspects of the life of men and Occidental societies. The values of this system then spread their materiality to all the individuals of these societies.

Certain factors promoted a century and a half ago, a general craze in the people for this materialistic system that would change everything, from education to the ways of thinking of man, powers and societies. Let us study why so many people have become infatuated with these ideologies which were at one time new, and why this craze still persists today. Why have these ideologies spellbound people and modern societies so much that, after a century, they continue to fascinate them just as much while at the same time materializing and dehumanizing them without them even noticing? To better understand let us retrace the main lines of these forces which activated this worship everywhere in the occidental world.

Plan

We will fist try to understand the great philosophical movements of thought which led scientists to develop a "magic thought" to avoid being criticized. Afterwards, we will try to understand what made Karl Marx devise a whole new system of society. This system was called the "the scientific and social materialism" and has deeply transformed man and society. We will then see the great stages of this gigantic saga of the institutionalization of this scientific and social materialism which followed and which brought men of science, intellectuals, technocrats, governments and societies to transform all domains of human action into science. Karl Marx's new social ideologies and concepts upset the people and the societies of the twentieth century and even continue their progression today, at the beginning of the twenty-first century, with very few modifications. Thereafter we will demonstrate how, along the way, these ideologies invaded education and the training of people to a point where it wove a canvas of the numerous new movements that still deeply mark all our contemporary Occidental civilization.

We will finally talk about the dreadful effects of this scientific and social materialistic system that delighted scientists, intellectuals, governments and occidental institutions because they became the foundation of the supremacy of their powers, and how this system has recently "lost the legitimacy of its power".

A. New ways of thinking that give rise to scientific and social materialism

The progress of knowledge, the desire to monopolize the powers of religion, the obsession of scientists and intellectuals to change society, the desires of scientists to assure the physical immortality of man and to believe themselves to be as powerful as God, were powerful motors that allowed the explosion of these new ideologies and societies and largely contributed to mark the end of the nineteenth century and the whole twentieth century.

The desire to monopolize the power of the Church

At the beginning of nineteenth century, the 1789 French Revolution swept the occidental world with a wind of new ideologies on the freedom of the people and the States. Ideological changes arousing during the French Revolution are not at all foreign to this excitement. France is still trying by "reason" (logos), to release itself from the hold of the monarchy and the great medieval Lords, but the real power still remains in the hands of the Church. For decades now, intellectuals have wanted to monopolize the religious powers and separate them from those of the States. This ideological culture has marked the nineteenth century in Europe and in America. The occidental countries will soon try to adopt the new ideologies of this scientific revolution, so anxious are they to help their scientists and intellectuals separate the religious powers from the civil ones. **Nobody knew until Hegel how to achieve this legitimately with elegance.**

At the beginning of the nineteenth century, the supremacy of the powers still belonged to the Church and the monarchy. Armed with these secular powers, religions were bound to the civil powers and showed them the way. The religious powers greatly influenced the centers of State political decisions. The Church representatives even occupied royal palaces and centres of civil decisions. From there, they influenced the powers of these administrations on civil and State affairs and imposed their ideologies and their thoughts on leaders and statesmen. They also controlled the people, their education, their faiths and more often than not, those of the kings and the civil leaders. At that time, authorities and religions all sat at the same table and were almost always fed the same ideologies. With ramifications in all areas of society, these religious powers had the mission to make the people accept their ideologies as well as the civil decisions taken at higher levels.

If kings held their civil powers through their royal blood, the Church held theirs through philosophy. Since Aristotle, in fact, philosophic systems advocated that in order to exist, human beings as well as things had to be inevitably constituted of two "beings of reason" in the philosophical sense. These two "beings of reason" constitute the necessary and essential entities to the existence of beings and things, that is the essence and the existence as Aristotle's cosmology defined it. Other philosophical systems, such as the one of Thomas of Aquinas in his *"Somme théologique"*, described this philosophical conception of the world by speaking of the matter and the form, while Paul Claudel called these entities animus and anima to express the same cosmologic concept.

Secure in these great secular fundamental principles, almost common to all chapters of cosmology of the philosophic systems for over 2000 years of human history, all religions justified the existence of God on this fundamental and sacred philosophical principle. And so, the essence or the matter of things was respectively considered as "liable to putrefy and finite" beings, while the "existence or the shape of things" was considered as not liable to putrefy and immaterial. The "existence", the "form" or the "anima" of the existing beings explained and gave, always in the philosophical sense, the qualities of "non putrescent" or "non corruptible", that is the qualities of the Supreme Being, "God".

All occidental or Christian religions were secure in these fundamental philosophic principles of their cosmology system, and the majority justified the existence of God by these philosophical reasoning. This conception of the immaterial and material beings of the existence of things had never been denied by any philosophical system which respected itself, in spite of numerous attempts. The powers of religions, and thus of the States, were legitimized by these fundamental cosmology principles of the philosophies which defined the existing beings.

Monotheism had been strengthened by the Jewish tradition and by the Greek thought. The latter had even arrived at the conception of the existence of a unique God, by the only force of the human thought in *Le*

Timé de Platon (Timé, a student of Plato's), over three centuries before the Revelation. The revelation confirmed the existence of this unique God that the Greeks had already philosophically demonstrated by the only power of their thought.

Philosophic systems thus assured their permanence and legitimized the secular powers of the churches as well as their supremacy over the State civil powers and from there, the rules of deontology and morality which ensued to control the behaviour of the people and the societies. Until Hegel and Marx, morality always ensued from these philosophic systems. It was practised, legitimized and confirmed by still "human" State laws. States and governments were thus consequently legitimized by the same principles of the existence of God and by the same systems of philosophy.

Regrettably, History teaches us that religions and especially those that held the powers of religions were abusers. **It has been obvious for a long time now that more and more intellectuals were looking for an elegant way of separating the civil powers of the States from the religious powers, and to seize them by all possible means.**

One day Hegel came

One day, Hegel came. The German philosopher presented, around 1830, an all new philosophy, based on the eternity of the <u>matter</u>. Hegel's philosophy was built on the fundamental hypothesis that it was the matter that was eternal. This philosophy thus allowed for the first time, for all those oppressed and dissatisfied with the religious powers, to finally escape elegantly the power of the Church. **It is needless to say that the Hegelian philosophy would become, for these reasons, one of the most powerful ideological ferments to explain the boundless craze of scientists and intellectuals for the Hegelian philosophy.** This philosophy finally supplied the ideal philosophical justification to legally monopolize the powers of religions and the Church and to transfer them to themselves... And so, by ways of a philosophic ideology, secular intellectuals and scientists seized the powers that the Church had over the people and the State. **<u>From this moment on, scientists and intellectuals will define themselves as the only legitimate holders of absolute power over the people. We shall see further on how they, also, would abuse it.</u>**

Nobody had ever previously dared claim that matter was eternal and not biodegradable, even though there was no proof of this. The primary principle of Hegel's philosophy was simply a philosophic assumption, as were the cosmologic principles of the philosophies of Aristotle and of

Thomas of Aquinas, respectively on the essence and the existence or on the matter and the form, and which philosophically justified the existence of God.

Hegel's philosophy of the eternal matter seriously upset the secular principles of philosophy, which were based on the essence and the existence of human beings and things **because, if it is the matter that is eternal, it is inevitably the matter that is God.** By supposing that the matter was eternal, Hegel was thus making the matter "his" god. This idea was not lost on intellectuals and scientists.

By supporting these ideologies, scientists and intellectuals brushed aside the principles of the matter and the form or the essence and the existence to explain the existence of human beings and things. **At the same time, the secular foundations of the religious powers and the Church were shaken in their foundations,** and they have never recovered.

The concept of the "eternal matter" brings scientists and intellectuals **to devise new ideologies on how it would be advisable to conceive this new world, when it is the matter that is God.** Many of the ideologies which will arise from this concept will, effectively, deeply transform the world and allow scientists and intellectuals, (after a few decades of procrastination with the Church), **to dictate to the people and to the societies their "new" behaviour,** based on reason only (logos) and in a completely new world, where the matter is God.

And so the philosophical "tour de force" was done. Intellectuals and scientists finally had in hand a philosophy which allowed them to redesign the world and to legitimize at the same time their new powers. And so they began by explaining to the people how they must now behave, how they must think, live, evolve and be in this new world where the matter is God.

Hegel's philosophy not only allows scientists and intellectuals to remove the powers from the hands of religion, it also allows them to change in depth the people, the world, the ideologies, the governments, the institutions, the philosophies of life, the societies, the education, the professions, the occupations, the morality, etc. **Everything had to be redone according to the principle that the matter was God.** A new world was being born. The thinkers that followed Hegel and Marx were enthused at getting to work **to imagine the foundation of this new world.**

Loads of new ideologies called scientific but that have never been proven scientifically, because they are only the fruit of pure reason and not science, will be used by **pseudo scientists and intellectuals.** Some of these ideologies will completely transform the societies of the occidental world. Scientists will take advantage of this to take on all the powers formerly possessed by the religions and the States over people and society. Henceforth, it is out of the question to search for the truths of our world in philosophical or biblical ideologies. Only the knowledge acquired by science and "reason" will be accepted and imposed on the people as the only provable truth by the "reason". **But the fruit of the reason (logos) are in no way scientific truths.**

Scientists of physical science were convinced that the knowledge acquired on the matter would one day enable to feed billions of human beings, get them more essential goods, relieve their suffering, facilitate their ways of life and their environment, cure their diseases and even to push back the frontiers of death… The world of the matter being their new god, science and scientists knew no limits. And, as children breastfeeding, they profusely drank these new ideologies, even thought they were not confirmed with scientific methodology and experiments, and without even questioning it. Still today, they use like gluttons all the technological applications in order to facilitate their material life, and their material aspirations have not stopped progressing so far. **The previous chapters have demonstrated that by using only their "reason", the matter and the uncompromising technologies for over a century, people have managed to pollute their environment to the point where it threatens the existence of any life on earth, including theirs. They mutually materialized and destroyed each other to the point where all humanity in them has disappeared.** We are entitled to ask ourselves today if the so numerous side effects stemming from science, technology and this "reason" did not lead man straight to his own destruction. By using only science and reason, scientists and intellectuals forgot that the greatest part of human history was guided by instinct and feelings, and not by this intelligence or the "reason". If intelligence is responsible for these side effects, it is it that it has behaved and continues to behave as "this poisoned gift" about which spoke the ancients. Let us see how all this took place.

The canvas of the historic film of events which marks the framework of this scientific and rational revolution, and which would upset today's occidental world, enables us to distinguish the main motors which, at the beginning of the nineteenth century, gave wings to science and to the new ideologies which emerged from the intellectuals' brains. Everything really began with Hegel's philosophy which allows scientists to conceive a new world and a certain "magic thought" which only awaited legitimacy to burst.

At the end of the seventeenth century, science had already begun to bubble. **Scientists** are starting to better uncover the secrets of the matter.

By studying these secrets, they bring to light treasures unknown to this day. In the eyes of the people of the time, they are transformed into magicians. New technological applications appear everywhere and it isn't long before they are implemented in every day material life. They gradually invade all areas of the lives of people and societies. Minds are turning at full speed and amplify the appearance of new ideologies, based on scientific reasoning and on the scientific and social materialism which arises from this effervescence in the middle of the nineteenth century.

Numerous technological applications stemming from this new scientific knowledge allow men to hope to be able, one day, to break through all the secrets of the matter; these secrets of the philosopher's stone that the Alchemists of the Middle-Ages had so searched for... **The era of knowledge and technology thus really begins at this turning point of history.**

The prodigious increase of knowledge revived an old dream in man: the one of becoming immortal one day. Man has always been haunted by this desire to become immortal like his Creator. This dream goes back to the mists of time. Knowledge is already accumulating to the point where people think that scientists are some sort of demigods. **Does not the apple referred to in the Bible symbolize the knowledge that should lead man to immortality?** Religions have settled this human problem by teaching man that reproduction was his way of immortalizing himself and resembling the Immortal Divine Duplicate.

But by multiplying in order to look like the "Immortal Divine Copy", man has only satisfied part of his desires of immortality. He strived for so much more, like to prolong his material physical life eternally. With the transplantation of organs and the animal and human cloning, is man not entering today in the eye of this evolution towards the immortality of his physical body?

This desire for immortality which is haunting man has never stopped filling his dreams and disturbing his spirit. The acquisition of scientific knowledge on the matter has revived this old dream of immortality and scientists were seen as these demigods who could best insure him this opportunity. If science has already managed to facilitate his physical way of life, it could also possibly prolong his life and assure his immortality. The speed at which scientific knowledge and technological discoveries has accumulated, was enough to convince man that one day he would reach this greater well-being and even reach the immortality of his physical body. The man of this still recent era is thus more and more convinced that science is powerful enough to allow him to cure his diseases and to prolong his life "ad infinitum". Hopes are great, even unlimited...

Today's people are still engulfed in this whirlwind of aspirations of human madness. Is this craze without limits of the people for this gold mine of science, technology, data processing, communications, biotechnology, genetics and especially the cloning of the human being, organ transplant, etc., not enough proof? These aspirations, aroused by the acquisition of new knowledge on the matter, are going to deeply modify the objectives of the lives of the people and therefore their societies.

And so, science gets the mandate to realize these great human dreams. Scientists were themselves convinced that by studying the functioning of the material world, they would succeed one day in immortalising man's physical life. And so they got to work.

Again the new philosophy

Hegel's philosophy, which made the eternal matter its basic principle, has more that seduced the sorcerer's apprentices and the intellectuals of the nineteenth century: **first because it gives them the chance to monopolize the powers of religions and secondly it gives them the leisure to work at the discovery of their god, the matter.** Brought up over a century ago, the philosophy of the matter, God, quickly makes its way with scientists and intellectuals. It carved itself a place of choice among all the other ideologies which will guide the societies of the twentieth century, because it legitimizes, on a philosophical basis, all the new ideologies which will mark the explosion of a new era, **that of scientists and intellectuals that now possess the powers of religions, monarchies and governments.**

And so intellectuals, scientists and pseudo scientists (by opposition to the scientists of the exact physical sciences) began to elaborate ways and principles that people and societies had to accept in order to live in their new world, where the matter is God... Imagine the novelty in the middle of the nineteenth century! As science explodes, the mission of scientists becomes clearer. The sorcerer's apprentice can finally search for the secrets of his immortality, and the intellectuals (pseudo scientists) for the principles that would guide society in this new world where the matter is God! What can be greater and nobler for a scientist and an intellectual than to work at perfecting his god, the matter? What can be better for a society than to live at the rhythm of this god, the matter? **<u>Hegel's ideology thus becomes the central philosophic ferment which will give birth to scientific and social materialism.</u>**

The simplicity and the legitimacy that Hegel's philosophy gives to scientists are at their apogee. This philosophy finally allows scientists to demonstrate to the world that they are the only ones that have the power to lead the people and the societies. At the same time, it gives them the legitimacy of the powers and allows them to hope that one day they will find all the solutions to the secrets of the matter, that they will own them and have the greatest imaginable powers: **those to manage the people, the societies and the governments of the planet.** We hope to have demonstrated in the previous chapters that it is exactly what science is achieving at this beginning of the twenty-first century.

The choice of a methodology to search for the truth

To avoid wandering in its quest for the truth, science has chosen a rigorous methodology of research, a method allowing few errors and that will shield scientists from any criticism. René Descartes' method of research, published at the beginning of the seventeenth century in the *Discours sur la méthode*, lends itself admirably well to this elaborate objective. It only applies, however, to the study of material things.

The Cartesian method of research first imposes to the one that uses it to doubt of its premises. It teaches him that there is no effect without cause and that each effect has its own cause. The Cartesian methodology lends itself admirably well to the research objectives of scientists working on the matter. It always imposes upon them to build their research on concrete facts that can be verified with the senses. So far, all research hypotheses to this day have always been supported by concrete, tangible and true facts verifiable by man's senses.

Armed with this Cartesian method of research and of reason, and a philosophy that teaches him that the matter is his god, the researcher has come to believe that he can no longer make mistakes. He feels he is really on the right track, confident that in applying the reasoning of this method, he can certainly find the causes responsible for the effects he is looking for, that is the truth about material things. In other words, **scientists believe that the Hegelian philosophy together with the Cartesian research method of reasoning can but allow them to have access to the only truth on things...** This method of research thus appears as the "ideal method" to use for the search of the truth on the matter, God.

To find the cause of an observable effect, the Cartesian method of research always begins with doubt and what is more, only accepts as premises facts that are observable, verifiable and perceived with the senses, that is, only the data and facts that concern the senses. Scientists now limit themselves voluntarily to finding only the truths about material things, because they always start their research only with material and verifiable data. If scientists working on the physical matter always respect more easily these limits of the method, pseudo scientists will too often accept as the truth the lucubration of their reasoning, without experimenting it.

Since only the material things are observable and verifiable with the senses, the scientific method of research inevitably limits itself and can only be applied to study the matter. Pseudo scientists forget this basic principle of the method. No one, for instance, can use this method for studying facts which cannot be verified with senses, such as the great human questionings, the intuitions, the feelings, the love, the respect, the kindness, the empathy, etc. In its actual form, the scientists' method of research is and will always be deficient and incomplete to study data that have nothing to do with the senses.

Pseudo scientists and intellectuals should never have made the mistake of rejecting all that they could not study with their method of material research, **declaring them as non existing entities...** as they did too often. **They also rejected, without any scientific proof to support their lucubration, the existence of all non material entities.** What is more, these pseudo intellectuals should have never considered their intellectual lucubration as scientifically demonstrated truths!

With all the accumulated knowledge and their linear and rigid way of reasoning which does not allow for any mistake, or so they think, these pseudo scientists were even more certain of being the true holders of material powers. They became in a sense the high priests of science and the instruments of the search for truth. Scientists were blissfully happy because they were less subject to criticism. **The craze for the "reason" and the "logical thought" has so rapidly earned the trust of scientists, intellectuals and governmental powers that they consider "this magic thought" as the philosophical Stone,** so sought after by the men of science of the Middle Ages... In his *Apologia of the Knowledge*, had not Socrates come to the conclusion that he was the most intelligent because he knew that he did not know everything?

Knowledge acquired with the methodology of scientific research was enough to persuade the world that scientists were almighty persons. People even came to substitute them for God himself. So, in the eyes of the people, scientists were considered for a long time as the ones that would one day share the powers of the people and of the world. Who cannot be in favour of this method of logic and "reason" to search for the truth?

Believing that they were the sole holders of the truth and of the "reason", scientists and intellectuals took advantage of this to slip into positions of power, mainly within the governments. Governments were the first to hire scientists and intellectuals recently formed at this school of sciences and technology. They are called the experts. **Today, there are**

thousands of such scientists and pseudo scientists, experts and non elected technicians that control the fate of most of the countries of the planet. They are the "éminence grise" of yesterday, the power behind the throne. These experts control science and technologies and hold the keys to all the powers. Many elected members have not yet realized the deception!

Pseudo scientists are so delighted with their knowledge and their realizations that they have come to believe **and to teach that all the pieces of the game identified during their research are in the natural order of things, that they generate only truths and that the elements can only fall automatically into place by themselves and always at the right place!**

Societies are quickly becoming saturated with this scientific materialism

"Reason" and the scientific method of research, the acquired knowledge and the technical applications, all have literally invaded the lives of the people in our Occidental societies. No one could be against the search for the truth. All philosophers, sociologists, thinkers, scientists and intellectuals were the first to become soaked in this new scientific materialism and its values. The attraction to science and knowledge becomes a tremendous stimulus which greatly enthuses a great deal of people as well as governments for all that is scientific.

Scientists and intellectuals in power make the laws and dictate people's behaviour. They have become so full of themselves because they hold these powers that they think that scientific and social materialism and their ideologies are the "summit" of the "reason". The science of the "reason" is at its apogee.

Let us now analyze how this science and its methodology of research and of "reason" have been implemented with such force by the powers in the training and the education of the people, that they remain to this day ferments that maintain the scientific flame, immortalize this materialism and kill the humanity of man everywhere, in schools, universities, governments, institutions, etc.

Craze without limits for science

Science acquired its first letters of nobility with the acquisition of new knowledge on the matter. The technological realizations ensuing from it allowed the people to free themselves from unrewarding material tasks. Now armed with the "magic thought" which confers them so much power, scientists and intellectuals took the liberty of thinking "bigger". They promote material values as being the "summit" of the human thought.

We have demonstrated in the previous chapters that technology, knowledge and education with science shape the ways of thinking, acting and working of the people, educate them and integrate all spheres of human activity (cf. Chapters 4 and 5). The Hegelian philosophy of god the matter gives new "electrifying" arguments to the scientists and the intellectuals of the time. It allows them amongst other things to confirm that they are the true new owners of the powers of the people, as religion was before them, and that they really possess the right to redo this new world according to the image of their new God and of their "reason". **As soon as the philosophy of the matter god appeared as the Philosopher's Stone that we had been looking for since the mists of time, scientists and intellectuals started to imagine new ideologies on ways to conceive their new societies, and the new powers that will manage this future social and material Hegelian world.**

Some of these ideologies were unfortunately too often built without any scientific proof by too many pseudo-scientists. Of this craze without limits for new ideologies, intellectuals, scientists, thinkers, philosophers and a multitude of people, resulted the creation of new domains of sciences and pseudo social sciences some of which still exist today.

The pseudo-domains of science such as psychology, theology, history, economy, etc. by opposition to the physical sciences, do not lend themselves very well to the game of scientific methodology with its linear and mechanical thoughts of "reason" because they cannot be proven by the senses as physical sciences always can. And so scientists attempted to train the people in less tangible ideologies, ideologies that could not be proven with scientific experiments. Many of these pseudo scientific experts rapidly settled down in all governmental spheres and universities and proclaimed themselves and their ideologies as the holders of the only truths accessible to man. People were forced to accept these pseudo-experts, their ideologies and their arguments of "reason" as absolute truths. Several ideological legends were engendered by the scientists and intellectuals that had just seized the powers.

These experts elaborated great currents of social, economic, and political thoughts which marked the twentieth century. Some of these currents of thought still continue influencing the people at the beginning of this new millennium. All the social structures will conform to this new ideological whirlwind. All are built according to Hegel's philosophy of the eternal matter and try to conform to his methodology of research. Everything will change, from the conception of work to the university faculties and administration schools, to psychology, theology, etc. and all branches of human activity. These new conceptions of the world were all marked by science and the scientific reasoning according to the principle that the matter is God and of the limitless faith in science. So new currents of thought took shape and numerous models of society appeared everywhere in the occidental world. Certain concepts originating from this craze were so impressive and revolutionary that they still manage to this day to influence people and societies. **Even though these new concepts and system were presented as scientific truths, many were never demonstrated scientifically by scientific methods and remain wishful thinking.** Some of these currents of thoughts still dominate education, governmental powers, institutions and people, even at this beginning of the twenty-first century, without anybody daring to question their scientific legitimacy.

The social ideologies of the future socialist totalitarian communist societies were the first to have the wind in their sails. Very soon afterwards, not to be left behind, numerous other domains and societies, as well as capitalist, democratic, neo-liberal, liberal or other political systems underwent and still continue to undergo the influence of such currents of thoughts and ideologies, some of which only stem from pure invention without any scientific base.

B. Some of the great materialistic ideological tendencies

Karl Marx, Hegel's friend, **was the first to imagine the great principles which would guide people and societies to live and to be administered according to the principles of god the matter.** Published in 1850, Marx's *Social Capital* is a sort of catechism on social materialism as a system. The "scientific and social materialism" as he would call it and of which we have often demonstrated the evil effects, was the first of a series of ways of thinking which would mark the new societies. **Marx's "scientific and social materialism" spreads quickly. The work is new and attractive. It arouses the interest of many other thinkers by describing how people, governments and their leaders must govern, and how the people must agree to be governed in societies that accept that the matter is God.**

Numerous variants of these new types of societies develop according to the ideologies of Marx's thought. Depending on the State, they are called communists, Marxists, Marxist - Leninists, Bolcheviks, Mencheviks or are transformed into a multitude of socialist systems all based on Marx's social and scientific materialism. Certain will modify to form different socialist societies, some of which still exist today. Even liberal, conservative or capitalist governments of Occidental countries base their ideologies on this Marxist socialism. They modify their humanistic educations and religious societies to implant new forms of education with science and technology. They even force their citizens to accept these new ideologies and their pseudo scientific principles. The most powerful pseudo-experts and scientists in these governments imposed and still try today to impose their systems as a religion all over the planet.

Trotski and Lenine applied Hegel's ideologies and Karl Marx's socialist system in Russia since 1917 in order to be accepted. After a bloody revolution, this social system which feeds on pure Marxist ideologies will be known under the name of Marxism-Leninism. It will persist for about 75 years before becoming obsolete and finally ending in 1989, leaving behind desolate States and impoverished people without counting the dozens of millions who had died defending or contradicting the merits of this socialistic ideological lucubration.

Even before the middle of the twentieth century, in China, a very close variant of this Marxism-Leninism is called Maoism. In other countries, other variants will be called communism, Bolshevism, etc. All acquired their Marxist letters of nobility following revolutions before implementing socialist forms of governments in their societies.

Throughout the century that just ended, several countries of the world exchanged very often, and always during revolutions, their social and traditional governmental systems to become communist totalitarian States or variants of the Marxist socialism. Whether they are called today social democracy, democracy, socialism or other names, all these forms of powers make "Marxist socialist" **laws <u>which are laws that favour collective rights over people's individual rights and liberties</u>.**

From their first application in Russia at the beginning of the twentieth century, these social ideologies reached the Eastern countries, China under Mao, North Vietnam, North Korea, several African countries and many other small countries, often under left-wing communists or social democrats who will sometimes manage to be elected and to govern by implementing harsh and exaggerated social and collective measures. After the decolonization that followed the Second World War, several other occidental States will follow this path. Everywhere these totalitarian powers settle down, however, the leaders abolish the individual rights and liberties of their citizens to establish their new societies directed by collective laws. **<u>In order to realize this, however, these powers must inevitably reduce the humanity of their citizens by depriving them</u>**

of their rights and personal freedoms! The people of these socialist societies eventually learned, much later, of the sacrifices they had to make to establish Marx's ideologies in their societies.

Fortunately, these totalitarian societies almost completely disappeared from the surface of the earth during the last ten years of the twentieth century, with three exceptions: China, North Vietnam and Cuba. But several governments on the planet still continue to refer to these socialistic ideologies, including those that call themselves democracies and that boast at the same time of being liberals, republicans, democrats, capitalist "liberals", etc. Great contradictions still persist after years of being implemented in the heads of people and governments...

Today, many socialist countries of the world manage their people according to Marx's principles. These ideologies first captivated the intellectuals and successively the people of Occidental societies. Everywhere in these countries human values are put aside and quickly replaced by material values. **Pseudo-scientists and intellectuals still consider their systems as the greatest development of the human mind and as the only true existing values.** This revolution will continue to dominate and to spread the hegemony of the sciences and of this materialism in all societies. We find today that this socialistic (leftist) *modus vivendi* is very popular in several countries of the planet.

But in all these countries, when scientists take over the powers, they always seize the individual rights and liberties of their subjects. They change human laws by inhuman laws that crush individual rights. They teach science and its material values and dictate to the people their ways of life and their duty to obey the powers of the State (see previous chapters). Materialistic ideologies indelibly mark today's human activities, societies and institutions.

The modern world is still confronted with the same paradoxes of scientific materialism which marked the twentieth century: that of a science and a philosophical thought on the eternal matter, adulated by men of science and pseudo intellectuals who give everything with one hand and destroy everything with the other. "A dreadful materialistic evil" invades the people.

This disease which killed man's humanity arose from the materialism taught by these pseudo-scientists and took its source in the scientific reasoning, **that is in this other fault of the intelligence which considers only what comes from the senses and the matter.** Regrettably this materialism consciously rejects all human aspects of man. All the entities which are not material and all the values which do not answer to the logic of this reasoning and to the most primitive reptilian senses of man were rejected.

This is a mistake that man had already made by eating one day the symbolic apple which would make him similar to his Creator: immortal. The intelligence appears once more in the heart of this poisoned present about which spoke the ancients, because it rejects all the non material aspects of man.

Science becomes the fashion and the screen through which everything must pass

Intellectuals and experts educated in these ideologies have even come to believe, like the scientists, that it is sufficient to use this methodology of scientific reasoning for the results of that research to become irrefutable proofs. That is why all branches and university faculties have become "scientific" after having been subjected to the constraints of the scientific method of the search for the "truth". **Nobody would have dared claim that his domain was a Science, without having gone through the "mill of the reason" which, according to them, leads to the truth...**

Few today are the domains or radiuses of action which have not yet experienced this scientific methodology of this "self-satisfied reasoning". **What branch of human activity has not yet been studied with this linear, Cartesian and mechanical logic of the "reason"?** Where are the scientists' followers and experts that did not follow this scientific fashion? Scientists, governmental administrations, political powers, houses of education, social institutions, all were transformed into heaps of "techniques" or sciences after being scrutinized under the light of this linear reasoning. Less material pseudo-sciences have accepted without scientific experiments the lucubration of their pseudo-experts to create new currents of thoughts. Here are some examples of this.

Sigmund Freud, neurologist and psychiatrist, tries to make the science of the human psyche reasonable, scientific and mathematical. He suggests curing psychiatric diseases by pure psychic analysis. For over a century all kinds of psychoanalyses were tried to cure the diseases of the soul with such techniques. But is the world any better? "It is not" wrote James Hillman and Michael Ventura in their book *"We have done a Hundred Years of Psychotherapy – and the World's Getting Worse"* (*Ulmus Company Ltd*).

Wanting to promote psychology to the rank of an objective science, and not having found the reason for human behaviour in their laboratory rats, the fathers of behaviourism, **Watson and Skinner** found themselves obliged, after over 20 years of research, to invent a new psychology: **behaviourism**, based on constant laws which make interact the stimuli which activate the reactions in rats... Activated by the automatism of the behaviour after different stimuli on rats, this new materialistic psychology thus became the "study" of human behaviour. Researchers conclude that the reactions to stimuli condition the human behaviour.

In England, **Edward R. Pease** founds in 1883 the *Fabian Society* when trying to "reconstruct society in accord with the highest ideal moral principles" and Marx's ideologies. The members of this group were mostly intellectuals such as Sidney and Beatrice Webb, Bernard Shaw, Wells and many others. Closer to us, students of this school of thought were Montreal lawyer F.R. Scott and later, Pierre Elliot Trudeau, Prime Minister of Canada. These "Fabian" intellectuals will plead to implement the "State Providence" which we know very well today in democratic countries. The fabian slogan is noble: "Imagination at the Service of Society". These same fabians will help in the creation of the English Socialist Labour Party in 1906.

Herbert Marcuse, a German-American philosopher, criticizes the industrial civilization from the Freudian-Marxism of the "reason". He puts up front the principles of the superiority of scientific techniques which, beneath democratic appearances, manipulate the conscience.

John Kenneth Galbraith, an American economist, institutionalizes and tries to materialise the sciences of economy. He imagined a new science based <u>on the behaviour of the people in front of a given situation</u>. It is Galbraith who popularized the science of behaviour in the sciences of economy.

Frederick Winslow Taylor, an American engineer and economist, invents the taylorism. This science makes the promotion of the scientific management of work and becomes its promoter. Taylor was the first scientist to establish the practical measure of the time of execution of work. Timed work and production lines had begun.

Creation of education fields on a scientific basis

Everywhere, science dominates these great currents of thought, as much as the technologies running industries at full speed. Already, in the middle of the twentieth century, few university faculties had not yet accepted the constraints of the scientific methodology to restructure themselves according to the "reason" which is logical, linear and Cartesian. The previous chapters explained how all the university domains have passed through the mill of the scientific methodology.

Scientists have even come to teach the techniques of History. Although History is an ill-assorted conglomeration of facts which were never planned by science or reason, History has become a science! It has become possible to teach History, now that it has become a science called "the techniques of History". It is the same for many other non tangible and non physical domains of human activity such as theology, psychology, economy and human behaviour, etc. which are all moulded on this scientific reasoning.

Most processes of destruction in our societies have originated with the abusive and disorderly application of scientific knowledge and the materialistic technologies of Marx. The absence of human considerations is at the origin of several technological applications which have been proven fatal to both the environment and the people themselves. There is not the shadow of a doubt today that it is the careless applications of science and of technologies on the environment that caused most of the disastrous side-effects that undergo people and societies. I listened recently on the radio to **a song entitled "Side Effects"** which is well suited to the matter discussed here.

The explosion of materialistic and socialistic ideologies

Let us repeat once again that scientists and intellectuals were strongly driven by the desire to monopolize the powers of the religions which have oppressed people for centuries. To be quickly liberated from these religious powers, scientists and intellectuals settled down at the commands of society. In order to establish the supremacy of their powers, they went as far as creating the science of religion, the science of theology, etc. All that was "ideology" at the time was strongly tinted with materialism and the scientific method of research. New sciences have exploded everywhere. This is why there are so many fields of expertise today and so many new technicians and /or scientific specialties. More often than not, these new sciences were developed in reaction to the religious powers which formerly dominated human and social activities in all these areas.

Pressed to make all human ideologies to the image of Karl Marx's Social Capital, and, next to exact sciences such as physics and chemistry, much less exact and pseudo-scientific fields were developed, as we said earlier, such as the sciences of economy, administration, politics, psychology, human behaviour, etc.

Some of these new domains and currents of thought have spread at the speed of light, mainly in scientists and intellectuals and the new university faculties that like to boast that they are searching for the truth in the "reason". This craze for new scientific structures and new types of currents of thought similar to those of Karl Marx became limitless in many circles. Many pseudo scientists and intellectuals were happy to finally look for the truth by using a scientific methodology which no longer brings into account the religious faiths, the sacred and the ideologies based only on faith. The first planning of societies, built on knowledge and the new ways of thinking, are so stimulating that they allow researchers, intellectuals, thinkers, governments, journalists, writers, societies and institutions to build new ways of developing their respective domains, always according to the philosophy of the matter god. **But rarely have these currents of thoughts or new pseudo-sciences been really scientifically verified, even if these ideologies were presented as scientifically proven truths, which they were not.**

Sciences are now considered a religion, a way of thinking and a culture which is found in the actions of the people educated with these scientific principles. These new fields of sciences have thus become established in all societies and systems of thinking, leaving their imprint on writing, education, the way powers administer individuals and societies and especially by sowing everywhere their hegemony. People have learned how to live in a different way and now place all their faith in their new god, the almighty matter.

Major currents of political thoughts that continue to divide societies today

From the end of nineteenth century, two great political currents of thought began confronting each other on the planet: **the one that comes from the Marxist scientific socialism** which opposes **all other forms of already existing governments, the majority based on humanistic philosophies such as economic liberalism or capitalism**. These two ideological worlds are still in confrontation today after more than a century, but with much less power than at the beginning, because scientific and social materialism as well as science and technology have standardized education everywhere in the occidental countries.

Several forms of Marxist socialism have imprinted sciences and technologies, very often making their original identification difficult in the different forms of government which persist all the same in calling themselves liberal democracies, capitalists or neo-liberals. All the words are good to distinguish themselves from one another, whatever their resemblance or shared ideologies, the names they give themselves, the kinds of governmental powers they manage, etc. **Even current**

democracies are almost always combined, to different degrees, with socialist or Marxist ideologies. These two great ideological tendencies, liberal or capitalist on one hand and socialist on the other, also called right and left, have become standardized to the point where they have become difficult to differentiate in certain Occidental governments.

We attribute this confusing vocabulary to the fact that socialist ideologies resemble more the feminine ideologies, connected to the left part of the brain, while liberal and capitalist ideologies are especially dictated by the more rational, right part or mathematical brain of men. The leftists share the more the feminine ideologies of the left which have been very popular for half a century, while the right-wing people share the more male ideologies of liberalism or of capitalism. These last ones formerly characterized governments of the right. These fundamental distinctions are more and more subtle today in countries where people are formed by sciences and material values.

This simplistic vocabulary has transported, in a sense, the fundamental differences between the left-wing Marxist ideologies and the liberalism and capitalism of the right-wing ideologies. **The biggest difference is that the socialistic ideologies of the left used the collective rights and collective laws to dominate the individuals, while the ideologies of the right protected the democratic individual rights.** The true democracies of yesterday advocated the individual rights and liberties of the people and were embedded in the "Charters of Individual Rights and Liberties". Up to now, such charters of rights did not exist in the leftist societies. These fundamental distinctions are no longer respected today, because even the people from the most democratic countries have also lost many of their individual rights and liberties to the benefit of their socialistic governments. Hardly ever during their education with science and technology have people learned that other forms of thoughts had formerly existed.

Countries governed by liberal or neo-liberal ideologies

Neo-liberal countries and democracies (the rightists) had always been managed by governments elected by the people and by powers which differentiated the democratic societies from those managed by scientific and social materialism that were almost always obtained by force. But today, these so-called democratic societies only offer their citizens illusions of freedom and action, claiming not to enslave their citizens into the merciless slavery of the people who lived in the former totalitarian Marxist - Leninist countries. This semblance of freedom is only exercised today within limits narrower than ever, expressed in confusing and vague socialist vocabularies. **It is true that the citizens of these countries are free to work where they want, establish their own businesses as they wish, earn money according to their work, speak without inevitably having to undergo the torments of their powers and to change governments at more or less fixed periods… but under certain conditions.**

In return, the leaders of these pseudo democratic countries agree to redistribute their wealth according to standards and laws which are more and more binding and which are looking much more like socialist left-wing powers. The liberal and capitalist ideologies have more social and collective than liberal laws, that is more and more inhuman laws which stifle the individuals' rights and liberties and look more and more like the socialist laws which abolish the greatest part of the individual rights and liberties.

To summarize, right-wing ideologies feed on individual rights and liberties while socialistic left-wing ideologies always feed on collective rights and collective responsibilities. We will never repeat enough that these distinctions are more and more academic today, because the people of these two societies are both formed at the same school of scientific and social materialism with sciences, technologies, materialistic values and with less and less humanities. This education has standardized people and killed their humanity as in the socialist countries.

In so called liberal, neo-liberal or social democratic societies, the lure comes from the fact that the States are governed by elected powers, while in fact, the powers of these societies are in the hands of numerous scientists, technicians, intellectuals and mostly non elected pseudo-expert civil servants. They are the ones who create havoc and skilfully control the elected members and the Occidental societies, not the elected people.

Countries governed by Marxist or socialistic ideologies

The countries that adopt Marxist or socialist styles of government following revolutions exercise various forms of socialism. Their governments are generally not elected by the people. They came in power using arms and during revolutions. With morbid concern with visible equality and social justice, all these socialist governments, of whatever variances, **always administer by appropriating for themselves the civil and individual rights and liberties of their citizens in order to manage them,** to better control their activities, their liberties and their ways of thinking.

In return, their leaders supply them part of their essential and personal necessities in the form of popular social measures. In these countries, everything happens like in the fable "*The Dog and the Wolf*" of Jean de La Fontaine. The fable describes a fat dog well fed with household food but not free to run where he wants, and a thin, starving wolf free to go wherever he wants. It tells of the difference between being in prison and being free in a given country.

There are however few people in the former Marxist societies that lived like the dog of the fable. If these governments seem to give more social measures to their citizens, they are generally quite minimal. Work in these countries is not financially rewarding or stimulating and is always under the unique control of the centralizing powers of the state. Those that work must share with those that refuse to work…until the day no one wants to work anymore. No one is really interested to work for the benefit of those who won't. The loss of motivation to work inevitably brings a decrease in the wealth and in the goods of these countries. At the end, after a certain time, when all the wealth of the country has been wasted by the leaders and the social measures, when the elites have become rich with the people's goods, when no one wants to work to make these favoured elites live, the country goes bankrupt and its citizens remain in a state of extreme poverty and dependence. Examples of this are Russia, Cuba, North Chorea and many African countries after the Second World War.

The citizens of these countries come to behave like those birds fed near houses in the winter. They quickly forget how and where to find their food. Humans that are always fed by the State kill themselves doing nothing, as said Félix Leclerc in his song "*The best way to kill a man is to pay him for doing nothing*". Social bankruptcy always overcomes these left-wing States. And when it happens, these people beg other countries to help get them out of the mess that the social States put them in. There is another fable of La Fontaine that we can relate to in those States, that of "*The Cicada and the Ant*". As you will recall, the cicada "sang all summer long and did not have a piece of bread or even a worm to eat when winter came"…

France is an example of a former republican democracy which always claims to be a democratic republic because it preserves its right to vote. But the citizens deliberately rejected a few years ago their right liberal political system to replace it by a left socialist party. These governments call themselves social democracies and govern themselves with more socialist and Marxist laws than democratic laws. Their citizens are just like the governments they put in power: they think and work with the reason (logos), they train themselves with sciences, technologies, socialism and material values and they quickly become dehumanized.

Canada and the United States are also supposed to be democracies. The parties in place are elected and have no affirmed Marxist or leftist tendencies. They are presumed to refer to the liberal democratic principles of the right based on their previous history. But some of their non elected scientist leaders and experts in power, lawyers, politicians, technicians, administrators, computer scientists, high-ranking civil servants, economists, engineers and others, have all been educated for generations with sciences and technologies and as such all share more socialist and materialistic than human values. These states practise a sort of scientific socialism "of the center". Sometimes they manage with left-wing principles, sometime with right-wing ideologies, and so try to make everybody happy.

Certain Canadian provinces have already elected socialist parties to manage them. Bob Ray's socialist government got Ontario into debt in four years of dreadful social measures. Some socialistic governments, called neo democrats, have already swept the Canadian western provinces and several members of the government of the *Parti Québecois* lied for years by not calling themselves socialists or social democrats. Some do not dare give themselves a label, but when they vote for collective and more social laws, we knows which side they are on.

As demonstrated above, the true powers are in the hands of the scientists and intellectuals who set up the system, build up the political measures, write the laws, legalize them, advise the elected individuals, administer them, etc. All these civil servants in power are inevitably scientists and experts. All have received a scientific formation and are consequently followers of the ideologies of the scientific and social materialism. Their scientific education has made them materialists in the soul and all their actions are evidences of their ideological allegiance.

When these people come into power, they are quick to monopolize the rights and liberties of their citizens, in order to manage them in true left socialist ideologies. They legalize their actions with inhuman laws. The values that guide them are visible in the laws voted by the elected members. **Let us remember that, when a law makes the people lose their civil and individuals rights and liberties, imposes on the citizens their unique educational system, deprives them of their human, social or family values, controls their health, their schools, their medication and subjects them to the will of the leaders and the powers that enslave them, this law is inhuman.** It betrays the ideas of those that make it, because it models on socialist and collectivist principles. When a law of collective rights has priority on individual rights, this law stifles the individual rights, liberties and values and this law becomes inhuman in regards to individual rights.

Everything is a question of degree between left or right ideologies. It is nevertheless out of this degree of division that chaos was born. We were all witnesses to the monstrous demonstrations that socialists, syndicates, notorious communists and the numerous young people that united to prevent the extreme-right to take power in France in April and May of 2002. They betrayed themselves. It is reassuring to know that in this country all the socialist parties are well identified.

When it is not the case, it is not surprising that people have difficulty understanding the sense of such or such a law, to feel harmed and manipulated by their powers, to have no more interest in the *res publica*, to lose interest in their leaders and in their powers, to drop out of school and society, to become violent, etc. This happens when people have less and less individual rights and their powers impose more and more collective laws on them. A large percentage of disinterested people are no longer interested to work at contributing to the wealth of their country.

Socialist ideologies are beautiful on paper. But when time comes to set them up and to put them into practice, powers must inevitably deprive the individuals of their rights and liberties which are the motors of the human actions, and a general "demotivation" gains bit by bit the individuals. Work does not produce so much wealth, working hours are shorter, leisure activities increase and little by little, countries become impoverished, get into debt and anarchy settles in everywhere.

Vices of the scientific and social materialism

Other ideologies of Marx's socialism show their weaknesses. Not only have they been robbed of their rights and liberties and been imposed inhuman collective laws, the people must also conform to the dictates of the scientists and intellectuals who manage them. All those that oppose to the legalized dictates decreed by the powers are considered as being opposed to their material god. So they can easily be eliminated or legally executed or sent to expiation camps, as did in their time several communist countries. This is to what led this brilliant social Marxist ideology that justified millions of deaths under Stalin in Russia, Mao in China, as well as the atrocities in Korea, in Vietnam, in Cambodia and in a number of wars and exterminations of people during the twentieth century. These same social rules justified thereafter the multiple summary executions of their opponents on the planet. That those that have eyes to see look and that those that have ears to listen, listen.

The people must understand, before judging, the subtleties of their respective forms of governments and how science and scientific methods of research base their perceptions only on data accepted by their only most primitive senses! When the people of these societies lose their rights and liberties, it is always the scientists and the intellectuals in power who seize them in the name of society and of their god the matter. It is from these Marxist ideologies that powers acquire all their rights of life or death, their right to rob the rights and liberties of their subjects or to commit the atrocities described in the previous chapters. **These acts can only be justified by the new laws which legalize these powers and which no citizen can oppose without undergoing the consequences. The one that opposes to the expansion of "god the matter" or of society always deserves a punishment, even in democratic countries.**

The opium of the people of socialist countries, to borrow Marx's words and later Lenine's, is no longer religion, but the socialist ideologies of the Marxist-Leninists which refers to Mark's scientific and social materialism, the Hegelian philosophy and the "reason". The only legitimacy of this scientific and social materialism takes its source in the Hegelian philosophy of the "matter god". We shall go back on the facts demonstrating that matter is not eternal and that all socialist systems are sitting on the branch they themselves are cutting.

Wealth and financial means allow all kinds of models of socialist societies. They are spaced out from the extreme left, characterized by communism, Bolshevism, the totalitarian and totally inhuman Marxism-Leninism practised in several countries notably in Russia and in China and their satellite countries, passing by the "elected" socialist systems to the social neo liberalist and capitalist countries. The theft of the rights and liberties of the citizens is always a common denominator in all socialist States.

Voting at more or less fixed period is, strictly speaking, a democratic measure, the first in fact. The right to vote has often been diverted and no longer means that these governments are true democracies. Why? Because the true powers now belong to permanent scientists and powerful intellectuals who occupy key posts in governments. They are not elected and are permanently there to lead. We will also notice that, in several so-called right-wing liberal and democratic countries where voting still has its place - a sign that certain leaders would indeed like to consider as a mark of democracy and freedom -, the elected members are only puppets in the hands of these non-elected permanent scientific powers. The non elected members behind the elected powers are the ones that govern and <u>they</u> do not change after an election. **The elected members are often reduced to being representatives and doomed to remain behind these non elected scientists who hold the true power.** The elected members are cleverly driven by these more and more powerful scientists and intellectuals, these numerous "high-rank civil servants" of all sorts, who make the rules and who never have to answer of their acts.

The permanent military strategists, the deputy ministers of all the ministries and those that run hundreds of governmental companies such as Hydro-Québec (electricity company), financing societies, the CIA and the FBI in the United States to name but a few, already have policies and ideologies of functioning that can be very different from those of the newly elected powers. It is because of these parallel permanent powers that institutions as the American CIA or FBI can make poor decisions completely opposed to the ideologies of the elected political party.

Adapting our ways of thinking to the new ideologies

There are several variants of societies and ideologies within these two great tendencies. A priori, all have as objectives to scientifically justify their ideological truths and their right or left beliefs.

The ways of thinking of people and of powers are modified by the ideologies of their scientists and intellectuals, educated at the school of science and technology. Elected members often cannot do what they want because of these numerous permanent employees in place. A former Québec Prime Minister complained, over 30 years ago, of not being able to do as he wanted because of these permanent technocrats in power. Let us not forget the almost untouchable roles played by the leaders of institutions in place in the United States (FBI, CIA), when J.F. Kennedy took over at the beginning of the sixties. It was clear that the non elected leaders at the head of American institutions already imposed their own policies and their personal views on their institutions, on management abroad as well as within their own country, and often against the ideologies of the elected political party.

The twentieth century has known other political ideological system such as the existentialists and the ultra-liberals to name but these. These groupings tried to distinguish themselves from the main ideological currents by removing themselves from everything that did not please them. But there has never been a government openly ultra-liberal or existentialist who tested the ideologies of these types of government.

C. The role played by the teaching of sciences in the expansion of materialism

The important role played by the education with sciences in the perpetuation of the scientific and social materialism in the Occidental world was mentioned earlier. Let us now see how this materialism slowly took its place in education to eventually control it completely, as much in liberal and capitalist democracies as in Marxist-Leninist or socialist totalitarian countries.

The education with science and techniques without the humanities has become the common denominator that served at materializing and standardizing the people and their societies and to impose everywhere their materialistic ideologies. The forces that oppose each other today as they did yesterday are these materialistic forces on the one hand which oppose the traditional, religious, human or sacred forces on the other. Scientists in power in the Occidental governments were the first to choose to take away the humanities which characterized until then the education and imposed in its place the education with sciences. They chose to form their people at the school of science and technology rather than at the school of humanities. A union of these two forms of education would have been, in my opinion, a much better type of education to preserve the humanity of the people. It is important to understand that this education based on science and technology has completely replaced the traditional system of forming in all the Occidental societies, as well as in other societies of the planet. Scientists and intellectuals in power have imposed their materialistic system as being the most suited to the development of man and consumption societies. Regrettably, **being only material and based only on perceptions with the senses,** the education and values of this scientific system were essentially material. This Cartesian system of education has had dreadful shortcomings in education because it was incapable of fully forming man, that is the material as well as the spiritual man. This method cannot be used alone to form individuals, because it is incapable of giving at the same time a human training and a scientific training.

It has been demonstrated earlier that this method of current research can only be applied to the study of the physical thing of our universe. Science has always denied all the human realities of man. New research in neurobiology no longer supports the fact that the senses alone decide what is true and what is not. Comprehension by the senses alone only takes into account the decisions made by man's most primitive brain. This scientific method of research cannot take into account the data emanating from the superior levels of the human brain, such as feelings, arts, human solutions, etc.

A majority of intellectuals, technocrats, governments and people believed that science could never make a mistake because it searched for the truth. **It is unfortunate that few individuals have noticed that this materialistic methodology of forming can only take into account the material aspect of the training of people and that it has totally put aside the human and traditional values of which it could not confirm the existence with the senses alone.** Claiming that intellectuals did not realize that science and knowledge were only material is going a bit too far...To claim that science thought it could answer all the spiritual needs of the people is still going too far. If these aspirations were very legitimate and justifiable on the material plan, they were totally powerless to take into consideration the human side of man.

Trusting only the facts verifiable with the senses, science has totally rejected all human material entities in its educational system with sciences and technology. Let us remember that when these ideologies were put forward, scientists and intellectuals believed that the matter was God. Was not the Hegelian "*Deus ex machina*" the symbol of "god the matter" on which all scientists were working? People in power deliberately chose to teach and to form their citizens essentially with these material sciences, depriving individuals from a whole side of themselves. **Nevertheless scientists are now aware that this is no longer the case and that matter is not eternal, nor is it God.** Must we believe that these scientists have never been aware that their educations and their scientific values were only material...?

We believe that the powers deliberately chose this scientific and materialistic system of education to demonstrate that they were the true holders of power and that they were the ones responsible for the education and the forming of the people! It was thus clear from the beginning that this materialistic model of research was not a system for the individual and complete training of the people. One can only give and teach what it possesses, **and this system only possesses the knowledge on the material and physical world.** "*Nemo dat quod not habet*", says the proverb ("We do not give what we do not possess"). Even though feelings, art, introspection, the sacred, dignity, love, kindness, empathy, are not entities perceptible with the primitive senses, they are inherent parts of man. By selecting the scientific model for educating people, materialistic powers have forcefully excluded from education the human parts of the individuals as if they did not exist, and this, without any scientific demonstration that these entities were nonexistent. It is certain that these entities could never be verified by the scientific methodology of the "reason" guided only by the senses. Descartes had already given warnings on the use we could make of his method of research. **The evil comes from the fact that intellectuals and scientists should not have denied the existence of the things which they could not study with their incomplete method of teaching.** Extremely valid for research in science and to obtain information on the matter, **this method was wrongly and inappropriately used by scientists and powers to educate the people.** This method was perfectly adequate for training researchers to study the physical aspects of their world, **but was totally inappropriate to insure the complete training of man.**

It is and has been clear from the beginning that this materialistic system of thought was defective in itself and powerless to serve in the training of people. The error is then inherent to the choice of this method of education and of its use by the powers to train of the people. The method being defective, the system of education with this method was defective. Is this why it so badly marked of extreme materialism the generations of people that received this incomplete formation?

It is thus this lack of understanding of the powers that drove humanity to its actual state of dehumanisation which eventually killed the humanity in man. Scientists and atheist intellectuals wanted it this way, and so powers that decided to impose this system have committed a crime against humanity, or else they were perfectly total idiots.

Where are we at with this education of scientific materialism at the beginning of this twenty-first century?

The roles played by these material values, without their counterpart of human values in the education with sciences and the materialistic way of training, had numerous unhealthy repercussions on people and societies. This social disease is still in full expansion. In these countries everything begins with technique and science and ends with technique and science. It is the same from one country to the next because education with sciences was given and planned according only to the material necessities and the techniques necessary for consumer societies.

Nevertheless, in spite of all the evidences they have in front of their eyes, powers still keep on forcing professors to teach the same scientific reasoning from primary school to university, starting even before the children learn to read, write and count. In the occidental societies, the children literally learn the scientific methodology of thinking and of reasoning in elementary grades. Powers have introduced this Cartesian linear and mechanical methodology of reasoning in childhood. The scientific reasoning is very fashionable. Even though it is not bad in itself, it is abnormal that it should remain the only kind of reasoning that children will learn and keep all their lives. **The child is already an aspiring apprentice technician from the time he starts kindergarten and primary school.**

We are in awe of the reasoning of these children who seem to be prodigies. **They are formed this way to meet the requests of consumer societies.** It is necessary to know that learning the techniques is not learning how to think or how to learn and to be a conscious human being with all that it entails. The learning of human entities and of the human and traditional values that come from the superior levels of the human brain is not on the educational menu. **It is abnormal that today's young people learn only the material things that science can teach. There are other ways of thinking and of arguing in the human forming.** One cannot teach how "to argue" using only material and tangible things perceptible with the senses. If you never lean that respect or love is a human quality you will never respect or love others. A scientific system of formation that cannot teach the fundamental qualities in man has serious shortcomings. We also notice too often that with this education, the young people have not learned to structure their minds. There are holes in their ways of thinking. It is quite possible that all children are not made to become engineers or scientists. Certain may prefer to develop their artistic talents or others human attributes. Not all individuals are made to understand sciences or the mathematical linear scientific reasoning.

I have personally noticed that several students enter university and even commit to a master's degree or a doctorate, without having learned how to think in a logical way, besides not knowing how to write. These students are not less intelligent. They can very well structure and perform a research. But they do not have the necessary knowledge to analyze or see the relations between the data stemming from their own research. **A lack of general knowledge and humanistic thinking is clearly missing in these young people.** They know loads of things when they come out of university, but these are disorganized, in spite of their scientific formation and successful reasoning. Their problems begin not when they do their research, but when they have to put their work on paper or discuss the results of their observations. Many have great difficulty connecting together the data they nevertheless observed during their own research.

Mastering the language is another problem. We can understand why a study published in the popular press at the end of February 2000, showed that 20 % of the professors did not pass their French tests to enter university and that half of the pupils of secondary schools do not receive their diploma.

The consequences of a scientific education without humanism

When the education with sciences and technologies began, the systems and programs of training that had been well in place during the previous centuries were literally eliminated from the school training programs. None of the people in charge of education seemed to realize the possible impacts of their actions in the long run. The entire space was given to the education by sciences and to scientific training. This type of formation could explain why over 30 % of the students of secondary schools and college graduate today with a technical formation but without **any elements of general training.** These students are ready to earn their living by practising the techniques they learned to master. This lack of possibility to do anything but techniques must be frustrating enough to explain numerous suicides or dropping out of school or society. These individuals are extremely limited in their human relations with others and in their possibilities to understand the "news" and the values of society other than through their techniques. This lack of human formation does not allow them to get out of their totally materialistic universe. So they swim in the material values that they know: rave parties, beer, bars, sex, money, etc.

Since about three decades now, occidental students have lost contact with traditional values and have only scattered rudiments of human values acquired at home, if there are any. The only notions of traditional and human values come from their parents or from popular culture, which is very little. We know how much the parents are exceeded with the school affairs these days! **Many parents thought, wrongly, that the education of their young people was the responsibility of the school and of the professors.** Many young parents have also received the same dehumanized scientific education as their children. The professors also educated in the system of formation without the humanities, thought that the human and family education was the parents' business and that their tasks were limited to the scientific academic forming.

This is even truer when both parents work. Many must leave early in the morning to be at work on time. Children sometimes get to school with their lunchbox two hours ahead of time. Both parents and children are exhausted when they come home in the evening. Some even ask school authorities that their children have no homework or lessons at home. In any case, it is they, the children, who decide if they will do their homework or not... The teachers will tell you that they have more and more difficulty in asking that homework be done, that lessons be learned, and that they cannot penalize the children when it would be necessary. They have lost all authority, just as the parents have. After decades of such scientific forming, all the individuals who work in industries or society management know only about materialistic values. The industry needs high level technicians to do precise work in order to increase the production. We must increase the number of technicians by 50 % and this only to answer the current demand of the industry. Society needs employees to pay income taxes and taxes of all sorts to make the economy turn. Powers didn't think students should waste their time on a general formation or on learning how to think. And so, occidental societies are terribly short of thinkers! If the young people could think, besides using their technical training at work, would that be so bad? If, after college, the young could just solve the problems of man and the problem of "God", societies would change completely. **In societies educated only with the sciences without the humanities, these human questionings are not even listed in the school programs.**

Scientific education and current sciences are based on time management. Education is also "taylorised" in a sense. The objective is to "maximize the time and to learn to manage that time", had proclaimed the taylorists. And so, professors learn class management in a scientific way, according to time and the techniques to be learned. To make individuals understand what they are doing and how to do it does not seem a priority of this education system. Whether the students understand or not what they are learning is not the task of today's teachers. It is not even the task of the Ministries of Education. These prefer to inculcate to their employees a monolithic, uniform thought without feeling. **After decades of this type of education, people are standardized, hyper specialized and must refer to the powers to decide and to think in their place.** Is this not the way socialism was practised in Marxist – Leninist countries, where only science was taught without the humanities and the human values?

What can be said about the people in power who plan this education? **The important thing for them is that the children master techniques and a trade to earn their living.** And so these people just went and threw away centuries of humanistic education and traditional values. All that was not science was rejected: religion, the sacred, family and human values, ethnicity, deontology, public-spiritedness, allegiance, generosity and, let us say, the problem of man and the problem of God which will never be settled in this materialistic education: science simply ignores non material values. It was wrong of the people in power in these education ministries to believe that the young people of our societies of consumption did not need to learn other things than what was purely material. They anchored these materialistic principles so deeply in the minds of the young people that it is impossible even for those who would want to, to teach matters other than sciences and techniques. Students formed with these only material values are however free to learn all the techniques they want: the techniques of sport, of leisure activities, of learning, of reading, of mathematics, of computers, of administration, of social work, of nursing, of medicine, of law, of architecture, of agronomy, of aeronautics, of psychology, of theology, of history, etc. **In this culture of the technique, all that is not "technical" no longer interests students: the languages, history, morality, general forming, etc.**

Before this turning point of the forming with sciences, education was given with words and thoughts. The young learned how their ancestors had lived and how they thought. Today the young mostly learn through science the techniques that will help them earn a living. **So, modern pedagogy can be resumed to the management of teaching the techniques of science.** Even at university level, pedagogy amounts to learning more and more sophisticated techniques to practice sharper and better paying professions. **Medicine itself, formerly an art of treating both the human person and the disease, has been transformed into a faculty of learning the lines of conduct and specialized or hyper specialized techniques.** It is not rare today that a doctor specializes in a restricted field of knowledge to earn his living, such as the reading techniques of X Rays by nuclear magnetic resonance of the brain, only using hyper specialized techniques. It is out of the question for such specialized doctors to use their years of medicine to practise medicine. Authorities keep them from doing so. Very early on after having chosen surgery as a specialty, the future specialist will study for four, five, six years, sometimes more how to manipulate the techniques of a more specialized surgery, and must thereafter perform only these techniques such as surgery of the eyes, the ears bones, the replacement of heart valves, etc. All these techniques are learned in medicine as in several others professions, to earn a living by practising a particular technique or a group of restricted specialized techniques.

The art of practising medicine as did the doctors of yesterday now belongs to the past. Soon, general medicine may become a profession that "had" existed. Computer systems, programmed to take into account the "lines of conduct" when faced with such and such a situation, will one day replace the actual general practitioners. Modern hospitals have become places where loads of technologies, as different as day and night, are practised every day. Medicine is less and less a profession that treats diseases, listens to patients' complaints or takes care of them. At this level the human part of the individual is forgotten. **Modern hospitals**

are dehumanized places where very sophisticated technologies are applied to repair or replace damaged organs, rarely to take care of the human problems at the origin of the disease.

A book recently published in Quebec by teacher Gilles Gagné has a very revealing title regarding current education and on the almost complete takeover of education by the powers of the State: the book is called *"Mainmise sur l'Éducation"* (Takeover of Education). To meet the technicians' demands, the ministries of Education have invented a quick technical formation system to manage the classes. Students can no longer have a general vision of the subject learned. The techniques of the pedagogy of management do not allow it. In this automatic system of training and of education by sciences, teachers also have less general knowledge: they too are technicians. Teachers are often technicians whose only role is to teach techniques to the students. Culture and a global vision on the matter taught are considered a waste of time. This absence of humanism in education creates this dehumanised attitude and a lack of interest in teaching in the professors, and, in the student, a lack of interest and taste to learn. It is wearing for professors to teach without thinking and to only follow the lines of conduct imposed by those in power.

In front of such disinterest, is it necessary to ask why students drop out? The answer seems simple enough, simplistic even, to be understood by the analytical minds of the scientists in power. Students, like anyone else, have little or no interest in learning when they do not understand or have only a limited general vision of the matter they are studying. How can students who never learn history have a vision of what history is? How can a student who has never learned to think about what he is doing on this earth, what the humanities and human traditions are or that spirituality may be a part of him, suddenly become interested in the humanities, human traditions or spirituality? Those who did not learn to think about their origins, about whom they are and where they are going, have no feeling of belonging to society and reject it. This is simple to understand but impossible to do so with the only reasoning of the material scientists. It is the powers that make the decisions about education, that will teach them where they come from, what to think and how to solve their problems concerning education. Those people in power have judged, because they think themselves capable of judging, that the human aspect of man is not important to practise a technique, to earn a living and that it does not even exist.

Today, History is taught in museums, during journeys, by reading and on television. But only a small number of individuals will visit museums and only a small number will learn ill-assorted rudiments of history on scattered subjects and without continuance, so what they learn will never be classified at the right place in the individual's mind. There is little room for personal learning in the actual dehumanized forming. The learning of techniques passes inevitably before the learning of general and personal knowledge in these dehumanized consumer societies. The people in power within the ministries are probably already all experienced in the education of sciences and in the management of the scientific techniques. How could they have imagined teaching in a domain of which they ignored even the existence?

One day, when they are older, in adult educational courses or in lectures somewhere, some people will feel the need to learn some rudiments of human culture... The scientists, the scientific and social materialism, the material values, the Cartesian methodology of scientific research, the magic thinking of pseudo intellectuals, the knowledge acquired on the matter and the materialism of technological applications, have all joined hands to make man lose his humanity and to lead him towards unprecedented dehumanization and atheism which strike humanity today.

D. Matter is not eternal and the incomplete scientific methodology must be rethought to take into account new and non tangible facts

The saga of science and technology has been prodigious. The human and social changes which occurred in the last century in this dazzling technical and ideological revolution are impressive. Regrettably however, we must understand that this great evolution has produced numerous side effects. Among others, it has led people and societies to this materialistic and dehumanizing decadence which is at the origin of the chaos that we have been talking about.

Two fundamental errors were committed because of the lack of knowledge at the time of the elaboration of this historical turning point. Both were caused by the enthusiasm and the excitement which followed the elaboration of Hegel's philosophy and the infernal logic of this **only defective Cartesian method of research for forming individuals.** Maybe it was impossible to do otherwise at the time?

But today, more than one hundred and fifty years after Hegel, scientists know that matter is not eternal, neither is it God, as their fathers thought. So great for the study of material things, they also know that the Cartesian method of research they use has the deep incapacity of studying things other than material.

Matter is not eternal

The world learned one day that all this scientific madness that has upset the people for over a century and half, that all these new currents of thought, that all this training of the people with their only scientific method of research on the matter, that all the powers that scientists, intellectuals and governments have appropriated for themselves, are not justifiable any more. The fundamental principle which supported the legitimacy of this scientific legend is no longer valid. Scientists can no longer base themselves on Hegel's principle of the eternal matter: matter is not eternal. We had to wait 150 years after Hegel for science to demonstrate that the matter is not eternal, nor is it God.

The hypothesis of this theoretical assertion of the "eternal matter" will remain true until 1983. Scientific proof was presented that autumn in a congress of Physics in Utah, U.S.A. **Researchers of the IBM Company as well as others have demonstrated without a doubt that matter could not be eternal.** It degrades at the power level 10-32!

Since by definition God cannot degrade, the emotion has been enormous in the scientific community! What a disaster for the communist or socialist countries and for all those that share the Marxist ideologies! From the moment that matter is no longer eternal, all the teachings, the ideologies, all the euphoria, all the thoughts elaborated by the great thinkers who followed Karl Marx' hypothesis, are no longer justified. **Hegel falls and inevitably all that had found its legitimacy in this philosophy also falls. All the social ideologies elaborated by Karl Marx and those that came after him had to be revised.** These ideologies had lost their reason to be... the world, such as pseudo scientists and pseudo intellectuals imagined it without scientific data must also be revised... Imagine all those social ideologies that followed Hegel and Marx for over a century and a half and on which the occidental world legitimized its powers, including those of their governments: they have no more reason to be. Karl Marx's beautiful socialist ideologies and those that followed disappeared in thin air. Most of these ideologies were only lucubration of pseudo intellectuals and were never demonstrated scientifically with experiments. Can you imagine the amount of damage done during all those years?

The world changed during all those years because of the masterful influence of the Hegelian philosophy. Sciences had taught and shaped the world for over a century, as if Hegel were right. **The world, our world, must be redone and rethought because it was wrongly modeled on the philosophy of the eternal matter, his god.** Pseudo scientists and intellectuals lost all their legitimacy. They must revise the materialistic truths that they taught. Even their methodology of research, their ways of teaching and of reasoning that justified forming the people and the minds as if the matter was a god, are not valid any more. All the societies that received this education and that are now set in this atheistic materialism because of their scientific education must start all over ... The philosophy of the eternal matter which gave birth to and propagated Karl Marx's scientific and social materialism and engendered this multitude of new materialist ideological currents of thought, is no longer a legitimate truth. **The powers of the pseudo social scientists, intellectuals and the governments they manage are no longer legitimized!**

Even after being demonstrated for 20 years, this evil and deep materialism still continues to mark the people and their societies formed for decades by this false education... This time, it is materialism and science that step down from their pedestal, as religions did when scientists and intellectuals slipped into the posts of command. **From the day scientists themselves demonstration that matter is biodegradable and not a god, they lost all their legitimacy.** Scientists and intellectuals cannot teach any more or even consider all that they had elaborated for decades on this false principle of the "eternal" matter. If the foundations and ideologies of the Hegelian materialistic philosophy suddenly collapse like a house of cards, then Mark's socialism also collapses. They no longer have a reason to exist! All Marxist - Leninist societies or other forms of socialist societies based on the eternal matter, also lose their reason to be.

This finding is tragic not only for the people and the Marxist societies, but also for the socialists educated only in this scientific and social materialism, as well as for the whole occidental world, the pseudo scientists, the intellectuals and a great number of their ideological teachings. Let us hope that in the future less materialistic societies will help soften this materialism, help science to mature and to change its method of rigid and linear reasoning, especially in the forming of the people. **The intellectuals of the reason (logos), as well as all the ideologies from centuries of "reason", must immediately revise all that was wrong in all their no longer valid reasons of running people and societies.**

The fall of the socialist communist countries soon after the demonstration that the matter was not God, as well as the events of September 11th, 2001 must be considered as warnings that should engender strong awareness of the tragic "human emptiness" in our education system. Maybe everything is not lost, after all. It is understandable that we cannot make disappear with the stroke of a pen, decades of training and centuries of "false ideologies and thoughts" nor without breaking eggs or without all the people becoming aware of the wrongful teachings they received for a long time.

We must start all over again at "zero". But we cannot forget that millions of people died in the name of all these social and political ideologies which arose from Marx's scientific and social materialism. Millions of persons suffered for a century seeing their values trampled, their false governmental powers deceiving them with their inhuman and dehumanizing laws made in the name of these "false" ideologies. The harm is done and continues to eat away at the people and the societies. Important problems were created because a considerable number of individuals of this planet were formed to think and to live this materialism as if Hegel's and Karl Marx's ideologies were the truth.

We now know for sure that scientists and intellectuals set up a defective educational system and justified their hegemony and that of their powers with these same false truths! What can we do? Will it be necessary to start again at zero? If so, from where exactly do we start? The traditional and "ex cathedra" values of the scientists are shaken. The materialistic truths that denied the human being in man should be reviewed. No one can now legally use these materialistic dogmas, teach them, communicate them to others and much less use them as the ideal way to form people. As in the totalitarian countries that tried to experience this system, the lies were there, and lies are a much more material than human value.

Having undergone the torments of the Marxist totalitarian societies, some societies now turn towards a semblance of democracy based only on the rights of man. "Could yesterday's democracy still not be the least bad form of society to have existed?" asked Fukuyama.

The Americans are trying to export their social "democratized?" capitalism as the best model for the world. In his book and film "*Bowling for Colombine*", March, 2002, Michael Moore, actor and American film-maker, harshly questions his materialistic American society. He doesn't mince his words when he says that "the white human being", his fellowman, is materialized to the marrow of his bones. He qualifies his fellow Americans as the worst product of humanity. **They are the greediest and the most gluttonous for material things and money, and they still believe they are superior to everyone.** "It is this white American", he continues, "who has the most faults, commits the most crimes, is the most poorly educated, the one that reads the least and knows the least about the world he lives in". Is it this not a rather dreadful description of the democracy that American politicians want to export all over the world?

The limits of the method

Something must be done: we must completely review the scientific method of research to study entities other than the material ones of our universe. If this scientific method of research and of reasoning gave us so much physical knowledge on the matter, it also has its weaknesses and its limits. **We knew that the knowledge acquired with this methodology of "reason" was always and inevitably of a material order and answered only to the most primitive senses.** The need to always verify data by the senses is the Cartesian method of reasoning. **It can only be confirmed by using the more primitive senses of the human beings: the reptilian senses.**

Already limited with the linearity of the "reason", it seems that this same method of scientific research cannot, in its current state, take into account the more subtle data registered later on in the superior levels of the human brain. In other words, this method cannot take into account the data accumulated and registered in the part of the brain where is found the taste for arts, the emotional brain on which reason has no influence, the bran of the instinct which guided evolution during several thousands of years. The senses cannot recognize these higher levels of knowledge in the brain **that is the less tangible data that the reptilian senses cannot perceive.**

In neurobiological studies, such data have been found to be stored in the superior levels of the brain. These new discoveries add weight to the deficiencies of the Cartesian methodology which rely only on what can be analysed with the primitive senses. The scientists of the nineteenth century as well as Descartes did not know at the time that there were more evolved levels of perception in the brain. **These other levels of consciousness take into account more recent and subtle data of the evolution of man,** I mean data that have been stored there probably because they were of a different order and important to keep in mind for the survival of the species. Researchers even think that one day they will be able to localize the centre of the human consciousness in the human brain... We are far from the material perceptions that can be taken into account by the senses only!

However effective for the accumulation of data on the matter, the current method of research of scientists regrettably relies only on the simplest perceptions of the brain to accept or to deny a given fact or to discriminate finer and less tangible subtleties acquired during the more recent stages of evolution.

Scientists, pseudo scientists and intellectuals can no longer ignore these noticeable entities that the primitive senses can not perceive. It may be possible to one day correct the errors made by scientists and their followers educated only with the reason of the primitive senses and that drive them and our societies so easily to increasing atheism. **It is also possible that errors of reasoning were made simply because the method of research could not take account of all these new data.** We are aware that numerous hypotheses were so far rejected only because of the limitations imposed by this method. Several data should be reanalyzed in order to take into account a number of these less tangible and perceptible facts. Several new observations in quantum physics, astrophysics, for example, that appear as non material strictly speaking, but which cannot be studied at the present time because of this defective method of research. **And so, there is an urgent necessity to modify the actual method of research, in order to take into account entities that are not tangible with the senses and that have been rejected as nonexistent for that reason. One must urgently modify this method or develop an entirely new method that could and would take into account material facts as well as less material entities.**

What has become of man and society?

The societies that lived under Marx's scientific and social materialism applied in the totalitarian countries are dismantled today. They are searching for a new legitimacy that intellectuals and researchers have not yet been able to offer them. The Marxist - Leninist socialism does not exist any more: it lost its philosophical foundations and was never demonstrated with sound experiments. Societies and the people formed at that school must now purge themselves from these false ideologies and start on new bases. The task will be difficult. People cannot be brushed aside, as we mentioned earlier, like their ideological system when it lost its legitimacy! These people are now living in their respective societies and have even lost their reason for being! What will they do? No one knows!

Some of these societies are trying to restructure politically, to find other social models and to modify their ideologies on principles other than that of Hegel's "eternal matter". Some of these societies try to turn towards other forms of existing societies somewhere between socialists, neo liberals or capitalists, but have not yet gotten there. **Some thought of turning to the Human Rights of the democracies and thought they had found a new reason to exist. But Human Rights oppose the spreading of collective rights and the laws now in place in all our socialist, liberal or capitalist societies.** One thing is certain, the mutant individuals of the socialist materialism are in mutation towards something else, still ill defined and which could be qualified as democracy's primitive rudiment.

It seems so far that there are only vague and unclear ideologies put forward to get closer to the economic liberalism of the democracies. Socialist societies are trying to adopt the Human Rights in their new basic principles because in spite of their defects, as will say Fukuyama, democracies still seem to be the best form of government that have so far existed on earth. And while we talk, we keep on teaching and forming citizens as if nothing had happened. Until when will political powers dare refer to these socialist ideologies, knowing full well now that they are in the wrong? Hegel's fall and the fall of socialism left a lot of confusion in its trail.

In *"La Démocratie contre elle-même"* (Democracy against Itself), Marcel Gauchet, Gallimard, 2002, believes that all ex-communist countries and their various interpretations of democracy, **now confront the democracies with their greatest adversaries: themselves.** Democracy becomes undone as it progresses, said Gauchet. Is it for that reason that we think democracy is just a lure? Even though more and more people and politicians profusely use the world democracy, this type of government is no longer what it once was. **Democracy is the power of the people by the people, not by groups of people and even less by governments that forget the people once election time is over.** This means much more than a vote in more or less fixed periods of time. It must be redefined.

I have trouble imagining how a society can be at the same time a democracy and a socialist state which constantly appropriates all the individual rights and liberties of its citizens. A society cannot be at the same time a democracy and a socialist State. There is an improbable semantic paradox in all modern democratic states and mainly in social democracies. To rebuild the world on new democratic individual values, one will have to return to the former conception of the essence a democracy. It will be necessary for science to consider that the human brain cannot rely only on its material primitive senses, since it is also made of feelings, arts, respect, instinct, etc. Where are the neutral philosophers today? They have all been killed by the materialistic scientific education. **Where are the scientists and the intellectuals capable of freeing us from this "dreadful evil of our modern societies", where scientific and social materialism has almost swallowed us all?**

Generations of people have already been lost: we must acknowledge this. If the powers that intellectuals and pseudo scientists took from religions by referring to the Hegelian philosophy are not justifiable any more, **who will possess this power tomorrow? We believe that the**

ball is now in the camp of each individual and that each of us should become aware of this and own it.

People educated in materialism are at a loss and do not know what to do about it. Nobody has any plans or ideas to replace the absence of humanity in their socialist societies. They still continue to cling to the last vestiges of their fallen powers. Nobody really knows how to reorient the societies on a reasonable ideological foundation. All the education based on science, technologies and material values given for decades must be rethought. Governments and their technocrats have no other legitimacy today than that of an election to resume their powers. Ideologies of replacement have so far been disconcertingly poor. **Must the people continue to feed on the "dehumanized reason" (logos),** even though the ideologies of the scientific materialism have lost all their meaning and that the true democracies are thinking about their development? The horizon seems a rather empty space to gather people around one valid thought...

Will we assist at a comeback of the religions...in this flight towards the sacred where we could at least turn towards more human and sacred values that are so missed by the people since such a long time?

Maybe it is time for each individual to take himself in charge and to take the reins of his own fate and so become aware that the generations of atheistic materialists that controlled his fate can no longer continue to drag him in this absolute negativism. Their administrations are illegitimate and obsolete, because they cannot base themselves any longer on powers other than science and "reason" which have lost their power and the legitimacy to continue to managing governments.

We hope to have demonstrated so far that scientific and social materialism and their materialistic values are no longer justified in education and that the scientific model of forming is incomplete and inadequate to form the people without including the humanities in education.

The task will be difficult. Scientists and intellectuals should immerse themselves in Alexis Carrel's book *"L'Homme cet inconnu"* (Man, that Stranger) which says that "Men accept with great difficulty the things they do not wish to change from the bottom of their hearts, for it would oblige them to question absolutely everything".

A return to more fundamental values, to more human choices of life, to the human as a being, to both matter and heart and to respect for the human person that has been denied with the sciences, is absolutely necessary for the future of humanity. And this, in spite of all the challenges it might represent for our level 0 sectarian and egoistic civilization. (See above).

In the proposed solutions (cf. Chapter 9), we will stress the fact that every individual of this planet should succeed in becoming individually aware of his own powers. Every man on earth owes himself to decide what is good and what is bad for him, his humanity and his world, after knowing all the options, not only part of them. **The taking over of people's powers by governments, religions, alleged intellectuals or any other group who are finally no better than the individual himself, should be systematically revised.**

Will the twenty-first century be religious or not?

It is while looking at the speed of expansion of the social materialism invading our world, at the frightening dehumanization of humanity and at the predominance of material values over traditional ones, **that Malraux had the vision that our world could not survive beyond the twenty-first century to the growing materialism which already struck him at the end of the fifties.**

Materialized people so badly need to autodestruct! There was no need for science and "reason", completely disconnected from the instinctive and human values, to further confuse man and humanity! It has always been evident to Malraux that the history of the "reason" of the twentieth century with its perpetual wars and new atrocities on a planetary scale, had already demonstrated fifty year ago that material values put forward in the teaching of science, technology, and the linear and rigid reasoning of scientists and intellectuals, have commanded a return to more humanity. Added to man's innate desires to use the most sophisticated weapons to destroy **himself it became clear to Malraux that in front of this enormous dehumanisation, and without a return to more humanity, the sacred and the religious, man would not survive the twenty-first century.**

I also refuse to believe that the French philosopher Gabriel Marcel, Malraux' contemporary, was able to write *"Les hommes contre l'humain"* (Men against the Humane), without thinking about the great cupidity of human nature which materialized through technology without reacting. The philosopher was not without having noticed also, as his contemporaries André Gide and Léon Blum, that materialism and Marx's social ideologies were doomed to failure because they disconnected man from his human reality. We are right in the middle of this inhuman world described by George Orwell 50 years ago, and we are unaware of it!

E. Conclusion

Saturated by centuries of abuse by religious powers, spellbound for years by science, its mechanics of "reason" and the bubbling of new social ideologies, scientists and intellectuals of the nineteenth century threw themselves head first in the scientific and socialist adventure. They took advantage of the quickly expanding scientific knowledge to seize the powers of religion over the people, crying high and loud that Hegel's new philosophy on the eternity of the matter, his god, conferred them all this legitimacy. They adopted the Cartesian scientific methodology of research to explore our material universe and took advantage of the new accumulated knowledge to slip into the commands of the States. For nearly a century these scientists and intellectuals have completely governed the occidental world.

The new material knowledge acquired at the speed of light on the matter and the plethora of technological applications coming from this knowledge has deeply changed the conditions of life of the inhabitants of the planet all along this scientific excursion. This dominance of sciences still continues to expand today at a vertiginous rhythm. Many of the materialistic and social ideologies that have been put forward refuse to die and greatly help spread the evil of materialism among men. Most currents thoughts and ideologies developed at that time continue to grow through education with science and technology, and are taught as being the only real truth. Reinforced by the natural thirst of man for all that is material, the new technologies have changed the world as well as man's instinctive taste for this materiality that continues to stand out pitifully everywhere in man and society. The world is today literally eaten away by this materialistic evil which turns people to asocial levels and dehumanizes societies.

Born of the enthusiasm of Hegel followers and Marx's scientific materialism, all these social ideologies have left a great number of disastrous side effects on our physical world as well as on people and society. The social ideologies are still taught by the defenders of the reason (logos) as being the only real truth, although they were never demonstrated with scientific experiments. Moreover, **scientists and holders of the reason have always refused to take into account both the true material and spiritual nature of the human being.** So, when they were applied without humanity, the unproven lucubration of the pseudo scientists of the reason and of intellectuals have often been catastrophic to the environment, to life on earth, and the human values were lost in these deeply dehumanized individuals and societies.

However, 150 years after Hegel's philosophy of god the matter and the strong domination of Karl Marx's scientific and social materialism in the Occident, physicians have demonstrated that the matter is biodegradable and thus cannot be eternal, nor can it be considered as God. **And so, suddenly, all the social systems developed for men to live in societies where the matter is God and all the powers built on these false social ideologies lost all their legitimacy.**

The people educated in this materialistic system of education without the humanities were then transformed into mechanical robots without humanity. **So the air, the water, the earth and the seas, these elements necessary to life on earth were inhumanly damaged by the great pressure of the technological applications and the absence of humanity.** This is well demonstrated by the disastrous effects of pollution on our world, the destruction of life on earth, the materialization of man and his societies and the endless chaos, wars, extinctions of people, destructions of humanity, etc. that marked the twentieth century as the century of technological explosion and of the reason (logos) applied without the humanities.

There is a great imbalance between man and nature and between materialism and the humanity of man. Confrontations appear everywhere between these protagonists. They are tearing our world apart.

Through education, people communicate this materiality like an epidemic into all fields of human activity and spread its evil and morbidity everywhere. The supremacy of the sciences and their material values in education dethrones the former humanities and the traditional values. In its place, they transmit their materialism to the soul of the people and of the societies. After a few decades of such scientific forming without the humanities, sacred social symbols disappear and people are transformed into materialized and atheistic robots... **These humanoid robots forget that they were former human beings made with matter and spirit.**

The world is now invaded with this dehumanized and increasing atheism of the individuals using mainly material values in their everyday lives. They engender indescribable chaos and transform into science everything they touch: economy, politics, administration, psychology, the matters taught, the traditional professions, etc. All ranges of human activity are undergoing a similar indigestion of this materiality carried by sciences, technologies and their "sacred reason".

CHAPTER 8

Materialism and the downfall into despair

Introduction

Our low level of civilisation, man still being a wolf towards his fellowman and the past being guarantor of the future, this actual "unbridled materialism" can only continue to invade man and society. We will soon be living in an Orwellian universe if nothing changes. **"If a drastic turnaround is not made, the world will burst into a succession of conflicts and crises"** wrote Ervin Laszlo, in *Virage Global*: L'*effondrement de notre monde est-il inévitable*? (Is the Collapse of our World Inevitable?), Éditions de l'Homme, 2000. The author attributed the need for this "Global Turnaround" to the massive destruction of the environment, to the fast and unilateral globalization, to societies' thirst for money and to an unstable world economy.

The data presented in this work localize the origin of this materiality of our civilisation to the imposition of material values in education with sciences and technology without the humanities at the centre of the current degradation of our world. This scientific, technological and ideological evolution generates everywhere a multitude of perverse effects so fraught with consequences for life on earth and for man's future **that a "turnaround must take place within a generation. If not", wrote in substance Laszlo, "side effects will provoke such destruction in our world that the damages to the environment will no longer be repairable".** Unfortunately, Laszlo did not mention in his work the dehumanisation of man and society. He neither links these disastrous side effects of the environment to science and technology. But, what Laszlo describes indicates to me that the people are soon going to live in that closed, controlled and dehumanized universe, in many points similar to the one described by George Orwell in *Big Brother* over fifty years ago, if a rapid turnaround in our living habits does not occur.

We are progressing towards "tomorrow" and we are all responsible for this tomorrow. The building of our future should not left to destiny and to fate! The task of changing things is so great that certain pseudo scientists prefer to believe that man is only a material machine that can be changed at will. But it not true: this Greek mechanical conception of the universe is no longer acceptable.

Every man should keep in mind the terrible living situations of the people formed in the socialist automated totalitarian modern states which abolish liberties and individual rights, intercept phone conversations, filter e-mails, degrade the people to the rank of animals, etc. Reporters without Borders and Amnesty International even state that this type of doubtful socialist practice allows countries such as United States, China, Vietnam and many others to even put in jail cyber dissidents. We must keep in mind the failure of this scientific and materialistic conception of man and societies.

A century of "reason" or of nonsense...?

A century of materialism has already spread desolation everywhere. From the war of the Boers at the end of the nineteenth century, the Russian Revolution (1913), the First World War (1914/1918) and the Second World War (1939/1945), the extermination of the people to the crazy materialistic ideologies of the Nazi concentration camps or of the Marxist countries in Russia, China, etc., millions of human beings have been sacrificed at the altar of unreasonableness by supporters of the "reason" (logos). Dreadful attempts to exterminate the Armenian people in 1915 and successively in Cambodia, in Uganda, in Rwanda or in Kosovo afterward, as well as the unthinkable and systematic slaughters caused by uncontrolled racism burst all around the world, **at the same time that environment was being destroyed and that man materialized.** Yesterday's wars were legitimized in the name of a myth, to spread a culture, a religion or to conquer territories. The atrocities of the twentieth century were all planned by men of science, tyrants or intellectuals in power and always methodically, for scientific reasons and with all the newest technologies.

Materialistic principles and social ideologies continue to perpetuate the disasters of the twentieth century. People and societies continue to materialize throughout the world and to dehumanize at an incredible rhythm; powers continue to deprive their citizens of their human values and citizens become "atheists" due to this continuous living in material values.

Scientists and intellectuals continue to impose their science and their values as a religion in order to preserve their powers over the people. They keep on claiming the supremacy of their "reason" and administer the people and the societies according to their laws and dehumanized values, as if possessing the absolute truth. **And without humanity, the people pollute their planet and unconsciously destroy**

their habitat, threatening more and more the existence of life on earth and their own destruction ...

Chaos continues to devastate the planet as in the twentieth century. Barbaric acts planned by the "reason" of the people in power are constant witnesses of this materialization in the Middle- East, in Palestine, in Israel, in Afghanistan, in India, in China, in Colombia, to name only these places of almost permanent conflicts. For over a century, these ideological battles are always expressed by the manifestation of indescribable destructions imagined by the scientists and intellectuals in power. Just like the unprecedented horrors which swept humanity during the century of technology and science, current horrors continue to justify their "reason for being" as they let scientists and their ideologies exercise their supremacy over the other people and societies of the planet. **One can almost assert with certainty that all the conflicts across the planet can be attributed to this superior thought of scientists and intellectuals that place science and "reason" (logos) at the top of human achievements.**

People continue to lose, in the hands of these pseudo-scientists and intellectuals in power, their civil and individual rights and liberties and the power to manage their own lives. Regrettably, the dictates of this materialism continue to drive the inhabitants of this planet, rich or poor, archaic, traditional or evolved societies, to an unconditional attraction for tangible assets, wealth, the lure of money, lust, etc. Such chaotic behaviours succeed one another and threaten the world with suffocation. **Let us see how.**

In the absence of humanity and human morality, the natural resources of the earth are and will keep on being exploited and wasted in a shameless way. The physical environment of the planet is degrading so quickly, that certain researchers think that the fragile achievements essential to life will not be able to survive, for more than a generation, **the assaults and human socioeconomic pressures exerted by these materialistic individuals.** Several populations are already reduced to not being able to satisfy their most basic and fundamental needs such as drinking water, non polluted food or clean air. And things are not improving...

The scientific thought should stop continually comparing the human brain to a computer and considering man a machine. These pseudo scientific ideologies of the cognitive movement must be strongly fought, because pseudo scientists and intellectuals want to make people believe that man is threatened with extinction... What will we do of the human rights if there are no more individuals left? If man is just a computer, where is the difference between killing a man and shutting down a computer, asked Jean-Claude Guillebaud in *"L'homme est-il en* *voie de disparition?"* (Is Man in the Process of Disappearing?), Éditions Fides, Montreal, 2004.

How will tomorrow be if nothing changes?

In the past, people could take centuries to adapt to the new changes in living. But the speed with which sciences and technologies develop today no longer allows the people to rectify the rapid degradation of the world before it falls into a state of non viability.

Even though the natural selector of AIDS has already decimated the young populations of South Sahara's African countries and will decimate millions of other persons in Asia and in China during the first decades of this century, (medicine not yet having found the way to fight this mass disease), the global population is still increasing in an almost geometrical way. So, to improve their conditions of living or to avoid the impoverishment of their populations, **people rush towards the large urban and industrial centres** of their country, or else try to immigrate to other countries better off than theirs. **These large movements of populations contribute at destabilizing the cities and the countries where these people gather.** Over 85 large cities of the world now possess shantytowns, where the poorer individuals pile up like animals to survive in incredibly unhealthy conglomerations.

Huge industrial centers are growing in these large cities and further contribute at increasing the pollution of air, water and the environment. Socioeconomic tensions are always more acute in these overpopulated cities.

The need for food and services of all kinds increase the pressure on the agricultural production of these countries. An agricultural overexploitation becomes necessary in order to produce and feed these new masses of people. The summers of 2002 and 2003, with their dreadful floods in Europe and in Asia especially, were once again witness to the climatic changes caused by all the pollution that keeps on accumulating.

Besides increasing pollution, **the overexploitation of agricultural land contributes to the draining of the earth's reserves and to literally empty them of their nourishing constituents.** So, the productive lands of the planet continue being impoverished and the culture of food becomes more and more difficult in our formerly rich earth suitable for cultivation. Added to the problems caused by the expansion of agriculture, industries and the services needed by the population movements, pollution and degradation of the environment will only continue and the elements essential to life will rarefy.

Pressure is continually being exerted on the elements essential to life: the water, the air and the earth. The reserves of drinkable water are the number one problem of this century. Scientists foresee that in twenty-five years, half of the inhabitants of the planet will lack drinking water.

The industrial needs for the production of goods for all these people will further increase the production of toxic waste and pollutants and so the pollution of the environment and of home surroundings can only increase.

With factories continually emanating their toxic poisons in the environment, the expansion of agriculture and its use of fertilizers, weed-killers and pesticides, etc., cars spewing billions of tons of CO2 in the air each year, waste accumulation, the deficiency in sanitary services, the socioeconomic conditions and the environment can only continue to deteriorate in the world.

The sad situation around large cities is already unbearable. Sewers in the open air are multiplying. Polluting greenhouses gases continue to be thrown back into the environment. **Seas continue to warm up besides receiving more and more pollutants and being emptied of their fish and seafood. Health systems are deteriorating, diseases are increasing, medical doctors are few and socioeconomic costs are excessive everywhere.**

The waste produced by all this overpopulation is incalculable. It must be sorted out, recycled or buried in large pits and is becoming, as mentioned above, a constant threat of contamination to brooks, rivers, lakes, sources of water supply and/or subterranean waters. Studies on nature revealed in September 2003 that the North Pacific salmons that die after reproduction contaminate seven times more the waters of the Alaskan lakes with PCB's (polychlorinated biphenyls) than the natural pollution of earth and sea in the world. This is an example where marine pollution re-contaminates the sources of fresh drinking water.

The use of fossil fuel still needed to insure the production of electricity in consumer societies, added to the sanitary systems overflowing in several cities, can only worsen the pollution of the elements of life.

Without real and quick international intervention to correct all these perverse side effects of science, technology and thinking, the earth's climate can only continue to degrade, the air to warm up, to be emptied always more of its oxygen and be replaced by new toxic pollutants. The ozone layer will continue to thin down and climates will warm even more. Polar ices will continue to melt, flooding cities and residences and some will even disappear. Cataclysms will succeed one another under the shape of tornadoes, floods, earthquakes,

masses of fallen earth, etc. The sources of drinking water will dry up and more people will literally die from thirst and from famine.

If people's health and ecosystems continue to degrade at such a rhythm, new diseases and cancers will appear and grow. Health costs, already difficult to support by the working populations, will only continue to increase. More medication and health products will be needed only try to counter these side effects, etc. Thousands of forms of life will gradually disappear from the surface of the earth and desertification will become emphasized everywhere, leaving complete populations hungry and terrified.

Instead of the long awaited evolution by scientists and intellectuals of the "reason" who lead the world today, it is the human shortcomings that will have the upper hand in the struggle for survival.

No one can distinguish any longer the linear and mechanical order expected by the upholders of the "reason" and the instigators of the scientific and social materialism. On the contrary, even in the midst of a supposed physical well-being and an unprecedented evolution of the order of material things, people are more and more confronted to a complete regression of their humanity. These socioeconomic troubles can only create deeper social crises. It is obvious that the powers of the scientists and intellectuals, **which were to lead the world to truth and order, did not do their homework. The disorder that came with the materiality they scattered everywhere will continue to progress, if a rapid turnaround is not done.**

At this point, we should ask ourselves if this lucubration of the socialist ideologies of the pseudo-scientists and intellectuals was not a monumental trickery for humanity, since they used scientists and intellectual powers to rob the rights and liberties of the individuals, to control them, impose on them their values and make slaves out of them.

Without the major awareness of all the inhabitants of this planet and the Global Turnaround suggested by Laszlo, chaos will only continue to progress everywhere and leaders will continue to enslave their people.

Disproportions among the rich and the poor will increase. Unusual and new kinds of violence, more important than those we know, will appear everywhere. Violence and crime will only increase and shake much more intensely the individuals who do not have the advantages of the rich, when the tyranny of the dehumanized "reason" reaches its heights.

Certain richer and more materialised countries will continue to arm themselves and dominate the world, always to speak about their "reason" to less fortunate countries. Forms of destitution different from those that humanity has known will appear. Other people will be oppressed.

330

Other sophisticated wars will invade the planet of the muted humanoids and other ethnic exterminations will still affect the people. Avalanches of lies, different from those the world already knows, are already replacing and will continue to replace the forgotten human attitudes...

It is paradoxical, for instance, that television and the Internet, to name but these two technological benefactions, serve as much to contaminate the mind as to educate the people. **It is also paradoxical that governmental powers can become distorted to the point of using all the new technologies to create disorder, wars, to exploit the people, their instincts and their vilest passions, to multiply lotteries, gambling areas, casinos, etc.** Dehumanized powers will soon try to control the commerce of humans for money (their master), just as they control gambling, bars, lotteries, drugs, etc., by abusing human weaknesses to force people even more into slavery.

<u>**This is the boat on which humanity has now embarked**</u>. People will continue to struggle in this gutter if they do not make a quick "Global Turnaround".

But when one considers the slowness of the negotiations and interventions of the nations only to realize the Kyoto agreements taken 20 years ago and their possible realizations only in 2010, we wonder how scientists will handle the other 20 or 30 major agreements **needed to arrive at solving the other great problems of the destruction of our planet.** One can only feel uneasiness. Especially that few people are thinking about why and how to alleviate the dehumanizing horrors, at least as serious as those that destroy the environment of people and animals.

Conclusions

Why must we wait for further proof demonstrating that the people are destroying themselves in their actual state of dehumanization and of material values to start to react? It is always the same unhealthy and inhuman ideologies from the same pseudo-scientist movements that have participated in designing the great lines of thought that lead the people at the edge of this bottomless pit suffocating humanity under the weight of its dehumanization.

Why wait to further destroy our environment or for the earth to be strewed with more deaths to finally become aware of the evil that awaits humanity? <u>**A fault was committed somewhere, it is obvious.**</u> Instinct tells us. But in order to react, people must still be able to feel these lost remaining fragments of humanity that science has always denied us. Science cannot move forward without proof. The brain of the instinct however, that nevertheless guided the evolution of man for millenniums, feels illness well before science and the tenants of the "reason" can prove it, and thus reacts without having to think or to prove anything.... **Let us give our instincts a chance...**

There are now too many scientific proofs clearly demonstrating that the instinct has been using much more the superior levels of the brain than we would have believed to guide the people during their evolution. **It also appears evident that decisions advanced with the only dehumanized "reason", which scientists and intellectuals rose to the rank of a god, relies only on the primitive reptilian senses to decide what is good and what is bad for humanity.** Scientists and intellectuals rely only on their "reason" and their lower reptilian "senses" to prove their decisions, to rise above the others and to manage them! Since neurobiologists have now shown that there are other levels in the human brain where are recorded the results of former experiments lived during our evolution, these should now be taken into consideration when deciding what is good or bad for man and humanity. <u>**This cannot be worse than keeping only the primitive reptilian senses for making decisions, as scientists, the "reason" and the Cartesian method of research have been doing so far.**</u>

Regrettably, as demonstrated in this work, in a completely scientific and dehumanized world, proof always comes too late and we are forced to admit that what had to happen has happened.

CHAPTER 9

Are solutions beginnings to emerge?

Introduction

The objective of this work is to make individuals become aware that, along with the fantastic evolution of man's material life in the past century, science, technology, material values and the use of the only "reason" (logos) have produced much more perverse and disastrous side effects on our material world and on the people than religions previously did during centuries. This evolution has destroyed the environment, life on earth and has produced chaos all over the world. At the same time, it has engendered a terrible dehumanization of men and their societies and has killed the humanity in people.

No real solution has yet been presented to stop the destruction of our world. **It should belong to every individual and to every government on this planet to concretely try and correct the most perverse side effects created by science and the material technologies.** Let us try to imagine all the work that people and governments would have to plan, propose, discuss and accept to initiate the necessary turnaround to simply begin to stabilize the environment and to re-humanize man. **The solutions elaborated since a quarter of a century to simply set up and to try stabilizing the disastrous process of the warming of the climate are just beginning at international levels and will not be really perceptible on the environment for years.**

Firstly, it is imperative that individuals and world leaders become aware of this disease that consumes man and society. When men become conscious of the problems in controlling and preserving the environment, life biodiversities, reducing desertification, controlling climatic changes, etc. which keep on destroying life on earth, only then will it become possible to really set up studies and plans to try and solve the most urgent problems.

Think of the years of work needed to demonstrate and plan the most urgent solutions to try stopping the destruction of the ozone layer and to have the agreements signed by countries... over 20 years later, these processes are still far from being a reality...<u>**Only vacuous optimists think it is still possible to move any faster. To simply palliate to the current problems which affect our environment, we would need over 30 Kyoto agreements**</u>! Imagine the hugeness of the task and especially the necessity of obtaining assents and budgets from people and governments to simply undertake the usual actions to begin solving these other urgent problems…

Once aware of the problems, each one of us should do a self examination and act on our own. **"Think globally and act locally"** said the Green Peace slogan. We cannot wait any longer for the governments alone to solve the environmental and human problems identified in this work. **Each one of us must personally act now.**

Without a general mobilization and the participation of every individual and country of the world, all the solutions initiated by only part of the global population would be a wasted effort. This is not an easy task. We think that even a hypothetical level of civilization two or three (2 or 3), which would be, according to astrophysicists, one million years more advanced than our sectarian and uncoordinated level of civilization zero (0), would have a hard time correcting the current problems within acceptable delays. If we suppose that, according to astrophysicists, it takes three to five hundred years for our current sectarian civilization 0 to reach the necessary global consensus which should characterize the level of civilization one (1), it means that the world should immediately start to try solving these problems.

When we think of the stages that each man, scientist, intellectual and government must go through only to begin the necessary changes to modify the influence of materialism in education and integrate notions of humanity and traditional values, we have to worry. We will thus limit ourselves at suggesting a few avenues that powers should first take into consideration, before really tackling more concretely the study of the real solutions to the problem of the generalized dehumanisation that invades our planet.

Paradoxically, the main obstacle comes from the fact that our world is at odds with itself, while at the same time being in the middle of the most gigantic scientific material evolution of its history. If the world continues this way, it will collapse under our very eyes much more quickly than the former societies collapsed, for they had not reached the level of dehumanization and suffocation that is killing us today.

Several mistakes have been made by scientists and intellectuals, for instance, by deciding to reject with their only "reason" and their scientific incomplete method of research, the traditional human values! Their fault today is to continue to pretend, in the face of the numerous side effects left by this education, that the disasters which that strike humanity and societies today are not the fruit of science and of the **ideologies of**

"this civilization of the reason", sowed to the four winds on the world and accepted as inviolable truths.

Let us look back at the data presented in the previous chapters to be convinced. They cannot remind us enough why it is so distressing to notice that **without the reinsertion of the humanities in education,** the magnificent sciences and technological applications will only continue to destroy the environment and life on earth as well as deprive the people of their individual rights and liberties and eventually enslave them.

Possible solutions

In front of these hard facts, **every one of us has the duty to try and remedy to this suffocating disease which strikes humanity, society and the whole world, "if" we are able to control it** before being swallowed by the monstrous creatures that scientists and intellectuals have created.

The beginnings of solutions initiated so far by scientists and intellectuals to try and correct the situation **were all of material order.** Even the ecological solutions proposed by biologists do not take into account the effects of all this materialism on the human being and society. **The scientific, political or sociological solutions proposed so far have not yet taken into account that at least the consciousness of each individual is not material.**

Before the events of September 11, 2001, we had written that we should indeed honour the human values in the thinking of solutions. The greatest problem comes from the fact that the materialized and dehumanized individuals formed in this education system to manage the people today, do not even realize that they are only materialistic, atheistic and distorted human beings lacking humanity! Maybe only repeated barbaric and demonic actions like those committed by international terrorists, can succeed in activating in one each of us the appropriate reflection **which is necessary to realise that perhaps something is terribly lacking in individuals.** We will indeed one day need to resolve the situation that infects the hearts of the people and of the intellectual materialists that manage us. I refuse to believe that only such demonic actions can succeed in making us question our inflexible judgments devoid of human sense!

The elected as well as the non elected civil servants that manage our occidental governments will have to acknowledge the absence of humanity in their actions. They will have to stop hiding behind what they call the legality of their inhuman laws and of their "politically correct" comments. They will also have to stop parading around and shouting from the rooftops of all the countries of the world their dishonest lies about

335

preserving their own citizens' rights and liberties. **How can these people confuse legality and morality?**

To begin with, it **will be necessary to stop abusing nature and destroying the elements necessary for the life, as well as stopping the authorities'** dreadful intrusions in the people's private lives. We must remove from the command posts the irresponsible individuals who commit these degrading acts. How can we continue to rely on these powers which treat us this way?

<u>**One of numerous problems to be dealt with is the uniquely materialistic way of thinking of the men in power that belongs to a depraved culture; they have an amoral way of thinking, believing that scientific and social materialism are the only real human values**</u>. Those people in charge often behave like machines and robots that cannot think and that have no notion of humanity. The Ministry of the Human Resources of Canada, (see Chapters 4), should be completely abolished because of its actions. We will recall that the civil servants of that Ministry came to think that their ways of doing were right and fair. Even though the government tells us that the unique file they had made on each citizen and which contains their every move since 20 years is destroyed, we must not believe that these information have actually been really destroyed... Knowing the materialistic mentality of these dehumanized persons, no one can believe them any more. **The vast culture of deceit which prevails in this particular ministry, as well as in many others of our materialized and dehumanized governments, is only the tip of an iceberg which amplifies the dreadful materialising evil that eats away like a cancer at the people and at our modern socialist societies.** These people are leading us at the edge of the same abyss which has already swallowed up the countries which were the first to teach and to practise Karl Marx's scientific and social materialism!

There is no doubt in our minds that one of the perverse material effects of the technological applications was to facilitate all these enormous interventions of the powers in the private lives of the people. Machiavellian minds, responsible for these violations of individual rights and personal freedoms, shamelessly impute the fault to technology. **Does technology allow the destruction of the world with nuclear weapons because these weapons exist?** However rigid, logical and mechanical these reasoning may be, technology does not give such rights to those in power. Has such a simplistic materialistic reasoning justified the millions of gratuitous murders by the Stalin, Mao, Pol Pot or other leaders of the totalitarian socialist countries?

We will begin to witness a turnover in certain governments when the people finally become aware of the abuses made by their past and current non elected leaders in power.

Much time will be needed to right all the wrongs made to nature and to man since over a century by the all too socialistic, administrative, educational and governmental mentalities. **How can we proceed to quickly initiate such changes, while there are less and less individuals in our occidental societies who are not yet contaminated by these scientific and socialist ideologies?** Is it possible that only tragic events or humanitarian disasters could induce the necessary changes? This will take time, probably decades of collaboration on the part of all the individuals and governments of the planet and of all the concerned countries.

When the people start to attend the councils of the Ministers, when they participate in the decisions now taken by the ministers' personal staffs and pronounce themselves on the things that really and directly concern them, **maybe then we will have returned to a more humanely acceptable form of democracy.** Governments elected for 4 or 5 years should well understand that they have been elected to administer and not to take decisions as important as increasing the social measures of the State to infinity, to excessively tax their citizens, to waste in trivialities the taxpayers' money, to create gigantic debts that will burden future generations, to appropriate people's health and education, to merge cities, to scatter the wealth of their country in favour of private companies, etc., **without specifically consulting their citizens in a just and honest way and without lying?**

To avoid the atrocities that the governments of the "reason" too often induced against the will of their citizens, <u>**it would be advisable to establish international courts to judge "criminal decisions"**</u> taken by certain private or public powers, when politics go against the private lives of the people. What is the necessity of having "charters of rights" if nobody respects them? I am referring here to the ethnic purges which multiplied in a inconceivable way during the century of the "reason", to the declarations of war of a country against another by their powers, to the use of nuclear, biologic or chemical weapons that endanger the world, to the right to pollute and to destroy the planet that certain public or private powers take upon themselves, or still to the dreadful thefts committed by leaders of private companies such as of Enron, Anderson on the stock exchange market to the detriment of their shareholders, etc.

<u>**It would be advisable to begin with an individual awareness of the phenomenon of destruction of our world to analyze to what point we got ourselves misled.**</u> Only **an individual awareness at first**, then a

global one of the current situation of our world could allow us to consider Laszlo's "Global Turnaround", if "the world and the people that live in it do not want to disappear in the bottomless pit that awaits them". (*cf. Ervin Laszlo's Virage Global, L'effrondrement de notre monde est-il inévitable? Éditions de l'homme, 2002*) (Global Turaround: Is the Collapse of our World Inevitable?).

It may be possible to reach this goal by going back to some form of education with human sciences, associated to a return to the humanities or a new method of humanized research and technology. I have in mind a method of research that would take into account at the same time both the material and the less tangible human data. It is no longer acceptable to take into account only scientific judgments of "reason" accepted by man's most primitive reptilian senses to decide what is right and what is not.

Scientists and intellectuals can no longer rely only on the "scientific reasoning" (logos) as the only way to perceive human thought and reasoning. We might have to go back to the methods of education prevailing before the appropriation of education by scientists and intellectuals, **such as a return to nature, history, arts, the humanities, general formation, etc. It would also be advisable to include in the techniques of teaching, as well as in science itself, certain principles and traditional human values, discarded too quickly by pseudo scientists and intellectuals...**

Powers should stop urging people to modify their secular symbolism, simply because they did not know at the time that "reason" and the scientific data of their method of research only met the needs of their primitive brain. Now that neurobiology has demonstrated that the simple comprehension of things uniquely by **the senses does not allow integrating into the hypotheses of research the less tangible data nevertheless stored in the other levels of the human brain,** it is compulsory to even consider them? Let us think about the accumulated knowledge of situations, data, judgments, feelings, arts or even of the more instinctive solutions acquired during the history of humanity, by opposition to the data justified only by the senses or to the lucubration of the reasoning of the pseudo intellectuals.

This is a neurobiological discovery that pseudo scientists should henceforth take into consideration in their scientific reasoning, **more particularly at this time of the human existence, where humanity must rapidly become aware of the shortcomings of its scientific and social system originating from the related inexact pseudo scientific domains.** It must be removed from this deep materialism where the rigid and linear reasoning devoid of human notions have imprisoned man for such a long time.

Scientists have not yet been able to adhere to any kind of symbolism or to less tangible data justified by the superior levels of the human brain, because they have based their reasoning only on their rigid and incapable ways of taking into account human feelings, for instance... **Using only "reason and senses" to justify their hypotheses of research, scientists and intellectuals have always cloistered themselves behind walls which do not allow them to see any further.** Nothing will change as long as they do not accept that they have limited themselves at understanding only the material data confirmed by their senses and their archaic brain to try to understand man and his world.

An element in the scientific reasoning of pseudo scientists has always been missing in the elaboration of their ideologies on less material and perceptible entities by using only their "reason": **it is that element of humanity for which these people have often been blamed for not considering!**

All the scientists, the intellectuals, the politicians, the mutant robots, the humanoids, the atheists formed by this education system with sciences without the humanities, would not be able to legitimize without a deep modification of their Cartesian method of research and reasoning, in other words, **if they do not acknowledge their mistakes and do not initiate the necessary changes needed in their ways of thinking, so that finally education, ways of being, reasoning and the governance of the people can take into account both the material and the spiritual side of the man.**

Now that we are aware that more spiritual, finer, subtle and human data have been lost in the analyses because of the use of an defective method of research and of reasoning when applied to the study of entities not perceptible with the primitive senses, **it is compulsory to conceive an improvement of this method, in order to take into account data or entities verifiable with other levels of the human brain, rather than only with the reptilian senses, to find the real truth!**

It is compulsory to immediately modify the actual method of research, or to replace it by a new one that will take into consideration the tangible as well as the less perceptible data that the material senses alone cannot take into account. If such a method is applied, man will only be ennobled, and we can expect that science will finally be able to see the existence of the supernatural, the poetry, the sacred, the human values, the beauty, the love, the kindness, the empathy, the respect, etc., entities which are so lacking in atheistic people, formed only in the scientific and social materialism. **Let us hope that these are frames of mind will finally be studied by science as live and existing entities!**

The knowledge acquired with such a modified method of research would not only consider the matter, but also the human part of man. **We like to believe that a research taking into account the more imponderable data could only help to humanize knowledge, people, powers and societies. And at the same time, to <u>help understand the gap that always separated the believers of material knowledge from those using the more intuitive, sacred or spiritual knowledge</u>.** The inclusion of such less palpable data into the hypothesis of research could also include, in our stifling and dehumanized education, more depth on traditional values or on strictly human entities which are so lacking in the pure Cartesian method.

One thing is certain, it is no longer acceptable for scientists to try and conclude anything about the human in man, without proofs other than material. Hegel's philosophy was wrong: matter is not eternal, nor is God the matter. All the opinions taken so far by pseudo scientists on beliefs and on "human questionings" or on the existence of human values must be reanalyzed in the light of this new method of research. History is filled with inappropriate judgments based only on data and incomplete analyses! Let us think that all the errors made by the religions on Copernic's or Galileo's discoveries in the fifteenth and sixteenth centuries respectively, will be corrected in the face of the weaknesses of their reasoning and of their materialistic methodologies. Scientists and intellectuals can no longer impose or propagate in their education system the false truths as the ideal method of formation, without taking into account these more recent data.

<u>The current problems of the materialized and dehumanized humanity will be corrected only if scientists agree one day that the truth is not only in the "reason and the material values". And so they cannot exclude the existence of traditional human values, love, kindness, empathy and especially the respect of others</u>. The pseudo scientists of non exact sciences must also acknowledge **the mistakes they made in the name of their "reason".**

<u>Thus, to change and to form the people with a more universal system of education, powers should quickly modify the actual model that uses Karl Marx' unbearable social ideologies and reintroduce the humanities and the traditional values in education</u>. What kind of reputation can a science have when it stupidly denies the existence of what it cannot study or understand? Or a science that bases its reasoning on the only perceptions of man's reptilian senses, without taking into account the more noble data, stored somewhere in the brains of evolved human beings?

It is very important to teach people that there also exists, next to scientific education, other forms of thought and other ways of forming the minds than only with the material values taught in the training of people by sciences.

The new education should take into account at the same time both the natural laws of physics and chemistry and those less tangible laws belonging to other spheres of human activity.

A return to instinct, this form of non reasoned intelligence of man and of animals which, unlike the "raw reason", can intervene in the results of former experiences stored in the genes of the brain's superior levels, is essential in the development of a new hyper conscious man. Could it be for that reason that the "reason of the intelligence alone" has never been able to understand why instinct has always guided so well the people during the millions of years of their evolution?

The upholders of the sacred should also modify their concepts and take into account the material realities now proven by science. The people must understand that there is no more time to lose in sterile and futile discussions, or in unclear epic debates among evolutionists and bible believers, among religious and non religious individuals, especially when the existence of life on earth is threatened... It is time for the truth, the real truth which concerns the survival of life on this earth and we can no longer allow these dilettante lucubration. **We have reached the end of the road: the very survival of our world and of the humanity of man, or the end of man.**

If people would admit once and for all that they are not only "reason" or "spirit", but at the same time matter and spirit, a giant step would be made towards the urgency to unify and to eradicate this evil that overcomes our physical world and man's spiritual world. No proof and no scientific data have ever demonstrated that it could be otherwise. Nothing in science justifies the vacuous atheism that swallows up so many people educated only with science and scientific reasoning (logos). **Current knowledge enables us to assert that there is nothing opposing the mind to finally comprehend knowledge "per se", without having to pass by the senses.**

In an ideal world, the great decisions would now suppose the union of these two forces of man, the material force and the spiritual force of all those that still feel some notions of humanity. A common movement must quickly be started to stop at once the destructive threats on the environment and on the people. We almost need to take a pause, because **both the dogmatic education of science and the "ex cathedra" of the spiritual world are now unacceptable: they both have lost their legitimacy.** The errors of the past and the murderous deviations which have too often marked the actions of those that believed to possess the truth can no longer be explained by past arguments or by a science which has not yet found again its humanity.

Will we succeed in extricating ourselves from all of this?

Will we succeed in extricating ourselves from all of this? We don't know, but we should at least try. We will certainly have to raise questions about many achievements, so far accepted as infallible certainties by scientists and which they are not.

The people of this planet have been locked up for too long by former powers: **in the mythical world** during millenniums, **in the sacred or religious world** during hundreds of years afterwards and **into the sterilized world of the raw scientific reason without humanity for a few centuries.**

Alternately in the history of humanity these mythical, religious and scientific masters proclaimed they were the holders of people's powers, that they were the guardians of their fate and the only individuals capable of taking charge of their actions, their rights, their individual liberties and their spirit. To be able to act more freely, the people in power have always held people in the "obscurantism" of the illiterates, in an unbalanced education or in the obscure plans of their reasoning.

We are forced to admit today that neither the myths of yesterday, nor the religions of yesterday, nor the tenants of the current "reason" that seized man's rights, personal freedoms or personal powers were right. People who have managed to "solve" the problems of man or have pretended to answer the essential needs of humanity were mostly wrong. In spite of their good will, these individuals have always only contributed in bringing people to the edge of the abyss from where nobody knew how or even <u>if</u> he could go out of. People have lost faith in these impostors that seized their individual powers: yesterday's religions, scientists, intellectuals, governments, etc.

If the behaviour of societies was formerly established on human values and religious dictates, it is now too much in the hands of these ideological pseudo scientific powers and pseudo intellectuals who have lost all legitimacy. **Hegelian and Marxist ideologies have no more reason to be.** They are now part of History. It is unconceivable that, without any legitimacy, these scientific and intellectual powers can still continue to remain at the heads of governmental powers by using their repressive and inequitable laws to control human beings. **All these powers established themselves by relying on the simple apprehension of the primitive perceptions of the "reason" and the senses.**

Every individual in the world must henceforth turn now to his own internal physical and spiritual strength to get out of the gigantic mess where humanity is headed at an incredible speed.

The solution of the conscience

Forced so far by religions and scientists, **it would be advisable from now on for every individual of this planet to become aware of his owns individual strength and becomes an autonomous being.**

To survive tomorrow, every individual of this planet will have to develop his own individual consciousness, his own personal behaviour. Those that share this awareness will have to regroup within associations that share the same "universal consciousness".

Each individual must liberate himself from all these people who still claim deceitfully to have the power to manage them or to manage their conscience and their lives, including their governments. Only then will the people be freed from the slavery in which their powers have so far illegitimately maintained them.

But time is short and the hour of truth for humanity and the environment has come. There is not much time to react and to right the wrongs. The agreements of Kyoto and Rio, which are not yet realized after more than two decades, and the 20 or 30 other agreements necessary today to stop the pollution of the seas, the disappearance of 85 % of the fish (our food) from the oceans, the deforestation, the desertification, the extinction of whales and of other sea species, etc., are still like thorns in the sides of our devastated civilization.

This work would like to be added to all the cries of alarm sent to every man on the planet, so that each one can become aware of the dangers that threaten the environment and for the alarming effects of science and technology on the disappearance of his humanity. Whatever his race, religion, country, culture, the ideologies he shares or the excessive use of the soulless technologies that put his planet in danger, every individual must become aware of the deep evil of materialism that suffocates him and eats away at him.

Scientists, ideologists, intellectuals, governmental leaders and other leaders should be the first to become aware that they have all abused of the powers that they stole from the people at "sword's end" or by "reason".

Let us imagine that only an individual and collective consciousness of those problems would allow humanity to perhaps avoid the disasters ahead. Let us hope that measures will be quickly taken so that each one of us has the time to form this "individual consciousness" indispensable to save our world and to quickly enter the more conscious "level of civilization 1" that we so badly need, now.

We like to believe that the time has come to make Malraux's reflections our own: "If the people of the twenty-first century do not go back to the religious and to the sacred, they must accept not being at all".

www.ingramcontent.com/pod-product-compliance
Lightning Source LLC
Chambersburg PA
CBHW030002190526
45157CB00014B/96